26th Process Plant Symposium (PPSS 2024)

Held at the 2024 AIChE Spring Meeting and 20th Global Congress on Process Safety

New Orleans, Louisiana, USA
24-28 March 2024

ISBN: 979-8-3313-1698-3

Printed from e-media with permission by:

Curran Associates, Inc.
57 Morehouse Lane
Red Hook, NY 12571

Some format issues inherent in the e-media version may also appear in this print version.

AIChE
120 Wall Street, FL 23
New York, NY 10005-4020

Phone: (800) 242-4363
Fax: (203) 775-5177

www.aiche.org

Additional copies of this publication are available from:

Curran Associates, Inc.
57 Morehouse Lane
Red Hook, NY 12571 USA
Phone: 845-758-0400
Fax: 845-758-2633
Email: curran@proceedings.com
Web: www.proceedings.com

TABLE OF CONTENTS

SPRING24
+20TH GCPS
A Joint AIChE and CCPS Meeting

Observing Your LOPAs in the Wild

Field Validation of LOPA Assumptions in Operating Plants

A.M. (Tony) Downes
Downes Process Safety LLC
Morristown, NJ 07960 USA
Anthony.M.Downes@gmail.com

Prepared for Presentation at
American Institute of Chemical Engineers
2024 Spring Meeting and 20th Global Congress on Process Safety
New Orleans, LA
March 24-28, 2024

Keywords: LOPA Validation, SIL Validation, Layer of Protection Analysis (LOPA), Process Safety, Risk Assessment, Validation, Industrial Process.

Abstract

For many years, Process Hazards Analyses were the standard in the industry. Beginning in the 1990s some companies implemented Layer of Protection Analyses (LOPAs) for the high consequence scenarios from the PHAs. The assumptions from the CCPS LOPA book "Layer of Protection Analysis: Simplified Process Risk Assessment" have been widely used since then.

Organizations must now ask "what's the real experience like? Are the safeguards as identified during the LOPA going to be effective? Is the preventative maintenance schedule in place and is that schedule being followed properly? Is there a test procedure for credited safeguards, and does that procedure make sense? What do those tests reveal?"

In scenarios like loading/unloading, which have a significant human error/human factor component, are the frequency assumptions still valid? Are the procedures and practices credited during the LOPA study really being followed? Are the independent protection layers (IPLs)? If checklists are credited, are they being used properly? Do alarms make sense? Regarding the human element, do operators understand the scenarios described in the LOPA? Do they understand the safeguards, how they work and why they are important?

Companies should ask themselves these questions. At least two large chemical operating companies began doing exactly that some time ago. A small group of corporate process safety engineers began putting their companies' most significant scenarios "under a microscope". They performed a deep dive on each site's LOPA scenarios, looking carefully at every aspect - from the consequence analysis to the maintenance records.

The results were interesting, and sometimes a little frightening. It may be instructive to other companies wondering how well their risks are actually managed, or to those looking to conduct the validations required by recognized and generally accepted good engineering practices (RAGAGEP) like ISA/IEC-61511. This paper will discuss methods used and real (though anonymized) patterns found.

1 Introduction

This paper focuses on the results from the LOPA for several reasons. The primary reason is that it focuses efforts for the validation. In the industry, LOPAs are only done for potentially catastrophic scenarios, so there are (hopefully) a limited set versus the myriad possibilities identified by the PHA.

The second reason is that the LOPA process specifically breaks the risk into its component parts. An example scenario is one where a pressure-storage vessel might rupture due to a control system fault. In this example, in order for this scenario to occur, a specific control valve must stick or be running in manual mode. A specific alarm must fail or be missed. A specific relief valve must fail to open. Personnel are assumed (in the LOPA) to be in the tank farm area less than 10% of the time. The LOPA process has already peeled apart the layers so they can be examined one by one.

Third, is that the LOPA sets out the expectations for the scenario in (semi) quantitative form. Instead of looking for something that should only happen less than once per 10,000 – or 100,000, or 1,000,000 years, a company can look at things expected to happen once per 10 years. They can see how often faults are being found in safeguards – faults that are expected to occur less often than once per 100 demands. They can check whether those safeguards are actually on-line and working more than 90% or 99% or 99.9% of the time, as they are intended to be. In other words, a company can see how well these components of the scenario have performed over the past 5-10 years to see if that performance matches with the LOPA expectations.

The following are real-world examples:

Example 1: A filtration system for hydrogen peroxide was protected by an active interlock. It was designed to open purge valve XZV-G101 in case of loss of flow detected by FZLL-101. However, this interlock was not put into the computerized maintenance management system (CMMS) for inspection. Approximately five years later, the interlock was called on to operate, but one of the valves stuck closed. The filter housing was damaged by the decomposition reaction and overpressure. Relief valve PSV-101 partially relieved the pressure but was not sufficient by itself. Fortunately, no one was hurt. A validation of the inspection, testing and preventive maintenance (ITPM) process would have discovered the FZLL-101 interlock was not being executed, so its risk reduction factor (RRF) had fallen well below the target.

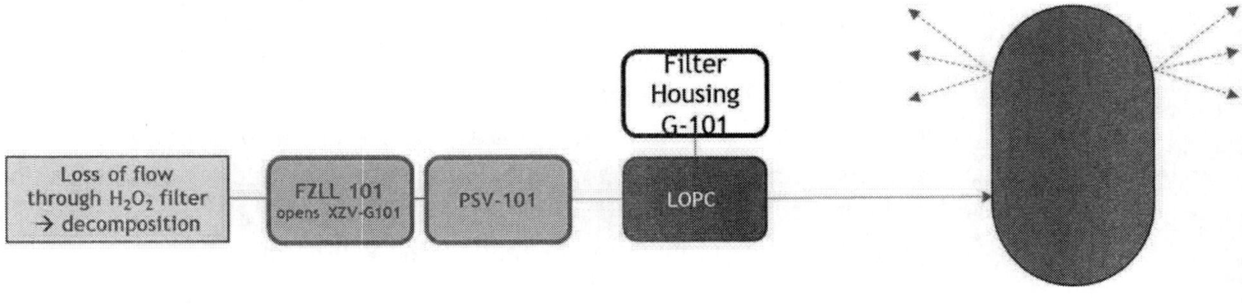

Figure 1. Simple "bowtie" diagram for a filter (one cause/one consequence)

Example 2: A reactor had the potential for an exothermic runaway that could burst the vessel leading to one or more fatalities if it were to occur. In the LOPA, it was assumed to have a temperature control loop fault about once per ten years (standard LOPA number per CCPS[i]). There was also an independent high-high-temperature interlock intended to prevent runaway reactions. It was found to have activated three times in one year. In other words, the initiating event frequency (IEF) appeared to be about 30 times higher than estimated by the LOPA. Clearly some further work was appropriate.

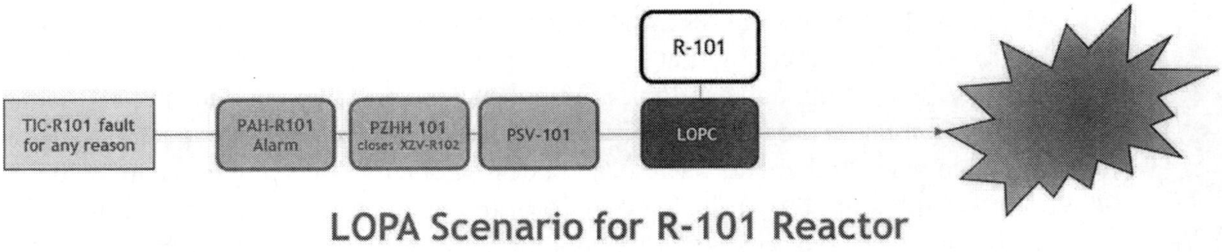

Figure 2. Reactor LOPA

Example 3: When looking across multiple scenarios, one company found that nearly half of their high hazard scenarios had been incorrectly evaluated for consequences by the PHA teams. Some had exaggerated the potential consequences, and some had underestimated them. Another found that the best starting place for validating LOPAs was to ensure that the estimated consequences made sense – perhaps the most logical approach to take.

The introduction outlined why the LOPA report is the right starting point. The rest of this paper will talk about the process that several companies have followed to validate their own LOPAs and sum up with conclusions and lessons learned. It will touch on each of the following topics:

- The input data sources (incidents, maintenance records and automation/controls data) used in the validation
- The rudiments of LOPA validation
- A method to validate procedural safeguards, including alarm responses
- A suggested sequence, including practical advice on how and where to start doing LOPA validations
- Skills, procedure and organization to conduct the validation.

2 Input data sources used in the validation

The LOPA results will be the key input used in the validation. The owner of the LOPA – usually the risk analyst - can outline and explain the scenario and its safeguards, ensuring the company understands the issues and will be invested in fixing them. Therefore, it is important that process safety information (PSI) match and document the cause and subsequent safeguards; the Piping & instrumentation diagrams (P&IDs), loop sheets, safe operating limits tables, cause and effect matrices and safety requirement specifications (SRS) should all align with the specifics of the scenario. The likelihood that the risk will be properly understood and kept sustainably low through the life of the unit will inherently increase.

The next sources of data are the site's incident and near-miss history, followed by the ITPM program and results which are usually found in the CMMS. Meetings with operators will confirm they understand the hazards as well as the administrative safeguards and/or any applicable emergency response protocols that they may have to execute. Their input into how and why the scenario could – or perhaps couldn't – occur is important.

The next source will come from the automation staff. It is a data dump from the alarm and event historian for the pertinent tags going back as far as they have data.

Table 1. LOPA/HAZOP assumptions and data sources to validate

Data source	Consequence	IEF	Safeguards	CM/EC
PSI/PHA/LOPA	X	X	X	?
Risk Analyst interviews	X	X	X	X
Incident/Near Miss reports		X		
ITPM / CMMS		X	X	
Operator Interviews		X	X	X
A&E History		X	X	?

3 The rudiments of LOPA validation

Lopa validation is about quantitatively confirming (or not) each of the assumptions that went into the LOPA for a specific scenario. LOPAs have four main components:

A. Consequence analysis (or a summary of the findings from the analysis)
B. Initiating event frequency (including the specific cause)
C. Identified independent protection layers (IPLs) and their availability/effectiveness
D. Enabling conditions and conditional modifiers (if any)

LOPA validation puts each of these under the microscope.

A. Consequence analysis (or a summary of the findings from the analysis)

It is worth the effort to validate the general scenario including its consequences. In one instance, a company found that teams had estimated overly conservative consequences for a substantial fraction of its scenarios. In some ways, this may seem like it is failing "safe". If teams are going to get it wrong, it's better that they overestimate the consequences of scenarios. In fact, several companies have PHA and LOPA "rules" that are deliberately conservative so PHA teams are unlikely to miss scenarios.

There are benefits to this approach, but it should not be overlooked that there is a natural tendency to be conservative. If an organization has any skeptics, they may point to these examples as "exaggerating the risk". It can undermine the credibility of the entire risk management process which is necessary to get continued support from the organizations. To deal with this conservative-bias, it is helpful to have a corporate-oversight process for high-consequence scenarios. Another consideration is that resources are always limited. Spending valuable time and money on unnecessary safeguards is wasteful.

This may be as simple as an A-E categorization of the possible consequences before considering active safeguards. Key categories include:

1. Potential catastrophes, such as multiple fatalities [1]
2. Single fatality
3-5 Less severe consequences (Injury etc.)

The first question should be "is this scenario really going to have catastrophic[2] consequences?" It may be necessary to use expert judgement and/or more sophisticated consequence analysis. The consequences can be validated using two steps. The first is qualitative - does the site's analysis "feel" correct? Some process safety professionals have been doing quantitative analysis of gas releases using computer tools since 1988, so their experience may be slightly different than others', but it is possible to develop a feel for whether a release could be large and whether it could affect large or sensitive receptors. A safety analyst visited the proposed location for a chlorine railcar unloading facility many years ago. Looking through fence they could see a daycare across the highway. There have been times when the PHA team was being overly conservative. If the consequence analysis "feels" right to an experienced risk analyst, it may be best to accept it and move on. In the case of the chlorine railcar unloading facility, the analyst and company moved on to step 2 – quantitative analysis.

For any consequence analysis that fails the qualitative screening, it is worth developing a computer model of both release rate for the chemical of interest, as well as its state, followed by dispersion analysis of the cloud. In an analysis of one company, more than half the "catastrophic" scenarios found could not have been so bad in reality. Simple models like ALOHA, as well as sophisticated engineering tools like SAFER, PHAST and others can be used. Specialized companies who evaluate occupied structures to see how they would fare if an incident were to occur can also be hired. These tools can more than pay for themselves as a company decides how to either reinforce or relocate control rooms or protect workers in maintenance shops.

B. Initiating event frequency (and cause)

Once the potential consequences of a scenario are found to be significant enough – usually one or more fatalities – it is time to validate the cause and the likelihood/frequency of the initiating event that could lead to that consequence. For example, if a control system fault could cause the incident, CCPS' LOPA book[ii] suggests this would be expected to occur about once per 10 years.

There are several ways to validate that a fault rate is similar to this general industry estimate. The first is to check the incident history. Most operating companies have good computer records that include "near miss" reporting. Recall Example 2 cited earlier in this paper. Too

[1] or some roughly equivalent environmental or business consequence such as devastating and long term environmental damage, or possible great economic/business loss.
[2] Most sophisticated operating companies have a risk matrix to focus their efforts on the highest consequence scenarios.

much weight should not be given to an absence of "near miss" reports though. A better source of data is to select an instrument with an alarm or interlock to protect against the scenario in question. Then look for activations of those alarms/interlocks in the site's alarm and event history record. While there is sophisticated software to monitor these, a lot can be done by exporting these from the control system, migrating them into Excel and filtering. When a list of activations has been compiled, the site should review them. Sometimes there are "innocent" explanations like testing, or other operations that could not have led to catastrophe. It is also important to look in the alarm and event history for "events" like mode changes (MD) where a control loop is taken out of automatic and placed into manual mode. The probability of a parameter going outside of its safe operating envelope can be much higher when its control loop is in manual mode. At the least, it should be noted if the event happens often or for long periods of time.

Three notes of caution on over-interpreting alarm and event history data:

1. In the case of more than one "instrumented" layer of protection – for example, if there is a high-pressure alarm that should activate before a high-high pressure interlock, then the focus should be on whichever should activate first.
2. The analysis usually looks for factor-of-10 differences in frequency.
3. This analysis only helps identify frequencies that are on the high side. Finding that there has only been one activation in the past 20 years is not statistically significant.

If a control system fault could cause a serious or catastrophic event, then its sensor should be in the site's calibration program, and that program should be managed through the computerized maintenance management system (CMMS), with records showing that calibrations are done on schedule. It is important to validate that the site is not finding big calibration issues. CMMS can also provide information on the number of repair work-orders being written for these transmitters. Both repair and preventive maintenance work orders should show up.

In the same CCPS table, "operator error" is expected to occur about once per 100 opportunities. Close examination of a company's procedures can determine how often a filter has to be changed or the tank-farm line-up changes. These data points can identify the number of opportunities per year for an operator to line something up incorrectly. If an operator error could lead to catastrophic consequences, then they'll have safeguards and memory-joggers like checklists or indicators. These instances should not be used in your "unmitigated" initiating event frequency. In some cases, it may be necessary to engage human factors specialists to evaluate cases where errors could lead to catastrophic consequences and the only safeguards are administrative.

Regarding operator error frequency, evaluators should ensure that the hazardous scenarios are described in the standard operating procedures (SOPs); check that operators are being trained; and interview some operators to make sure they know about the hazard and how their actions protect against it. There have been cases where operators were expected to take a specific action that newer employees were unaware of. Even when there are no issues identified, it is good reinforcement.

C. Safeguards and their availability/effectiveness

The first step in validating safeguard effectiveness is to verify to the extent possible that a particular safeguard could reliably protect against the scenario or its consequences. For example, a process safety professional investigated a batch makeup tank that had overflowed into its dike. It made a mess but was not particularly hazardous. During the investigation, it became clear that similar incidents had occurred recently. The high-level alarm was installed so that it was just a few inches below the overflow from the tank. The operators had less than 5 minutes to receive the alarm in the control room, relay it to the field, and for the field operator to stop the filling. After speaking with the operators, it became clear that this was not sufficient time to reliably[3] stop the pump.

Operator-based safeguards should be described in the SOP, and operators must be trained on them – to the point where they can describe them to the validation team. When validating administrative or human-based safeguards like checklists or secondary-signoff, analysts should go to the place where they are used and confirm that operators use them as intended. In several cases, analysts have observed paper checklists intended to prevent operation in the wrong order where the operators are in complete chemical-resistant suits. In each case, the operators filled in the checklist after the field task was over – defeating the safety purpose of the checklist.

For engineered safeguards, "effectiveness" is the probability that the given IPL will be effective in preventing the consequence of concern. Example scores would be: 9 times out of 10 or one LOPA order of magnitude credit, or better than 999 times out of 1000 for a SIL-3 safety instrumented function.

Beyond "actual functionality", there are two factors in effectiveness:

a. Does the IPL have an effective preventive maintenance program?
    a. Is the specific IPL in the site's scheduled ITPM program?
    b. Is the schedule being followed for that IPL?
    c. Do the results from the "as found" tests show that the IPL was working[4]?

    The following are some examples of findings at two large chemical companies when they began detailed validation of ITPM for engineered IPLs. In both companies, a corporate CMMS had been implemented in the late 1990's. About ten - twenty years later, the corporate process safety staff began working with the corporate reliability staff to ensure that PHA-credited safeguarding systems appeared in the CMMS and had a scheduled ITPM plan. Here's what they found:

    - More than 97% of the pressure safety valves (PSVs) appeared in the CMMS and had scheduled ITPMs. Where there were difficulties, they were related to PSV

[3] In fact, they were able to stop the transfer about 3 times out of 4. This data is available because the situation occurred every Friday when they were making a double batch of additive to get through the weekend.

[4] While occasionally failing an "as found" test is ok, this should be rare. For example, any specific SIL-2 interlock should fail the "as-found" test considerably less often than once per 100 tests.

tag-numbering as well as management of change (MOC) issues where valves had been added or changed after the site implemented the CMMS.
- At first, only a small fraction of the credited interlock "loops" appeared in the CMMS. In about half the cases, analog transmitters with only a "safety" function appeared in the maintenance calibration program, though rarely tag-by-tag which complicated validation.
- In many cases, safeguard layers had been added by PHA's once these began in the 1990's, but the information about these new items had not been passed to maintenance or was not entered into the CMMS.
- As the corporations became more sophisticated in their use and documentation of safety instrumented systems (SIS) for IPLs, ITPM procedures became more sophisticated in their use of test procedures and adjusting test intervals to achieve the target safety integrity level (SIL) value required by the PHA/LOPA. One upside of adopting SIS was that it was easier for maintenance to identify the associated sensor and final element devices as safety-critical.

At a few sites, the testing of interlocks was done by operations rather than maintenance. Only faults were reported to maintenance for repairs. As the number and sophistication of such interlocks increased, and as the corporate oversight of safety-critical ITPM increasingly relied on reporting via the CMMS, this became untenable.

b. Ensure that the IPL is NOT bypassed.
There are a couple of ways to survey this. The first way to check is to look through the unit's "bypass logbook". In the days of relay-based interlocks, these were called "jumper logs". Some companies manage their bypasses through their temporary MOC system. All sites should maintain a bypass log system to record, approve and manage their bypasses and to ensure the bypass is removed when it is no longer needed.

If the IPL is implemented via a programmable electronic system, the status is probably being recorded in the alarm and event (A&E) historian. The easiest way to check this is to look for the input or output "tags" of the safety-critical interlock loops. If they have been put into bypass mode (e.g. "BYP" in Honeywell systems) over the available history, explanations should be found for why each bypass was done. Implementing a bypass for a scheduled ITPM test is reassuring. Finding that a bypass was implemented after an input was activated but before final action could occur would be an indication[4] of some problem.

One company discovered an issue with bypassing of an interlock in a new unit, shortly after commissioning. When the corporate engineering staff reviewed the A&E history, they discovered dozens of activations of alarms that indicated a possible heat exchanger tube leak. In each case, the control board operator had bypassed the alarm. As it turned out, the issue was a combination of the setpoint for the analyzer being too "low" as well as the inherent drift and unreliability of process analyzers. This "unreliability" as well as a high spurious trip rate is a well-known attribute of analyzers. The answer was to adjust the setpoint "up" out of the signal noise and to move to a two-out-of-three voting system for the analyzer setup.

Bypassing a given interlock multiple times per year or for long periods of time would be an indication of another problem. After a serious near miss, one company implemented a project to find and eliminate situations where interlocks had to be bypassed in order to start processes. In some cases, this required automatic timers on bypasses. In some cases, it required process design changes.

The following are some examples of findings at several large chemical companies when they began checking for interlocks being in bypass-mode. While these numbers are much smaller than the issues identified via the ITPM analysis, they are at least as significant to the risk at the locations.

- When a company first started looking at bypasses on interlocks, they found more than 10% across the company.
- This varied widely from site to site, with at least one site having all more than twenty interlocks in bypass at the time of checking.
- About half the sites in in one company were found not to have robust bypass management systems.
- One company made it a project engineering objective to design processes to avoid the need for startup bypasses.
- In another company, one unit was found bypassing SIL-3 interlocks more than 200 times per year.

D. Enabling conditions and conditional modifiers

Consider modifiers like "occupancy" or the probability that someone will be in a remote part of the plant when an accident occurs, versus how often people are actually in that area, based on interviews or observations. Similarly, it should be noted if a particular regeneration process is expected to occur only once per year and last a couple of days (i.e. 2 days/year) but the validator finds that it takes 5 days and is done every quarter (i.e. 20 days/year).

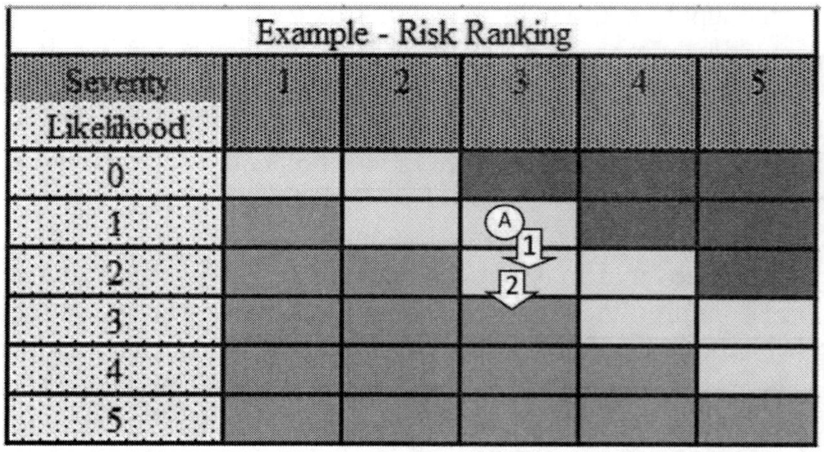

Figure 3. Risk ranking for a possible gasoline tank overfill

4 Suggested sequence, including practical advice on how and where to start doing LOPA validations

For those interested in trying LOPA validation in their company or at their site, here are some suggestions on how to get started:

- Start small. Pick three or four scenarios that already have LOPAs. Pick scenarios where management will care about the results – and be willing to fund further analysis if significant issues are found. It will take some effort to go through steps A-D above. Initially, it may take up to a week, spread across several weeks, to go through all aspects of a single LOPA scenario. With experience, this will come down to a day or two each, including time to document findings and recommendations.
- This should not be associated with an auditing process. Validating a risk assessment is still best done as a risk assessment process. It is too time-consuming to add onto the workload during an audit. Also, LOPA is a process of discovery. Operations and maintenance staff need to find value in it. Despite best efforts, audits can be viewed as nit-picky.
- Report the findings back to leadership. For example, "We looked at four high-consequence scenarios and found significant issues – significant enough that a risk gap may exist - in X percent of them". Don't be surprised if X is 50% or higher. That should build support for doing more such validations.

5 Skills, procedure and organization to conduct the validation

In previous companies, these validations have been a closely-held activity within the corporate or division level process safety team. To maintain the ability to compare results across sites, either a single experienced risk analyst is used, or a small team with tight technical oversight to ensure consistency. In several instances, delegating this to site staff or to larger groups of experienced risk analysts, led to such uneven results that it was undermining the credibility of the process. It also led to updating PHA/LOPA guidance based on these findings. This validation procedure and the feedback on its findings were shared with site-level PHA leaders and others.

Corporate reliability engineers can digitalize the ITPM validation efforts using the corporate CMMS with "PHA-credited safeguard" as a designator. This requires a good cross-functional team, excellent support/engagement from senior operations leadership and several years of steady effort to establish. Where this was done, results showed the value of the effort. Many critical IPLs were found to have preventive maintenance deficiencies: everything from higher-than-expected as-found faults to not being a part of the ITPM program at all. As a bonus, some devices/loops were found marked "PHA" in the CMMS, even though the PHA no longer credited them. With the support of the process safety team, these were redesignated, enabling maintenance to reduce their workload.

A number of major operational companies in the process industries are now using process historian data in their validation efforts. This requires close cross-functional efforts between process safety and automation staff. Again, enough problems are being found – and fixed – to justify further efforts.

6 Conclusions

This experience – demonstrated now over many sites at several different companies – makes it clear that the HAZOPs and LOPAs "as written" only reflect the true site risk level accurately a fraction of the time. Sometimes it is higher than the estimate. Sometimes it is lower.

Consequence analysis is a specialized activity. It is not unusual for consequence estimates done by site staff to be one to two orders of magnitude off – either overestimating or underestimating likely consequences – about half the time.

Many important safeguards have been found to either be bypassed or not tested as required by the ITPM program - or not be in the ITPM program at all. The fraction of active safeguards with problems varies from site to site as well as company to company. When process safety professionals validated these safeguards for the first time, they found 10-40% of SIL-rated safeguards had some problems. At a few sites, it was even higher than that.

Safeguards that rely on the actions of people have sometimes been found to be unlikely to work, either because operators were not following the requirements or had not been trained on them; or because there wasn't enough time for the operator to accomplish the required action.

PSM compliance audits very rarely have the time to get to the required level of detail to find the sort of problems described in this paper. Instead, one or more experienced process safety engineers had to do this validation as an additional activity. Once problems have been identified, leadership will be more willing to dedicate resources to finding where the real-world deviates from the LOPA and to fixing problems.

Remember, you don't reliably get what you *expect* – you get what you *inspect*.

7 References

Journal Article:
[1] Downes, A., Twarowski, A. *The benefits of comparing similar hazards across 'sister' plants*. Process Safety Progress, April 2009

[2] Downes, A., Goteti, P., Lindsay, S., Loseto, M. *Use Process Historian Data to Verify Safeguards.* Chemical Engineering Progress, February 2023

Book:
[3] "Layer of Protection Analysis" CCPS, 2001

Online Reference:
[4] Jamison C., Mentzer, R, *Study of Tank Overfill Incidents.* Purdue University, 2019

---

[i] CCPS "Layer of Protection Analysis" AIChE, 2001
[ii] CCPS "Layer of Protection Analysis" AIChE, 2001 pp 71

Independent Protective Layers and Human Response: Using Human Factors Principles for Verification of Effectiveness of Safeguards

Sarah Acton
Ascend Performance Materials
1010 Travis St #900, Houston, TX 77002
sacton@ascendmaterials.com

Keywords: Human Factors, Administrative Controls, Safeguard Verification

Abstract

Process Safety Management strives to protect people, the environment, and our business from process hazards. Designing an inherently safer process is our top priority. We sometimes rely on administrative controls (alarm response, critical procedure) as part of our layers of protection, especially when engineering controls are not practical.

However, it can be difficult to determine accurate verification criteria or how to improve the robustness of administrative controls. Using human factors principles to understand what shapes human performance is essential in determining when to use and how to improve administrative safeguards. This paper will discuss how to create a safeguard verification system using human factors to improve situation awareness and corrective action response. First, we will examine the importance that situation awareness plays in an operator's ability to detect, decide, and/or deflect. Next, we will describe how to identify error traps that can reduce the effectiveness of safeguards. Lastly, we will discuss learning from normal work which can help improve procedure writing, preventative maintenance programs, and selection of independent protective layers.

By employing human factors principles, we can create a verification program for safeguards in our PHA process so that we can better support human response and decision-making.

1 Introduction

The goal of Process Safety Management is to prevent incidents and to fail safely when things do not go as planned. This is accomplished by identifying the hazards, assessing the risks, and then controlling the risks. This process includes recognizing the scenarios that can cause harm, putting safeguards in place to prevent or mitigate harm, and verifying that those safeguards are effective. When building new facilities, it is easy to design inherently safer processes from the start. Hazards can more easily be eliminated or substituted, and safeguards can be designed into the process.

When choosing safeguards for chemical processes, Independent Protection Layers (IPLs) are safeguards that meet the criteria of detect, decide, and deflect [1]. Robust engineering controls, such as Safety Instrumented Systems (SIS), are preferred when designing a system because they can more quickly and accurately detect changes in the system and determine and carryout the actions need to keep the system from entering a dangerous state (deflect). SIS also must be independent from other control systems and not share a final element. Figure 1 below shows an example of the different elements in an SIS and how they relate to the detect, decide, deflect criteria.

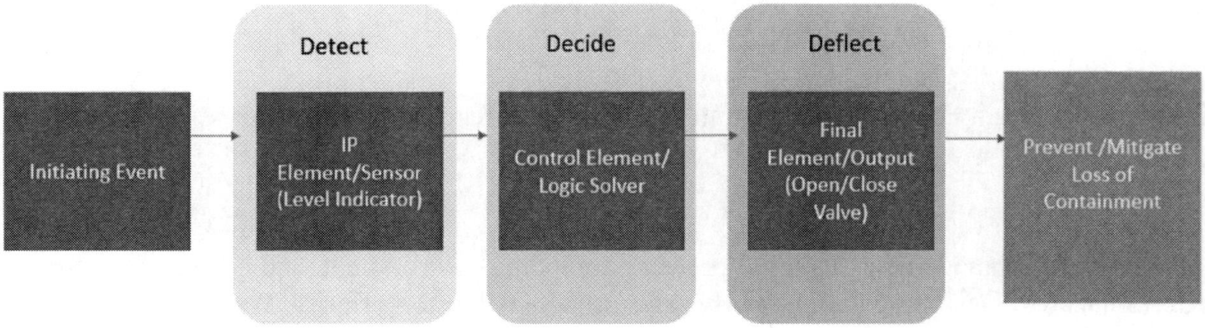

Figure 1. Engineering Control: Detect, Decide, and Deflect in an SIS

The verification of an SIS system is straightforward and would include checking the maintenance and inspection program and ensuring the independence in the system. However, a common struggle is to choose and verify appropriate safeguards in situations where more robust engineering controls cannot be applied. In these situations, we may turn to Non-SIS layers of protection such as Operator Response to Alarms or Critical Procedures (i.e., Start-Up). A Non-SIS Layer of Protection is defined as "any Independent Protection Layer that prevents the impact event." [2]. When a person is required to complete tasks as part of the safeguard, we categorize them as Administrative Controls or Human Protection Layer.

If no other options are available, administrative controls are better than no safeguard, but they are not always ideal. Humans can be inconsistent in their reactions, decisions making, and completion of tasks. The inconsistency can vary between person, shift, or even the same person may not perform the same tasks the same way each time. These inconsistencies in performance can be made greater given the type of task, the equipment, and work environment in which workers operate. The principles of Human Factors can be applied to understand how to select and verify administrative controls that have more consistent performance across various operators.

But what are Human Factors? While most people understand that Process Safety Management is about protecting people from the hazards associated with highly hazardous chemicals, the term Human Factors is not as readily understood. Human Factors is a cross disciplinary field including psychology, industrial engineering, biology, and some computer science. This field of study employs user-centered design to improve the interaction between humans and the system. Relatedly, human factors are also the name of a variety of aspects that shape how people perform their tasks. These factors can range from how well a procedure is written, to how much space is needed to perform the task, to the design of the equipment interface. When these factors increase the opportunity for error, we call them error traps or constraints. To reduce the opportunity for human error, we must influence the features that are associated with a worker's tools, tasks, and work environment [3]. Using Human Factors principles, we can identify and improve the factors that shape the performance of people in carrying out their part in Administrative Controls.

2 Human Factors in Verification of Safeguard Effectiveness

Understanding the role situation awareness and error traps play in the detect, decide, deflect criteria is important in being able to give these as IPL credit in the Process Hazard Analysis (PHA).

2.1 Situation Awareness

A simple definition of Situation Awareness (SA) is the ability to perceive changes in the environment, comprehend their meaning, and being able to project what the future status of the system will be [4]. Having good SA means that the operator can detect when the system is deviating from its desired state, understanding what the current state is, and what might need to be done to address the deviation or prevent it from getting worse. A lack of situation awareness has been shown to be a contributing factor in many incidents in safety critical fields [5]. Having good SA is important to control room operations and essential for Administrative Controls such as Operator Response to Alarms.

For administrative controls to be IPLs, they must meet the criteria for Detect, Decide and Deflect. When responding to an alarm, the safeguard has two components: the engineering side and the operator. The sensor detects a change and sends a signal to the operator. The operator then must be able to perceive the change, through alarm and visual ques, then make a decision based on procedure and carry out the actions needed to deflect.

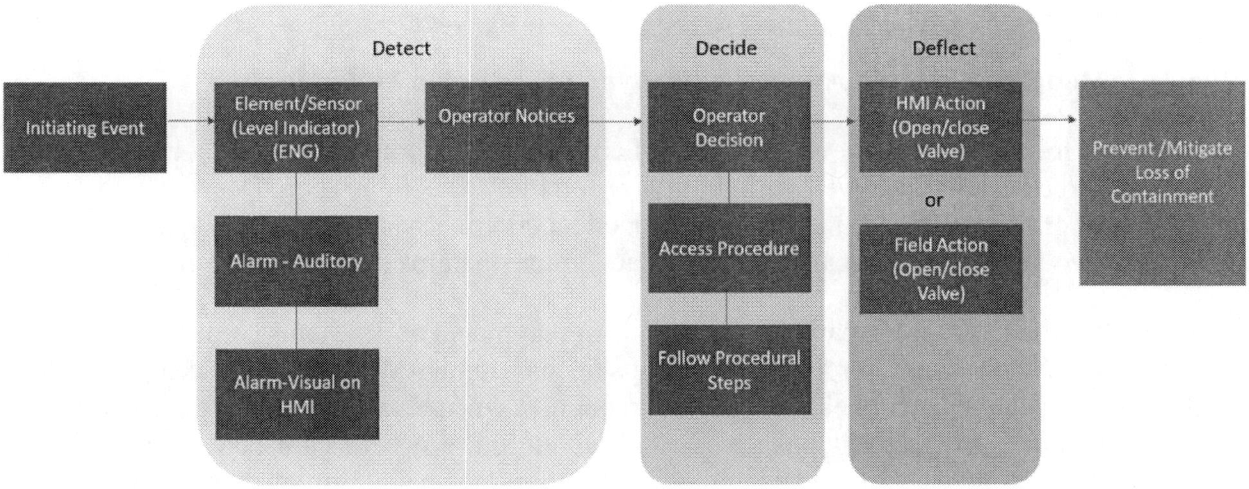

Figure 2. Administrative Control: Detect, Decide, Deflect in Operator Response to Alarm

Without being able to detect changes in the system and know what those changes are, an operator would not be able to respond or take the correct action to prevent a consequence. It is not the sole responsibility of the operator to be situationally aware, but the type and presentation of information from the system is what helps the operator with decision making and task completion. If provided with the right information at the right time, operators can make better decisions and reduce mental workload. Endsley's model of situation awareness shown in Figure 3 demonstrates the different task and system factors that can contribute to decision making and performance of action [6].

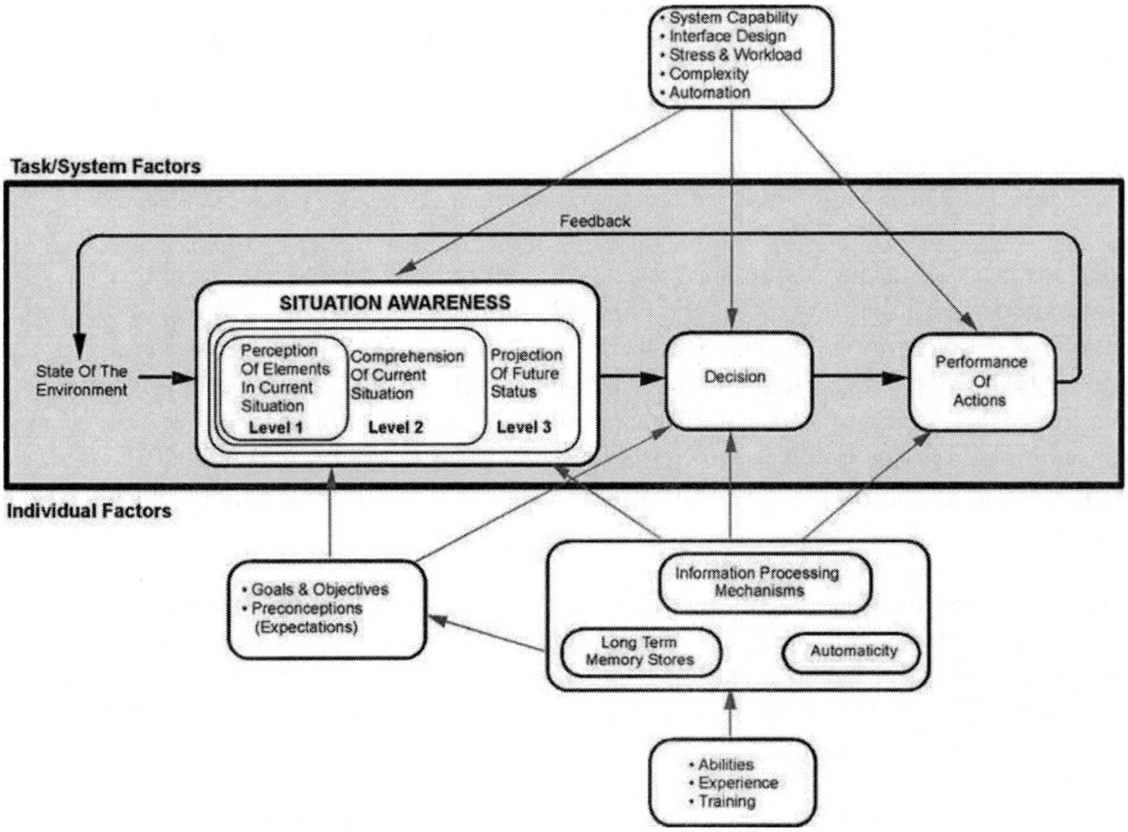

Figure 3. Model of Situation Awareness in Dynamic Decision Making [6]

In Endsley's model, the three levels of SA are Perception (1), Comprehension (2), and Projection (3). These three levels match well with the IPL criteria for Detection. Then, Decision and Performance of Action map to the Decide and Deflect criteria. System capability and interface design are two key inputs that can have an impact on the operator's SA.

In the control room, the information on the status of equipment and systems comes audibly through alarms and visually through the control screens. Human-Machine Interfaces (HMI) can be designed in such a way to help the operator recognize when changes occur in the system, comprehend those changes, and make predictions about future system status, covering all three levels of SA. It is important to understand what information is being provided and how so that we can verify that our administrative controls are functioning well.

The Abnormal Situation Management (ASM) Consortium is a group of companies that jointly conduct research in user-centered design to detect, prevent, and mitigate abnormal situations. The Consortium offers best practice guidance on alarm management, HMI design, procedures,

MOC and operations which all can improve situation awareness and decision making. When thinking about HMI design, the ASM Consortium has two fundamental design principles. The HMI should (1) allow the operator to develop and maintain a high level of real-time awareness of the process and (2) be consistent and easy to use, focusing on controlling the process [7]. The guidance on effective console design recommends four design practices to conduct these two principles:

- Display hierarchy supporting scope of work,
- Seamless integrated systems for operations,
- Uncluttered display layout with appropriate salience coding, and
- Human centered development lifecycle.

This paper cannot discuss all of these practices in detail, but Figure 4 gives examples of the display layout with poor design on the left and appropriate salience coding is shown on the right.

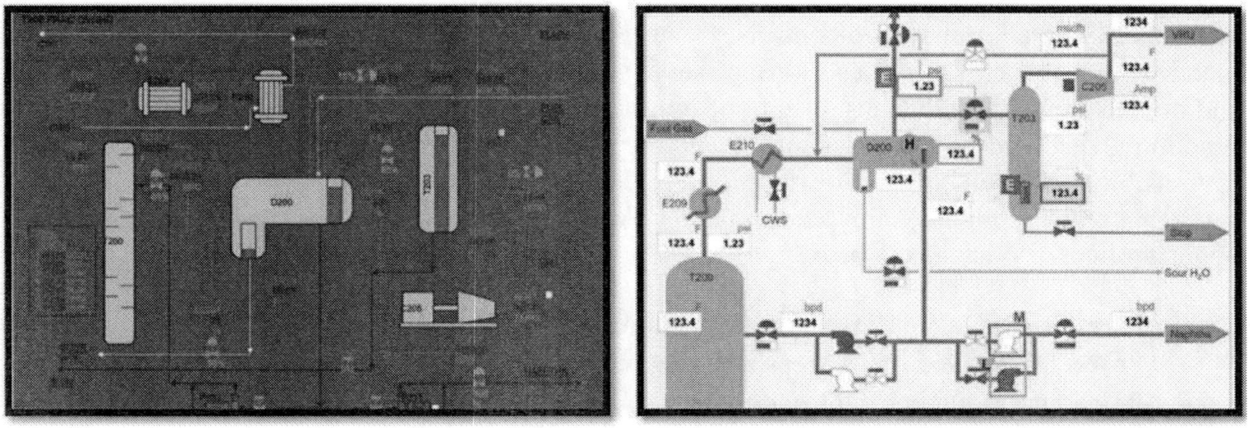

Figure 4. Examples from Abnormal Situation Management Consortium of displays that do not conform (left) and do conform (right) to guidelines for saliency coding [7].

Providing saliency is a key component that allows for detecting the changes in the system. Salience coding means that most important information, such as the cautions and warnings colors, are prominent and critical changes are readily seen. Meanwhile, non-critical information is shown in less prominent colors such as shades of grey. If reds and yellows are used for non-critical information, then it would appear that more parts of the system could be in trouble. This design (Figure 4, left) requires more mental energy to sort through to discover where the true critical changes in the system lie. Having a lighter background allows for critical colors to be prominent and easily seen and the operator response is quicker in detecting and processing the information leading to quicker, more accurate decision making.

2.2 Identification of Error Traps

Everybody makes mistakes. Any time a person conducts a task there is a potential for mistakes or failures. Error traps, or constraints, are circumstances that increase the likelihood of making a mistake. For instance, a poorly written procedure may be interpreted differently by each operator and the steps carried out differently by each operator, leading to performing the task incorrectly.

An example of poorly written procedures is in the Chernobyl disaster. During the investigation it was discovered that the operating procedures were so ambiguous that there is still uncertainty on whether the operators did or did not comply [8].

Table 1. Examples of Error Traps

Human Factor	Error Trap
Production Tempo	Rushed, behind schedule, setbacks
Tool Location	Locked in tool locker, not easy to find
Procedure	Difficult to find, poor document control

When verifying administrative controls, we need to identify and eliminate the circumstances that increase the chance of mistakes. Because some administrative controls have both an engineering and human element, both sides need to be assessed for error traps. Even though the engineering function is mechanical, the maintenance and inspections are done by people. People also are responsible for document control of procedures, training records, and other tools of the job that can help complete the task. For the human response side, there are many aspects that can create opportunities for mistakes to occur:

- Poor training or no training documentation,
- Procedure has too many steps and cannot be completed before consequence,
- Procedure requires troubleshooting,
- Procedure cannot be found,
- Ambiguous procedures,
- Not enough time to complete tasks before consequence,
- The same operator involved in the initiating event is also part of the safeguard (no independence).
    - Example: If the initiating event is "Valved not closed (procedure not followed)," then an operator following that same procedure cannot be the safeguard. However, if walking the line to check valves is a separate procedure with a different person, then that procedure could be a candidate for IPL.

Error traps in the system are a little more difficult to identify. Typical error traps or constraints that can be associated with the engineering side could be:

- Gaps in Inspection, Testing and Preventative Maintenance Programs,
    - No ITPM for alarms,
    - Low staffing on Maintenance teams,
    - Infrequent Inspections,
- No or poor document management systems,
- Inadequate or missing standards used for design of equipment,
- Poorly designed displays with visual clutter,

- IPL Alarms do not have proper warning/urgency color,
- IPL Alarms are missing and do not sound or appear on HMI,
- Control system only, no safety system (no independence),
- Shares final element with other control loop or with HMI controls,
- Response requires operator placing oneself in danger to carry out the required action,
- Cause of initiating event is also the safeguard,
    - Example: Interlock failing cannot be guarded against by same interlock.

This is not an exhaustive list, but being able to identify and verify these constraints can help remove the opportunities for error in the operator's tasks. Looking at these aspects can improve consistency of response no matter which operator, shift, or time of day.

2.3 Learning from Normal Work

Many people imagine how work takes place: managers, equipment designers, quality, safety professionals. We imagine work happening in a sterile room under perfect conditions. Procedures are written with this perfect condition world, perhaps without considering the availability of tools or distractions that may interfere with completing a step. Therefore, in reality, work is often carried out quite differently than imagined. When work cannot be done as imagined, it often moves away from the safeguards that are planned around how it would happen in ideal settings.

Despite work not being performed as imagined or planned, work is successful most of the time. The successful work is praised for getting it done on time, even though corners may have been cut to meet managerial expectations. When things go wrong, the same person is now blamed, and sometimes punished, for doing the same task the exact same way. The only difference may be the presence of the hazard that day. Therefore, a worker may be praised and punished for doing the same task the same way. This sends mixed messages to workers and creates mistrust and less likely to speak-up about hazards and problems. We need to take the time to understand why work is successful most of the time, even though there are constraints to following the procedure. Learning from normal work is a proactive way to learn about how we can improve work and add create safety before incidents happen [9].

Marcin Nazaruk describes learning from normal work as "proactively identifying those things that make the work difficult and challenging and addressing them to reduce risk." [9]. It is a way to understand the constraints, or error traps, which keep people from performing work as imagined and planned. It is not trying to refocus workers on following the procedure, but rather gaining an understanding why the procedure is not working. Understanding how work is normally done allows management to update procedures, update safeguards, and get worker buy-in.

How do we go about learning from normal work? It starts with being curious. If we are curious about the work and how success is made, then we will have better worker engagement. We can create a better speak-up culture by care and concern and looking to fix the system rather than assign blame to the worker. When asking about normal work, the focus should be on "What makes the task difficult." [9]. Do we have enough people for a separate person to walk the line? Is there enough time to complete this task safely? Does equipment have a confusing interface?

Let's take the example of a start-up procedure that we want to use as an IPL:

- Have open conversations about what makes each step difficult to complete or difficult to complete on time.
- Ask what they worry about most regarding safety and why.
- Get operator input into where they think an incident is most likely to happen. If they were training a new person, what would they warn them about?
- Looking always at how to improve the system and not how to make workers follow the rules. The rules (procedure) might be what needs to be updated!

These conversations are where error traps and constraints are discovered. If you want to use this procedure as an IPL, you need to be able to identify, mitigate or eliminate the error traps.

2.3 Verification in Practice

Verification of administrative controls is not as simple or straightforward as verifying engineering controls. The human element is complex and sometimes unpredictable. More complexity is added with operator response to alarms because of the combined engineering and human parts to the equation. Using human factors principles can set the worker up for success and help improve the effectiveness of the Non-SIS IPLs.

Ensuring that your company standards cover ITPM for alarms, Alarm Management, and HMI design with Human Factors best practices will go a long way to ensuring that your administration controls can meet the criteria of Detect, Decide, and Deflect. Figure 5 below shows an example of verification criteria used for Operator Response to Alarm.

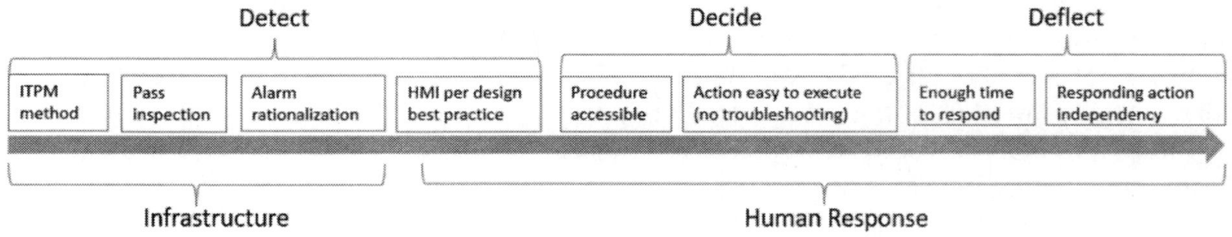

Figure 5. Verification Criteria of Alarm Response

This verification criteria covers the inspection and maintenance of the engineering assets and aspects for the human response. The alarm rationalization and HMI design standards should be based on ASM Consortium guidelines. The procedure should be easy to access and easy to execute. There should be enough time to complete the task and the responding action should be independent. Learning from normal work will help identify where the error traps may lie and expose your gaps in verification.

If the administrative control is following a critical procedure, the verification criteria would be similar in some respects. There would still need to be an accessible procedure and no troubleshooting. However, as seen in the example in Figure 6, the deflection portion would be more about the person who is following the procedure. To be considered a safeguard, the procedure should not put the operator in harm's way and should be a different person than in the initiating event.

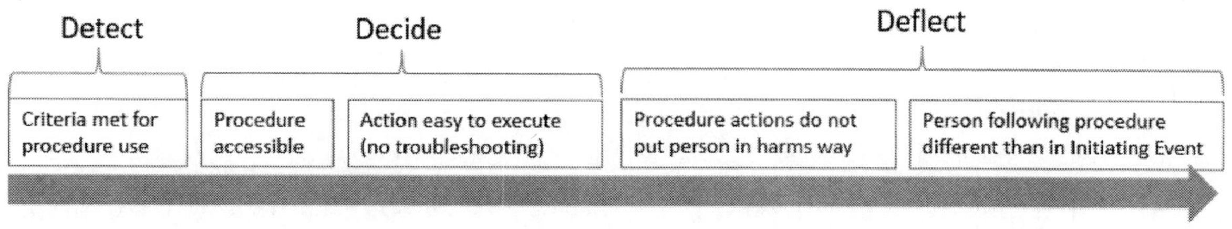

Figure 6. Verification Criteria of Critical Procedure

3 Conclusion

The main efforts to verify the effectiveness of administrative controls are to provide adequate situation awareness and identify and mitigate error traps. Taking time to learn from normal work, we can find the constraints that could downgrade the effectiveness of our administrative controls.

To reiterate from the introduction, engineering controls are always best. We also do not want to use the information in this paper to add more workload or stress to our operators by increasing the number of administrative controls. However, if you find that engineering controls are not practical, a robust verification of administrative controls can help ensure that we give the operator the right information at the right time to be able to detect, decide, and successfully deflect.

4 References

[1] Dowell, A. Layer of Protection Analysis: A New PHA Tool After Hazop, Before Fault Tree Analysis. *CCPS/AIChE International Conference and Workshop on Risk Analysis in Process Safety,* 1997.

[2] Dowell, A. Understanding IPL Boundaries. *Spring Meeting and 14th Global Congress on Process Safety Orlando*, 2018.

[3] Deker, S. *The field Guide to Human Error Investigation. 1st Edition*. Routledge; 2002.

[4] Mica R. Endsley, Betty Bolte, Debra G. Jones. Designing for Situation Awareness: An Approach to User-Centered Design. London and New York. Taylor and Francis. 2003.

[5] Jones, D. & Endsley, M. Sources of situation awareness errors in aviation. *Aviation, space, and environmental medicine*. 1996; **67**. 507-12.

[6] Endsley, M.R. Toward a theory of situation awareness in dynamic systems. *Human Factors*. 1995; **37 (1).** 32-64.

[7] Effective Console Operator HMI Design: Second Edition – Revised. *ASM Consortium Guidelines*. Editors P. Bullemer and Dal Bernon Reising. 2015; 165,170.

[8]*Chernobyl Accident - Appendix 1: Sequence of Events* https://www.world-nuclear.org/information-library/safety-and-security/safety-of-plants/appendices/chernobyl-accident-appendix-1-sequence-of-events.aspx updated June 2019.

[9] Nazaruk, M. Learning from Normal Work: How to Proactively Reduce Risk When Nothing Goes Wrong. *Professional Safety* .2023; **68 (11),** 14-21.

Ensuring Your Process Safeguards Stand Ready!
How to Make Sure You're Relying on the Right Safeguards in Process Hazards Analysis

Richard Carter, P.Eng., F.S. Eng. (TÜV Rheinland)
Watchmen Instrumented Safety Experts Ltd.
Calgary, AB, Canada
rcarter@watchmenise.com

Carsten Acker, P.L.(Eng.), F.S. Eng (TÜV Rheinland), CFSE, CSP
Watchmen Instrumented Safety Experts Ltd.
Calgary, AB, Canada
cacker@watchmenise.com

Keywords: Process Safety, Process Safety Management, PSM, Risk Analysis, Process Hazards Analysis, PHA, Hazard Identification and Risk Analysis, HIRA, Safeguard, Independent Layer of Protection, IPL, Conditional Modifier

Abstract

Process Hazards Analysis (PHA) is an essential part of an effective Process Safety Management system, but it is easy to end up with an inadequate analysis and a false sense of security. Identifying potentially hazardous events, and ensuring that there are adequate safeguards in place to prevent the unthinkable from happening, is a critical step in keeping people safe and preventing costly downtime and repairs. Unfortunately, however, these same safeguards contain traps that are easy to fall into in a PHA session, resulting in ineffective safeguards, lack of redundancy, and a much greater risk gap than intended. This can ultimately lead to catastrophic consequences.

Here are just a few examples of safeguarding traps commonly encountered within PHAs:

- Expecting superhuman perfection and speed from personnel when responding to alarms and hazardous situations
- Relying on personal gas monitors for rapidly-expanding hazardous gas clouds
- Pressure safety valves that are insufficiently sized for the scenario under review
- Over-reliance on instrumentation and controls, without adequate redundancy in the system
- Relying on non-critical procedures as a safeguard, or the failure of a procedure in question was the cause of the scenario under review
- Incorrect use of occupancy modifiers

This paper will identify potential pitfalls associated with the most common types of safeguards encountered in PHAs, and will discuss strategies for addressing these safeguards to ensure that the analysis is built on a strong foundation of effective, reliable safeguards.

Combined with adequate processes to ensure the continued reliability of these safeguards for the lifetime of the process, this approach will provide substantially greater safety in the design and operation of the facility. This will help to ensure the reliability and operability of the process, and help to prevent catastrophic incidents such as serious injuries and fatalities.

1 Introduction

Safeguards, also known as protection layers, are physical or administrative items that are included in a system in order to reduce the likelihood of occurrence of a certain hazardous event. Safeguards are a critical part of safe operation of any process. Every industrial system has the potential to exceed its safe operating limits, and safeguards are implemented in order to bring the system to a safe state if this happens.

However, there is often a misconception in Process Hazards Analysis (PHA) or Hazard Identification and Risk Analysis (HIRA), such as HAZOP or LOPA studies, regarding what constitutes an effective and reliable safeguard. This results in many such studies failing to properly identify the potential risk of a system, leading to operating the system with unidentified risk and therefore potentially allowing a hazardous event to occur more frequently than is tolerable.

This paper will discuss the characteristics and requirements of a safeguard in the context of PHA, and will address detailed considerations for the following common categories of safeguards that are relied upon in PHAs to keep people, the environment and equipment safe:

- Alarms and interlocks
- High-integrity interlocks (SIS and HIPPS)
- Pressure relief devices (PSVs, PVSVs, rupture discs)
- Operator rounds
- Fire and gas detection
- Occupancy factor modifiers (not considered a safeguard, but often utilized as such by PHA teams).

This paper will provide practical guidelines for each of these types of safeguards, and for safeguards as a whole, that can be used for any process or system to improve the quality of a PHA to better identify the true risk of the system. This will assist in making improved risk decisions that can reduce economic impacts, harm to environment, and save lives.

2 Identifying Effective and Dependable Safeguards

To ensure that the safeguards being relied upon in a PHA are effective and dependable, it is first important to discuss the required attributes of any safeguard. This section will discuss the requirements that every risk mitigation method must meet in order to be considered as a sufficiently reliable safeguard in a PHA.

2.1 Safeguards versus Normal Control

The first distinction to draw is the difference between "safeguards" and "normal controls". These are often conflated in PHA, which can cause confusion in a PHA and lead to a compromised safeguarding strategy.

Normal controls are those functions that are required to keep the system running in its normal operating state. These controls are usually frequently making adjustments to keep the process parameters within their intended operating limits. In many cases, they can be modified with little oversight, such as in the case of an operator changing the setpoint of a control loop or alarm. If there is a failure of one of these controls, it normally causes a process upset or puts the system in some potentially hazardous scenario situation, such as a level control valve malfunctioning closed and causing high level in a vessel.

Two examples of normal controls are shown in Figure 1. The left image shows a pneumatically-operated control valve, used to maintain the flow, level, pressure, or temperature of a process. The right image shows the accelerator and brake pedals of a car. Note that, as in the case of the car's pedals, it is not necessary for the control to be operating at all times to count as a normal control. If it is used frequently and is required to keep the system operating within its intended operating limit, it is considered a normal control.

Figure 1. Examples of normal controls: automated control valve, car accelerator and brake pedals (image credits: Rafał Rygielski via Wikimedia Commons, Richard Carter)

Safeguards, on the other hand, are not normally acting during normal operation. They are only required to act in the event that a process parameter is deviating outside of the intended operating limits. Because of this, the only times that a failure of a safeguard is identified is either during a proof test operation, or when the safeguard is required to act in a potentially unsafe situation.

Two examples of safeguards are shown in Figure 2. The pressure safety valve (PSV) shown in the left image is held closed by a spring, and is intended to open if the pressure in the system is dangerously high, but the rest of the time there is no indication of whether the PSV will work as intended unless it is tested. In the right image, a car airbag system is not intended to be in use except in the event of a collision, and in fact may cause a hazardous event if it activates during normal operation.

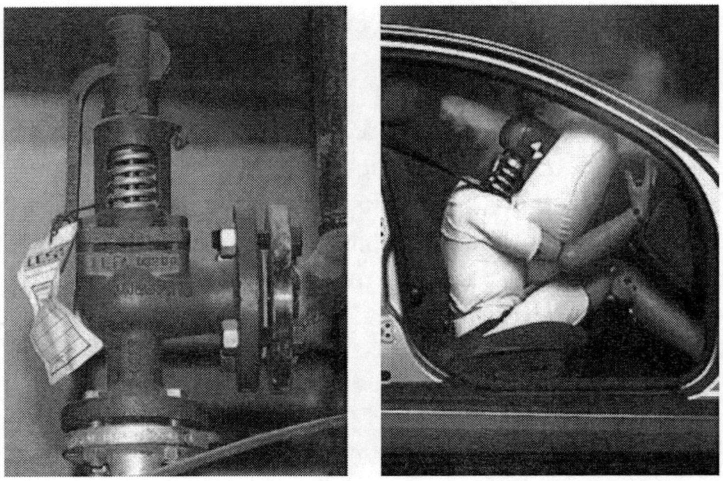

Figure 2. Examples of safeguards: pressure safety valve, car airbag (image credits: K. Krallis via Wikimedia Commons, Transport for NSW via Wikimedia Commons)

2.2 Factors Not Normally Considered PHA Safeguards

There are a number of factors which are not normally considered safeguards by themselves, but which support the normal controls and safeguard items [1]. These factors may affect the likelihood of a cause or the reliability of a safeguard, but by themselves they do not take action to prevent hazardous situations from occurring. They are normally assumed to be in place when describing a failure and a consequence, however any deficiency in these categories may adversely affect the likelihood of a hazardous event, or the efficacy of a safeguard, and therefore must be considered as part analysis of causes, consequences and safeguards.

- **Training and certification** – a certain level of training, certification and competency must be assumed for every individual that interacts with the process during design, construction, operation and decommissioning.
- **Design to codes and standards** – in most cases a baseline assumption can be made that the system is designed to applicable codes and standards. If any deficiencies or deviations are found, they must be addressed, but designing to codes and standards is not a safeguard by itself.

- **Operating and maintenance procedures** – As part of the difference between normal controls and safeguards, normal operating and maintenance procedures are assumed to be followed, unless it is known otherwise. A robust program for ensuring that these procedures are followed as intended can adjust the likelihood of a scenario caused by human factors, but is not a safeguard.
- **Normal testing and inspection** – Similar to operating and maintenance procedures, normal testing and inspection can influence the frequency or likelihood of scenario, or the reliability of a safeguard.
- **Preventive maintenance** – these are activities required to keep the equipment functioning as it is intended, and therefore will affect the likelihood that such equipment malfunctions. This may affect the likelihood of a scenario or the reliability of a safeguard.
- **Signage** – Signs are not considered safeguards, as they can easily be obscured, ignored, or misinterpreted. Furthermore, individuals will become familiar with signs they see often and will stop noticing them.
- **PPE (Personal Protective Equipment)** – The baseline set of PPE to be worn for a task should be considered in the potential hazard to personnel in a hazardous event. However, as the last line of defense against harm to a person, PPE is not intended to be relied upon to prevent injury. Furthermore, it is rarely highly reliable; for example, hazardous materials can enter the eyes around the edges of safety glasses, and serious injuries and fatalities from fires have been inflicted on workers wearing fire-retardant clothing.
- **Emergency response** – An emergency response plan (ERP) is required to be developed in most scenarios for safe operations, however as the last line of defense they are implemented once a significantly hazardous event is already underway and therefore it is not appropriate to consider it to be a safeguard. By the time an ERP is enacted, there is usually already significant damage or injury that has occurred, and the ERP is designed to limit the extent of this damage or injury and not prevent it outright.
- **Time** – The duration of the window between the inciting incident and the hazardous consequence may provide an opportunity for an automatic or manual response, but it does not prevent the event by itself and is therefore not a safeguard.

2.3 Safeguard Core Attributes

In order to be relied upon, each safeguard must meet certain criteria that are intended to confirm that the safeguard will be available and effective when required to act. The following four criteria are commonly used[1]:

- **Independent** of the cause and other safeguards.
- **Specific** to the hazardous scenario.
- **Dependable** to provide the risk reduction required and to take the system to a safe state.
- **Auditable** to verify that it is ready to act.

---

[1] Note that the CCPS book *Layer of Protection Analysis* [1] lists 3 criteria in section 6.3 and the CCPS book *Guidelines for Initiating Events and Independent Protection Layers in Layers of Protection Analysis* [2] lists seven core attributes in section 3. These sections are both used as reference for this paper.

These categories are explored further in this section.

2.3.1 Independent

One of the most commonly overlooked requirements of a safeguard is that it is independent of whatever may cause the scenario in which the safeguard is required to act, and independent of other safeguards that are intended to protect against the same scenario. In other words, there must be no common-cause event that would result in the scenario occurring and also prevent the safeguard acting, and no common-cause failure that would prevent the safeguard in question from acting correctly and also prevent any other safeguards acting correctly if they are intended to prevent the same hazardous scenario occurring.

Independence between cause and safeguard

If the event that causes the demand on the safeguard also prevents the safeguard from acting correctly, the safeguard is not adequately independent and therefore cannot be relied upon. For example, if a control function and an automated shutdown both take their input from the same sensing element then a malfunction of the sensing element, or error in the signal from the element, may cause an incorrect control action and, at the same time, prevent the safeguard from acting.

There are many industry incident examples of insufficient independence of a safeguard that has allowed a hazardous scenario to occur. As one example, the fatal liquid nitrogen release at the Foundation Food Group facility in Gainesville, GA was, in part, caused by insufficient independence between the control system measurement and the safeguard [3]. In this incident, a liquid-nitrogen bath freezer had a level control system that measured the level in the bath with a bubbler tube instrument. The high nitrogen level automatic shutdown also took its input reading from this instrument. At some point, the tube became bent such that the end of the tube (normally immersed in the liquid) was protruding above the maximum liquid nitrogen level height in the bath. This resulted in the control system detecting no level in the bath, which caused an overflow into the surrounding room and ultimately resulted in six fatalities. The high level shutdown action, which took its input from the same instrument, also did not activate as the instrument did not detect any level in the bath, and therefore the high level shutdown could not activate. The bent bubbler tube caused an incorrect control action and also prevented the safeguard from acting correctly.

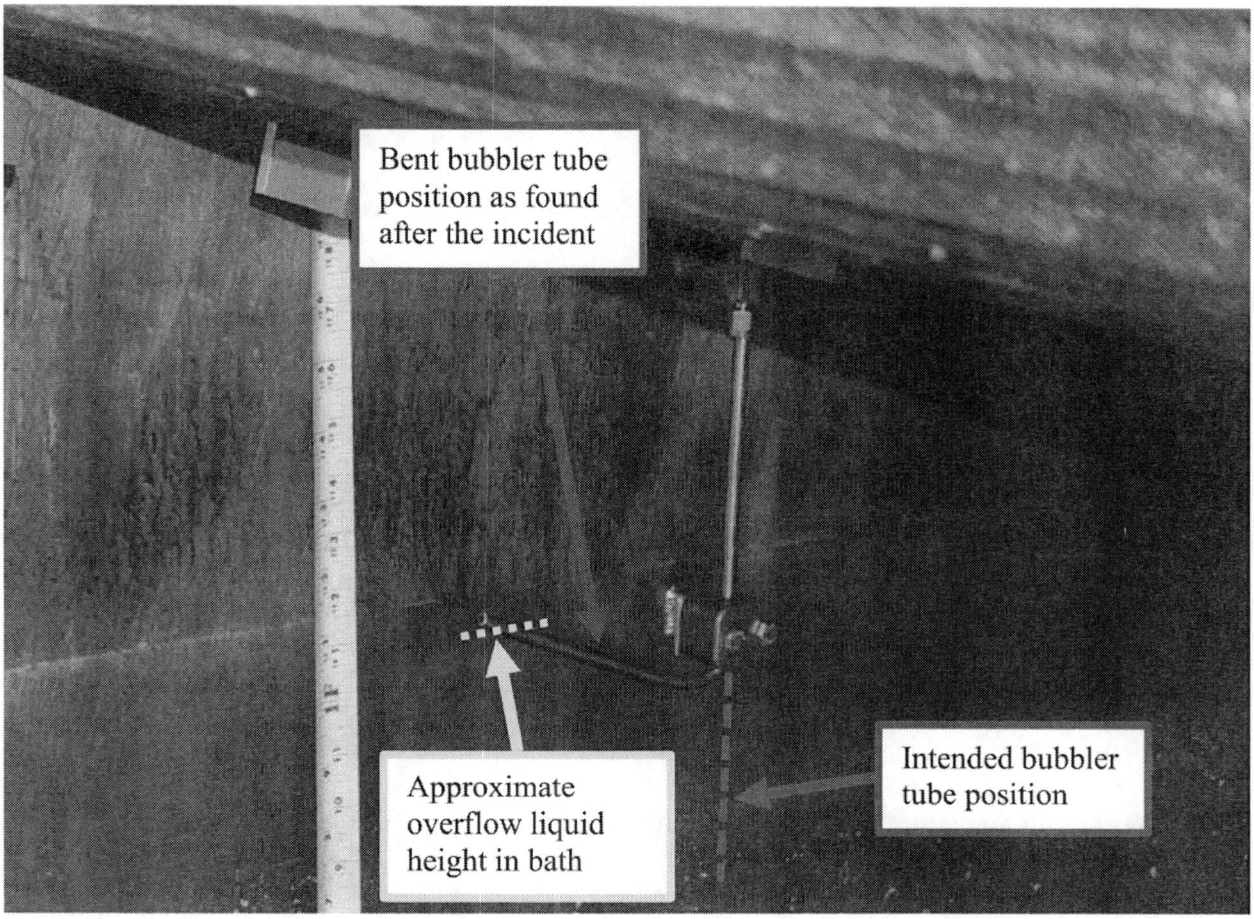

Figure 3. Bubbler tube instrument in Foundation Food Group incident (image credit: U.S. Chemical Safety and Hazard Investigation Board (CSB) [3]; annotations added)

Independence between safeguards

If there is a potential failure or event which could cause multiple safeguards to fail to act correctly for a scenario, only one of the safeguards can be considered to be adequately independent. For example, if there are two independent automated shutdown commands that are intended to prevent the same hazardous scenario, but they both take the same action such as closing the same emergency shutdown valve, there is likely to be insufficient independence because both actions rely on closing the same valve.[2] If the emergency shutdown valve fails to act as required, neither shutdown command will succeed. The same approach is used for alarms with human response; if a trained operator receives one alarm they may be able to adequately respond and prevent the hazardous scenario. If the same operator receives multiple alarms for the same event, however, they will not be able to respond to multiple alarms at the same time, and many alarms may even

[2] There are some cases where an analysis of the reliability of each component of a safeguard, and the overall reliability of the safeguard, is warranted to identify requirements for adequate independence to reach the desired risk reduction. However, for many processes it is sufficient for the operating company to sent standard minimum requirements for independence for safeguards instead of analyzing each on a case-by-case basis.

obscure the real process condition resulting in a higher likelihood of incorrect response from the operator.

2.3.2 Specific

A safeguard must be intended to prevent the hazardous scenario that is under review. This may seem obvious, however there are many cases where operators of a process must rely on indirect information to identify a potential hazardous scenario, or a safeguard is relied upon for a different reason than it was intended.

For example, in a distillation system the temperature and pressure of the process are closely interlinked, and therefore if there is a known issue with measurement of one of these variables the operators may rely on the other and infer the incorrect value. There are potential scenarios, however, when the process does not operate as expected and therefore this approach is insufficient, and therefore a high temperature action cannot be solely relied upon to identify a high pressure situation, and vice versa. When identifying safeguards in a PHA, it is important to ensure that safeguards are intended to identify and protect against the scenario that is in question.

2.3.3 Dependable

The safeguard must be dependable to perform its action and take the system to a safe state. This includes consideration that the safeguard will indeed take the system to a safe state when it functions correctly, and that it is sufficiently reliable that it can be reasonably expected to perform correctly.[3]

If the safeguard is later found to be less reliable or effective that it was previously expected to be, the relevant portion(s) of the PHA must be revalidated to ensure that adequate safeguarding is provided, given the reduced efficacy of the safeguard in question.

The Dependable criterion may be defeated in the event that either the safeguard will not take the system to a safe state, or the safeguard is not reliable enough for the demands of the scenario in question.

It is also important to consider whether the safeguard, performing its function as intended, will introduce any new hazards that would not otherwise be present. If a safeguard activating may cause another hazardous scenario, it is necessary to either ensure there is adequate protection against that scenario or, preferably, alter or replace the safeguard such that the system is taken to a safe state by the safeguard and no new hazards are created.

An example of a safeguard that introduces a new hazard is a pressure/vacuum relief valve (PVRV), also known as a pressure/vacuum safety valve, on an atmospheric-pressure storage tank that contains flammable gas. If the pressure in the tank falls below the ambient air pressure there is the potential for the tank to implode due to the vacuum created, and the PVRV prevents this vacuum by allowing atmospheric air to enter the tank. However, this causes a new hazard by introducing

[3] Note that the required reliability is dependent on the requirements of the process, the potential severity of a hazardous event if it fails, and the frequency that it is called upon to act.

oxygen to a flammable hydrocarbon gas environment, with the resultant potential for fire or explosion in the system. This new hazard must be also controlled for.

Finally, the potential for personnel working with the system to bypass or disable the safeguard must be considered. For example, an operator might disable an alarm because it activates often when not required to, creating an unnecessary distraction. While this may be a good decision for safe operation of the facility, it then means that the alarm cannot be considered a dependable safeguard in a PHA. If necessary controls must be put in place to prevent safeguards being bypassed or disabled, if they are considered necessary to control the risk, but also care must be taken to avoid introducing alarms and shutdowns that create nuisance interruptions for operators.

In qualitative assessments such as Hazard and Operability (HAZOP) studies, it is generally considered that a minimum threshold of dependability of 90% must be reached in order to consider the safeguard dependable. This means that the safeguard will take the system to a safe state at least nine out of ten times that it is required to act.

2.3.4 Auditable

All equipment and processes, whether equipment or administrative processes, degrade over time and require maintenance and investment to maintain an adequate reliability and functionality. A key part of this maintenance is auditing the safeguard to verify that, if it is required to act, it will perform its function effectively and with adequate reliability. This may include proof testing, calibration, verification, periodic replacement, and other checks or activities to provide assurance that the safeguard's reliability and function are maintained to an adequate level. The auditing process proves that the other safeguard criteria are being maintained. This process must be documented and verified to ensure it is meeting the requirements of the process in question.

3 Best Practices for Common Safeguard (and Modifier) Categories

This section will address some of the most common safeguard categories encountered in PHAs of all types, and will explore strategies for ensuring that the safeguards being relied on are providing the risk reduction assumed and expected.

3.1 Alarms and Interlocks

A lot of reliance is placed on alarms and interlocks in many PHAs. These safeguards are intended to detect when a process variable is exceeding its safe operating limits and either warn the operator so they can take action in the case of an alarm, or perform an automated action to prevent harm or damage occurring in the case of an interlock.

Image credit: Alexey Soucho via Unsplash

These safeguards must meet the below criteria, in addition to the standard safeguard criteria discussed in section 2.

Detection and Action

Regardless of the situation, every alarm and interlock needs to have a detection and an action component. The safeguard will not take the system to a safe state if the hazardous event is not detected. Similarly, if there is no action to be taken, or if the action is ineffective at preventing the scenario, then the alarm or interlock is not effective.

Independence

As will all other safeguards, these functions need to be adequately independent of the cause and of the other safeguards. With alarms and interlocks, however, it can be challenging to identify when there is inadequate independence. As discussed in section 2.3.1 with the Foundation Food Group incident, the bubbler tube instrument was a common point of failure between the liquid nitrogen bath level control and overflow shutoff interlock.

The level of independence required for a safeguard to be considered independent will depend on the specific scenario, control system architecture, and operating company standards and philosophy, however in most cases common-cause points of failure must be avoided. This means that for two interlocks to be considered safeguards for the same scenario, they must not share any components, or share them with the cause of the scenario, unless there is a sufficiently low probability of failure of the shared component that the safeguard is still considered to provide adequate reliability.

Access Control

The ability to modify the equipment, programming and setpoints of each alarm and interlock needs to be controlled as appropriate for that function, to reduce the likelihood of a person making a modification that (intentionally or unintentionally) prevents it from performing its required function with the reliability needed. For example, it is often desirable for the operator to be able to set the alarm setpoint to a value that they deem to be the best for their needs and the current process operation. However, there is a risk that the operator inadvertently sets the alarm at a value that will not provide adequate warning and therefore will not allow them time to respond to a hazardous event. Sometimes, alarms will activate regularly even if the system is performing normally, and this can lead operators to bypass alarms because they are distracting and not providing any useful information, known as a spurious alarm. If an alarm deemed to be safety-critical alarm is becoming a nuisance, a Management of Change (MOC) process must be followed to either fix the problem with the alarm, or review the safeguarding strategy for any associated hazardous events to ensure that adequate safeguarding is being provided if the alarm is removed.

Correct Setpoint

The setpoint of an alarm or interlock must consider the design and operation of the process. For example, a high temperature interlock must be set lower than the design temperature of the affected equipment in order for the safeguard to be adequate. Include consideration any other factors that may influence the required setpoint, such as the interaction of materials in the process, downstream or upstream systems, and any connected systems or utilities that may need to be protected by the safeguard.

In an overpressure protection system, it is common to have a high pressure shutdown action with a pressure relief device as a backup in case the shutdown fails to take the system to a safe state. In these cases, the setpoints of the shutdown and the relief device must be set such that the shutdown should always occur before the relief device is required to act. If the shutdown will not prevent the overpressure without the relief device acting, then the relief device is the sole safeguard against overpressure in that scenario. Many organizations will set high pressure shutdown setpoints to arbitrary, standardized value, such as 90% of the PSV setpoint. This standardized approach, however, does not consider the specifics of the scenario in question, and therefore each case should consider the system dynamics and process safety time available between the high pressure shutdown being initiated and the stop in the rise in pressure, to ensure that it will prevent the pressure from reaching the PSV setpoint.

Adequate Process Safety Time

The setpoint for the alarm or interlock must be set at a value that will provide adequate process safety time for the operator to respond and take action, or the interlock to activate and take action, such that the system will be protected. If there is insufficient time for a human or automated action, the incident will occur before it can be prevented, and therefore the action does not count as a safeguard.

In the case of an interlock it can usually be determined how quickly a certain system needs to act to prevent a hazardous scenario – although it is important to bear in mind that a quick activation of one interlock may cause an additional hazard, such as how an emergency shutdown valve on a liquid pipeline closing quickly may create a transient overpressure hazard.

For an alarm, the determination of "adequate" process safety time must consider:

- Staffing levels and operator availability;
- Time required for the operator to transfer their focus from any other activities they are working on (or potentially respond off-shift in the case of a remote response);
- Time required to identify the hazardous scenario that is occurring, and successfully determine the best course of action;
- Time required to reach the physical location, if an in-person response is required (consider the worst credible case, such as an off-shift response to a remote location in bad weather), and;
- Time required to implement the action from a remote control interface, or in-person.

3.1.1 Alarm-Specific Considerations

In addition to the considerations above, the following factors must be addressed for each alarm safeguard.

Operator Identification of Issue and Response

There must be a reasonable expectation that the operator will correctly identify what is going wrong in the process, and that the operator will correctly identify the action that must be taken to prevent an incident. This is another consideration in the process safety time allowed for in the safeguard design – the shorter the timeline that the operator has for response, the greater the stress placed on the operator, and therefore the greater the chance that the operator makes a mistake.

There must also be an action that the operator can take, otherwise the alarm is a notification only and not a safeguard. This action must not place personnel at additional risk, which was a contributing factor to the fatalities and injuries that resulted from the KMCO Chemical Facility in Crosby, Texas in 2019 [4]. In this incident, an release of flammable isobutane vapor could not be remotely isolated and therefore operators entered the hazardous area wearing SCBA (self-contained breathing apparatus) to control the situation, which contributed to the one fatality and two serious burn injuries that occurred when the isobutane cloud ignited.

Avoid Alarm Flooding

"**Alarm flooding**" is a common concern in large operating facilities, in which operators regularly receive too many alarms to adequately respond to each of them. In these cases, the important alarms that require operator response are difficult or impossible for the operator to identify among the frequent activation of many other alarms that do not require any action. The level of alarm flooding in the system in question will affect the maximum reliability that can be considered for any alarm safeguard in that system.

An alarm rationalization study should be conducted to ensure that only alarms that provide important information are relayed to operations, and to ensure each alarm is assigned an appropriate level of criticality.

Response May Exacerbate the Problem

Whenever relying on human analysis and response, it is important to remember that humans are fallible. It is possible for even very well-trained and highly-experienced individuals to make a mistake, make a decision with incomplete information, make an incorrect assumption, or come to an incorrect conclusion during problem-solving. The likelihood of a mistake increases if the individual is under stress, is inexperienced or inadequately trained, or does not have access to required information. In some cases, human response may even make a problem worse.

An incident that highlights the potential for operators to incorrectly respond to a scenario due to inadequate training is the ethylene fire at the Kuraray America ethylene and vinyl alcohol facility in Pasadena, Texas in May 2018. An ethylene release resulted in a large fire, during which 23 workers were injured, some extremely seriously. In its investigation, the CSB found that a contributing factor to the event was the lack of operator knowledge and training on how to respond to process alarms. The CSB concluded that with the correct procedures and training, the operators may have been able to avert the incident [5].

3.1.2 Interlock-Specific Considerations

In addition to the factors listed above for alarms and interlocks, it is also necessary to consider the specific requirements of an interlock.

Takes the System to a Safe State

In order to be considered a safeguard, the interlock must take the system to a safe state by itself. If the interlock only increases the amount of time before the hazardous event occurs then it is not, by itself, a safeguard. It may, however provide the time required for another action to bring the system to a safe state, in which case both the initial interlock and the follow-up action together must be considered as one combined safeguard.

If the interlock creates a new hazardous event then the new hazardous event must be protected against also.

Image credit: Bizzy Whizz via Wikimedia Commons

3.2 Higher-Integrity Interlocks: SIS and HIPPS

Two versions of interlocks that are designed to provide enhanced reliability are Safety Instrumented Functions (SIF) within Safety Instrumented Systems (SIS) and High Integrity Pressure Protection Systems (HIPPS). Each of these types of interlocks must meet all the requirements of an interlock from Section 3.1, but in addition must also meet the following requirements.

Design to the Required Reliability

Higher-reliability interlocks usually have a specific reliability requirement represented by Probability of Failure on Demand (PFD), which shows the probability that the safeguard will fail to adequately perform its function when it is demanded to do so. This reliability can also be represented using Safety Integrity Level (SIL), which is an order-of-magnitude representation of the reliability. In order to meet the required reliability, the entire interlock must be assessed as a whole in order to wholistically determine the PFD or SIL of the entire interlock from first detection to final action. This calculation must include all components that take part in the interlock, as well as how they work together.

Additional Design, Installation, Testing and Maintenance Requirements

Higher-reliability interlocks also have additional requirements in all phases of the lifecycle of the interlock, including design, installation, testing and maintenance [6].

A common issue with these interlocks is that over time, the reason behind the increased testing and maintenance requirements as compared with other interlocks is forgotten, or the documentation detailing these requirements is lost. If these interlocks are no longer tested and maintained to the standard required, they may not provide the reliability that was expected and therefore the system may be operating at unrecognized increased risk as compared to the intended risk.

Specialized Personnel Required to Install and Maintain

In many cases, personnel with specialized training and experience in functional safety are required to install, commission, and maintain these higher-reliability interlocks. This, in addition to the other requirements above, prevents higher-reliability interlocks from simply being an easy alternative to an inherently safer design. While higher-reliability interlocks are the best solution in some cases, it is sometimes more effective to look for other design solutions that can reduce the inherent risk of the process.

3.3 Pressure Relief Devices: PSVs, PVSVs, and Rupture Discs

Pressure relief devices such as Pressure Safety Valves (PSVs), Pressure/Vacuum Safety Valves (PVSVs) and Rupture Discs (RDs) are all types of mechanical device that are designed to prevent excessive pressure, or in some cases vacuum, from causing damage to equipment and loss of containment.

In addition to the criteria in section 2, these safeguards must meet all the following considerations.

Sizing Basis

Each instance of potential overpressure or vacuum will result in a maximum required capacity of the relief device, based on how much gas or liquid must be allowed into or out of the system in order to prevent damage. In order to be considered a safeguard for a given scenario, the pressure relief device must have sufficient capacity for that specific scenario. For example, a vessel with a PSV installed that is designed for fire case may not be protected against excessive flow from an upstream source, unless the PSV sizing calculations specifically include this scenario as a consideration.

Image credit: FastLoe via Wikimedia Commons

Correct Setpoint

The setpoint of the relief device must consider the design pressure of all equipment and piping that the device protects, and any surplus pressure that the equipment will be exposed to at full relief. For example, most PSVs are designed such that they begin to open at the setpoint pressure, and are fully open when the pressure is 10% above the setpoint. This means that the connected system will be exposed to 110% of the setpoint pressure, which may exceed the safe limits of the system.

Testing and Maintenance

As with all safeguards, pressure relief devices must be tested and maintained to ensure that they can be relied upon to provide the required reliability when called upon to act. In addition, it is necessary for any components that are considered in the sizing basis calculations for the device to be maintained adequately also. For example, if a restriction orifice or check valve is considered in the sizing basis calculations for the relief device, then the restriction orifice is a critical part of ensuring that the relief device has adequate capacity to protect the system, and the orifice must therefore be maintained adequately based on that criticality.

Image credit: Continental Disc Corporation

Potential Failure Mechanisms

Depending on the process characteristics, a pressure relief device may be susceptible to plugging, fouling, corrosion, freezing, or other mechanisms that can prevent the safeguard from performing its function correctly. This may reduce the amount of reliability that can be placed on the safeguard, and in some cases may prevent it from being considered a safeguard at all. Many organizations do not have specific criteria for what is considered “clean service” for a pressure relief device. The operating company must therefore confirm that the device is, or will be, clean and free of these failure mechanisms, or else take less credit for the device as a safeguard.

If a pressure relief device that was previously expected not to be affected by these mechanisms is later found to have been compromised by them, the relevant hazardous scenarios must be reviewed to ensure that adequate safeguarding is provided given the reduced reliability of the relief device.

Safe Relief Location

Each pressure relief device must direct gas or liquid away from the system that is experiencing overpressure, towards an alternate location. In some cases this destination location is a flare or vent system, in which case the system must be designed to accept the maximum flowrate that it can be expected to receive in order to be considered a safe location. If a vent is directed towards the atmosphere, it must be directed such that it does not pose a hazard to personnel, equipment or the environment from kinetic energy or toxic effects, and there are no credible concerns of ignition of flammable materials.

In the case of the previously-discussed 2018 Kuraray America fire, an overpressure scenario was relieved by an emergency pressure-relief system. This system directed the ethylene through vent piping that was horizontally directed toward an area where personnel were performing welding and other maintenance. The CSB concluded that, if the outlet piping had been directed to a safe location such as vertically upward, the impact of the scenario would likely have been limited [5].

3.4 Operator Rounds

Most facilities and operations rely to some extent on trained personnel regularly moving through the operating space, checking equipment and using their senses to detect any signs that the equipment and process is not operating as intended. These “operator rounds” activities are an important part of the safe operation of these facilities, however extreme care must be taken if a PHA team is considering relying on them as a safeguard.

Image credit: Khepster via Wikimedia Commons

Many operating companies consider operator rounds to be part of basic operations, similar to standard operating and maintenance procedures, and take a conservative approach by not allowing rounds to be considered a safeguard.

Rounds are also not appropriate to use as a safeguard against large-scale events, as they are not specific to the scenario in question. Potentially catastrophic impacts warrant safeguards that are designed specifically to prevent those scenarios.

Operator rounds are often not able to be considered a safeguard by themselves for the reasons below, although they may sometimes affect the likelihood of a cause or the reliability of another safeguard.

Scenario Must be Identifiable

In order for an operator to identify a potential hazardous situation during rounds, there must be something that the operator can detect to identify such a situation. For example, if there is no temperature measurement device installed, or there is no display for the operator to read, then the operator cannot identify if there is an abnormally high or low temperature during rounds. If it is unlikely that the operator would detect the evolving concern, then operator rounds are not appropriate for that scenario.

Adequate Time Considering Rounds Frequency

There must be adequate process safety time such that it is highly likely that the operator will conduct rounds between the moment that the problem becomes detectable and the moment that the hazardous impact occurs. In many cases it is necessary to allow for two or more sets of rounds in case one set of rounds is delayed or skipped due to other pressure issues, or in case a warning sign is overlooked the first time.

Scenario Must be Preventable on Discovery

If the operator detects the hazardous situation, they must then be able to take some action to prevent the hazardous impact. If the operator discovers that a toxic material has leaked from a storage tank into an adjacent river, it is too late by then to prevent the environmental impact and therefore the rounds would not count as a safeguard.

Must be Regularly Checked

Even a small processing facility includes many hundreds or thousands of potential sources of information, including instrument readouts, equipment sounds and vibration, and smells. In order to support considering rounds as a safeguard, there must be some assurance that the operator will routinely check the source of information that the PHA team is relying on. In many facilities this takes the form of paper or electronic "rounds sheets" in which the operator is required to record observations from the system or instruments to ensure that they are being checked.

3.5 Fire and Gas Detection

Image credit: Richard Carter

Fire and gas detection devices are intended to identify a hazardous situation that is already unfolding, and provide an opportunity to limit the amount of harm and damage that results from the event.

As with operator rounds, although these devices are an essential part of many systems that deal with hazardous materials, they are often not valid safeguards within preventative risk assessments due to their potential for failure and due to the fact that they provide mitigative protection only; they do not prevent an initial loss of containment and instead only provide an opportunity to try to avoid further harm.

Likelihood of Detection

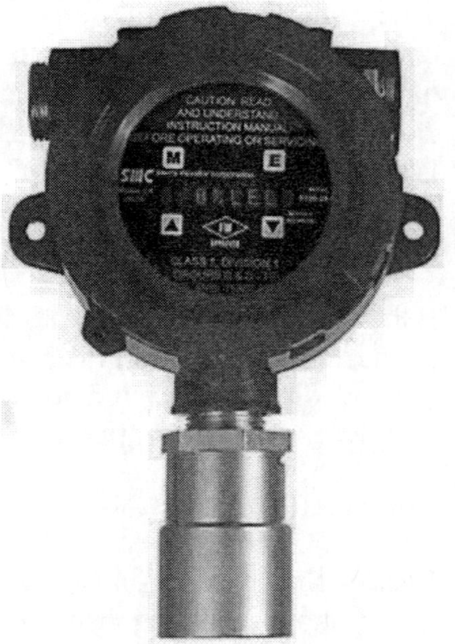

Image credit: MSA Safety

When considering fire and gas detection as a potential safeguard, the PHA team needs to consider the likelihood that the detector will identify the gas release or fire. The team must consider the type and location of detection devices, the limitations of the area in which gas or fire could be identified, and the dispersion and propagation characteristics of the gas cloud or fire. Without a detector coverage assessment, gas detection may not identify the release, and fire detectors may not provide adequate warning to prevent the scenario in question.

Must Reliably Alert Personnel

If part of the required response on detection is that personnel evacuate the area, or personnel are prevented from entering the hazardous area, there must be a reliable method of alerting personnel to the hazardous situation. This may include audible, visual, or other alarms. The efficacy of these alarms must be addressed considering the characteristics of the process; for example, some equipment is so loud that an audible alarm cannot be heard due to lack of decibel differential with background noise, and some flashing lights can be obscured or the bulb may be inoperable.

Must be Adequate Time and Route for Evacuation

If a hazardous situation is detected, there must be sufficient time for personnel to evacuate the area before the hazardous impact occurs, and an effective evacuation route must be available. In some cases the situation develops too quickly for personnel to reliably evacuate before they are impacted by the incident. For example, a flammable gas cloud may expand too quickly for personnel to evacuate, and therefore those people may still be harmed if and when the cloud finds an ignition source.

Note on Personal Gas Monitors

In many facilities where hazardous atmospheres may be present, personnel are required to carry monitoring devices that can warn them of a hazardous atmosphere. It is important to consider that these monitors are worn on the person's body, and therefore the person is exposed to hazardous atmosphere at the same time as the monitor. In many cases, this does not provide the individual with adequate time to evacuate and therefore in most cases these should not be considered safeguards in and of themselves.

Note on Fire Detection Defeat

There are several ways that fire detection can be defeated and fail to operate as intended. Historically, there have been issues with other heat sources such as hot work (e.g., welding), or even the sun reflecting off surfaces, causing spurious activation of these devices, which has sometimes led to operators bypassing the devices to prevent spurious shutdown of the system. This can mean that the detectors are not in service at the time a fire forms. This "normalization of deviance" can lead to operations bypassing these whenever work is being done on-site, leaving those personnel unprotected exactly when the highest risk activities are taking place

Furthermore, Watchmen personnel have found many incidents of fire detection devices that are obscured by temporary storage of equipment, or new construction that blocks the line of sight necessary for the device to identify a fire. In these cases the detector likely was working as intended, but would not be able to detect a fire.

3.6 Occupancy Factor Modifiers

Although they are not considered safeguards, a modifier of "personnel in the area less than X% of the time" or similar statement (known as an occupancy factor) is sometimes used in a PHA to represent a reduced likelihood that a person would be harmed by a hazardous event, should it occur. However extreme care is required if the PHA team wishes to consider using an occupancy modifier in a PHA, and the following points must be carefully considered.

Remember that an occupancy factor is relying on "luck," saying that if a hazardous event happens, the team is hoping that no one will be there to be hurt.

Consider All Personnel

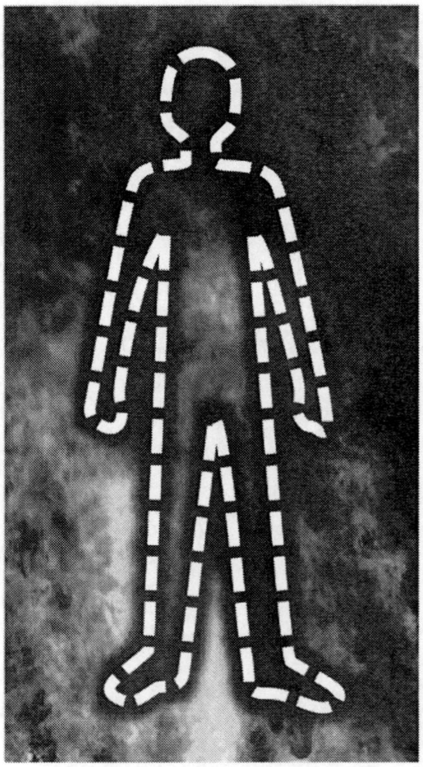

Fire image credit: LoggaWiggler via Pixabay

The occupancy factor must include consideration of **all** personnel who may be present in the hazardous area at any given time. This includes operators, maintenance personnel, contractors, construction personnel, technical personnel, site tours, nearby members of the public, and any other people who might be in the area at some time.

For example, in one PHA facilitated by the author for a remote facility, an occupancy factor was being considered since there were no people anywhere near the process normally, other than the operator visiting the remote site once or twice a week for an hour or less at a time. However, it was discovered that there would be construction taking place in the area nearly continuously for two years after this facility was started up, and therefore the occupancy modifier was not valid as there would be personnel in that area often for those two years.

Restricted Areas and Access Controls

Consider whether access controls to certain areas are necessary in order to ensure that the occupancy in that area is going to indeed be as low as the team is relying on. This might include access controls to fenced-off areas or administrative controls to ensure that people only enter the area if they are required to do so. It might also include active systems, such as the access control systems to the enclosures around many turbines, which shut down the turbine if the door is opened to prevent people being inside the turbine enclosure while the turbine is operating.

Note that in these scenarios, the access controls should be considered a safeguard, rather than listing an occupancy factor.

Scenario Caused by a Person in the Area

If the hazardous event is caused by a person being present, such as inadvertent operation of a manual valve, an occupancy factor is not valid because the scenario will only occur with someone present in the area.

Note that often the causation of a hazard is generalized and non-specific. In many cases this is due to human intervention. For example, an automated valve may move to the limit of travel because someone is conducting maintenance without proper maintenance bypasses in place. People might also be diagnosing a failure which hasn't manifested into a large incident yet.

During startup operations, or changes in operating modes, it is common to have additional personnel onsite, and so occupancy may be greater that the average during normal operations. Although these modes of operation only occur during a small percentage of the time in many processes, it is still a relatively frequent occurrence and so should be considered as such.

Operator Drawn to Area During Incident

When operations receives an alarm of a problem with a system, or a potential problem is identified, the first response is often to send an operator to the area to investigate the issue. This puts the operator directly in the hazardous area if a hazardous event occurs. This alone is a reason that, for many scenarios, occupancy factors are not reliable.

This is a frequently-overlooked aspect of these modifier considerations. However, three catastrophic incidents in different industries show this potential for personnel to be drawn to a hazardous situation:

- A worker received a remote alarm from the Aghorn Operating Waterflood Station in Odessa TX in 2019 and responded by going to the facility, where he was fatally injured by a toxic hydrogen sulfide (H_2S) release. It is not known whether the release occurred before the worker arrived, or while he was already in the incident area. What is known, however, is that after several hours his spouse, concerned that she had not heard from him, drove to the facility to find him and was also fatally injured by the same toxic atmosphere [7]. In this scenario, a latent hazardous atmosphere resulted in either one or two fatalities, including that of a member of the public.

- During a fatal incident at the AB Specialty Silicones facility in Waukegan, IL in 2019, an operator witnessed an unexpected fog formation inside, and overflow of, a reactor. The operator shouted for help, which drew two additional people to the hazardous area prior to the explosion [8].

- When a leak was discovered at the Chevron Richmond refinery in 2012, many personnel congregated in the area of the leak to attempt to determine the cause and solution. Although there were luckily no fatalities in this particular incident, there were many people in the area including operators, supervisors, technical staff and firefighters, when the pipe suddenly catastrophically failed and soon caught fire [9].

Latent Impacts

A latent impact is one that does not occur immediately, but only happens after a period of time. This could be, for example, a nitrogen release inside a building; although it may be a low likelihood that someone is there when the release first occurs, the nitrogen gas can fill the building and then the hazardous atmosphere is present indefinitely, potentially resulting in a fatality by asphyxiation when someone inevitably enters the building later.

If latent impact is a possibility, occupancy modifiers are not valid because someone will eventually enter the hazardous area and potentially be harmed. Any methods of detection of such a situation, such as low oxygen detection alerting personnel not to enter the area, would be considered the safeguard against this situation instead.

An example of a latent impact is the incident previously discussed involving nitrogen release and personnel asphyxiation at the Foundation Food Group facility in Gainesville, GA. In that incident, there were fatal injuries to personnel close to the time of the release, but further fatal injuries and serious injuries occurred at a later time as a result of people later entering the hazardous atmosphere [3]. In this scenario, the occupancy in the hazardous area at the time of the incident did not dictate the full extent of the catastrophic impact, because further personnel later entered the hazardous area.

Use Sparingly and Carefully

Remember, if the PHA team is relying on an occupancy modifier, they are relying on hoping that everyone gets lucky and no one is in the area when the hazardous situation is present. Occupancy factors are also more likely than most other types of safeguards to changing, for example due to staffing changes, operating mode changes, or facility configuration changes. If the occupancy in an area changes, any PHA scenarios which rely on occupancy modifiers for that area must be revalidated to determine if adequate risk mitigation is still in place.

Also remember that saying "personnel are in the area less than 10% of the time" is claiming that the PHA team is relying on people being there for just as much risk reduction as an alarm with operator response or an instrumented shutdown. If the occupancy modifier is "less than 1% of the time," the team is relying on that modifier just as much as they would be relying on a pressure safety valve or SIL 2 Safety Instrumented Function, which are both designed to be high-reliability safeguards.

4 Conclusions

This paper has discussed many specific items that must be considered in a Process Hazards Analysis when considering whether to rely on any given safeguard to prevent a hazardous event occurring. Although there are many considerations for each specific type of safeguard, the following questions can be used as a guide to consider the suitability and reliability of any type of safeguard:

- Is this safeguard designed to prevent this scenario?
- Will this safeguard take the system to a safe state in this scenario, without needing other safeguards to act also?
- Does it act independently of the cause and the other safeguards being relied upon to prevent this scenario?
- Is the safeguard well-maintained and expected to be available?
- Can you verify that it's working?
- Can you stand behind your justification for each of your safeguards and modifiers?

Remember, be honest with yourself and with your team about the realities of the system, facility, team and organization. For example, can you be sure that the area that was considered to have low occupancy will not become a higher traffic area in the future? If the team is discussing a potential critical operating procedure as a new safeguard but it is known that all operating procedures are generally not followed, a critical operating procedure will not be a valid safeguard because in reality it might not be followed to the extent it needs to be to provide that risk reduction.

By following the above questions, and the more detailed consideration for each safeguard type presented in the paper, your PHA will provide a more accurate assessment of the risk of your process. This will help with making effective risk decisions that can reduce equipment damage, minimize costly downtime, avoid environmental impact, and prevent harm to personnel.

5 References

[1] Center for Chemical Process Safety. *Layer of Protection Analysis: Simplified Process Risk Assessment*. American Institute of Chemical Engineers, 2001.

[2] Center for Chemical Process Safety. *Guidelines for Initiating Events and Independent Protection Layers in Layers of Protection Analysis.* John Wiley & Sons, 2015.

[3] US Chemical Safety and Hazard Investigation Board. *Fatal Liquid Nitrogen Release at Foundation Food Group: Investigation Report.* https://www.csb.gov/file.aspx?DocumentId=6268. December 2023. Accessed December 2023.

[4] US Chemical Safety and Hazard Investigation Board. *Fatal Equipment Rupture, Explosion, and Fire at the KMCO Chemical Facility: Investigation Report.* https://www.csb.gov/file.aspx?DocumentId=6265. December 2023. Accessed December 2023.

[5] US Chemical Safety and Hazard Investigation Board. Ethylene Release and Fire at Kuraray America, Inc. EVAL Plant: Investigation Report. https://www.csb.gov/file.aspx?DocumentId=6204. December 2022. Accessed February 2023.

[6] International Electrotechnical Commission. *IEC 61511-1 Functional Safety – Safety instrumented systems for the process industry sector – Part 1: Framework, definitions, system, hardware and application programming requirements.* Edition 2.1, August 2017.

[7] US Chemical Safety and Hazard Investigation Board. *Hydrogen Sulfide Release at Aghorn Operating Waterflood Station.* https://www.csb.gov/file.aspx?DocumentId=6155. May 2021. Accessed January 2024.

[8] US Chemical Safety and Hazard Investigation Board. *Chemical Reaction, Hydrogen Release, Explosion, and Fire at AB Specialty Silicones: Investigation Report.* https://www.csb.gov/file.aspx?DocumentId=6220. September 24, 2021. Accessed January 2024.

[9] US Chemical Safety and Hazard Investigation Board. *Chevron Richmond Refinery Pipe Rupture and Fire: Final Investigation Report.* https://www.csb.gov/file.aspx?DocumentId=5917. August 6, 2012. Accessed January 2024.

The Urgency of Risk-Based Facility Siting of CO_2 Pipelines

Russell Ogle
Exponent, Inc.
Warrenville, IL 60555
Russellogle12@gmail.com

Brenton Cox, Nicholas Schulman, Joanna Schneider
Exponent, Inc.

Keywords: carbon dioxide, pipeline failure, facility siting, dense gas dispersion

Abstract

The U.S. government has set two ambitious sustainability goals for the United States: carbon-free electricity generation by 2035 and net-zero carbon emissions by 2050. Other nations have set their own ambitious goals. Carbon capture with geological storage is an essential strategy for meeting these goals. However, carbon emitters are often not co-located with appropriate geological formations. Thus, carbon dioxide must be transported by pipelines from its point of generation to a suitable geological destination.

Pipeline transport of carbon dioxide is a demonstrated technology with lower failure rates and injury statistics than natural gas or hazardous liquid pipelines [1]. However, pipeline construction plans have been met with fierce public opposition. The release of carbon dioxide on February 22, 2020, in Satartia, Mississippi has become the focal point for many climate activists and organizations opposed to carbon dioxide pipelines.

The Department of Transportation (DOT) regulates pipeline design, operation, and maintenance in the US. However, DOT regulations are rather vague on facility siting issues, yielding siting decisions primarily to the states, but these rules may change in response to the Satartia release [2]. This presentation outlines a practical, sensible approach to implement a risk-based facility siting program to manage the risk of carbon dioxide releases to an acceptable level.

1 Introduction

To reduce and eventually minimize greenhouse gas emissions (GHG), various sectors of the economy must dramatically change their manufacturing technologies. This overarching strategy, called decarbonization, will require some level of carbon capture and geological storage. In the United States, a significant constraint is that the carbon emitters are not co-located with appropriate geological formations. Thus, captured carbon dioxide must be transported from its point of capture to its geological destination.

Carbon dioxide can be most conveniently transported via pipeline in a supercritical fluid state [3]. Similar to the design of natural gas transmission networks, a carbon dioxide transmission network will require compression stations to move the fluid from its source to its destination. At the destination the carbon dioxide will be injected into the target geological formation which may be a depleted oil and gas formation or a saline aquifer. With proper design, operation, maintenance, and monitoring, geological storage should provide secure sequestration of the stored carbon dioxide for periods of time approaching geological time.

Pipeline transport is by far the safest and most economical mode of transport. Alternative modes of transport such as truck or rail cannot compete with pipelines on either a safety or cost basis. The low risk of pipelines is demonstrated by the long history of pipeline transport and low failure rates of pipeline systems. Carbon dioxide pipelines have demonstrated lower failure rates and injury statistics compared to natural gas and hazardous liquid pipelines [4].

But proposals for the construction of carbon dioxide pipelines have met with strong public opposition. This negative perception has been largely driven by the February 22, 2020, failure of a carbon dioxide pipeline in Satartia, Mississippi. During this event, the release of approximately 5,000 m^3 of CO_2 from a pipeline sent 49 people to the hospital and required the evacuation of about 200 residents. The release was caused by an overload failure induced by impact from a landslide which resulted in the full fracture of a 24-inch pipeline. Furthermore, the unfortunate location of the pipeline failure released dense carbon dioxide vapor at an elevation higher than the Satartia community, allowing downward dispersion of the dense vapor to settle into the community and a local highway.

This single incident has become the focal point for many climate activists and organizations opposed to carbon dioxide pipelines. While this incident offers important lessons learned, it should not become the sole basis for risk management decisions. The risks posed by carbon dioxide pipelines must not be dismissed; instead, they should be responsibly managed. A risk-based facility siting program can provide a sound basis for risk management. In this paper we outline a practical, sensible approach for implementing a risk-based facility siting program to manage the risk of carbon dioxide releases to an acceptable level.

2 Carbon dioxide pipeline hazards

Carbon dioxide (CO_2) is a natural byproduct of respiration. The ambient air has a CO_2 concentration on the order of 400 parts per million (ppm), and human exhalation has a CO_2 concentration on the order of 40,000 ppm or 4.0%. Carbon dioxide is moderately toxic. In the United States the eight-hour exposure limit is 5,000 ppm and the immediately dangerous to life and health (IDLH) limit is 40,000 ppm. Carbon dioxide is heavier than air and large releases can displace the ambient air causing suffocation (physical asphyxiant) [5, 6].

Accidents involving CO_2 in nature and industry

A 2014 UK study performed at the Tyndall Centre for Climate Change Research at the University of Manchester revealed that while people did not generally oppose carbon capture & storage (CCS) technologies, they voiced a high degree of skepticism about pipeline construction [7]. Participants wondered whether appropriate health and safety measured could be successfully implemented for this new technology, questioned speed and accuracy in leak detection, and expressed concern regarding local environmental impacts of unintended CO_2 releases. In general, these points highlight broader concerns for safety associated with implementation of CO_2 pipelines.

High profile incidents, such as the natural releases of carbon dioxide in Pompeii, Italy and Lake Nyos, Cameroon may also contribute to the public's perception of risk associated with release of CO_2. The oldest historical record of fatalities produced by natural emissions of carbon dioxide was in 79 A.D. at the Pompeii eruption of Mount Vesuvius [8]. In this event and other subsequent volcanic releases of carbon dioxide, volcanic vents or openings, called fumaroles, emitted gases and vapors that raised local CO_2 levels fatally. Volcanic releases from fumaroles can raise local CO_2 levels by varying amounts, posing significant risks to human health. However, concentrations above 30% in the atmosphere can cause death [8]. As an order of magnitude of the scale of a volcanic release, the 1980 eruption of Mount St. Helens was estimated to release 10 million tonnes of CO_2 [9].

Two other noteworthy catastrophic environmental releases of carbon dioxide occurred in Cameroon in 1984 and 1986. In August 1984, a landslide caused a lethal gas burst of CO_2 from Lake Monoun, resulting in 37 deaths due to asphyxiation [10]. In August 1986, a larger-scale CO_2 release occurred at Lake Nyos. Despite the lack of direct volcanic activity there, magmatic CO_2 was released from a stagnant layer of water deep within the lake, resulting in the deaths of over 1,700 people due to asphyxiation [11]. This CO_2 release has been estimated at 1.6 million tonnes [12]. In addition to natural disasters, there have also been a number of accidental CO_2 releases from pipelines. The most high-profile CO_2 release from a pipeline to date occurred in Satartia, Mississippi in 2020. A case study of this event is detailed in Section 3.

CO_2 pipeline failure frequencies

Statistics documented by the Pipeline and Hazardous Materials Safety Administration (PHMSA) within the Department of Transportation reveal 61 reported incidents for CO_2 pipelines between 2012 and 2022 with no reported injuries or fatalities [13]. Figure 1 shows a comparison of the number of pipeline incidents reported by PHMSA between 2012 and 2022 for CO_2 pipelines. Further analysis of CO_2 pipeline failures for the time period between 2012 and 2022 delineates the primary cause of each failure, including material/weld/equipment failure, incorrect operation, pipeline corrosion, natural forces, and other causes, as shown in Figure 2. The primary cause of failure of CO_2 pipelines is documented as material/weld/equipment failure, with 42 of 61 CO_2 pipeline failures (69%) documented to be a result of this cause. Meanwhile, ten CO_2 pipeline failures (16%) are attributed to incorrect operation.

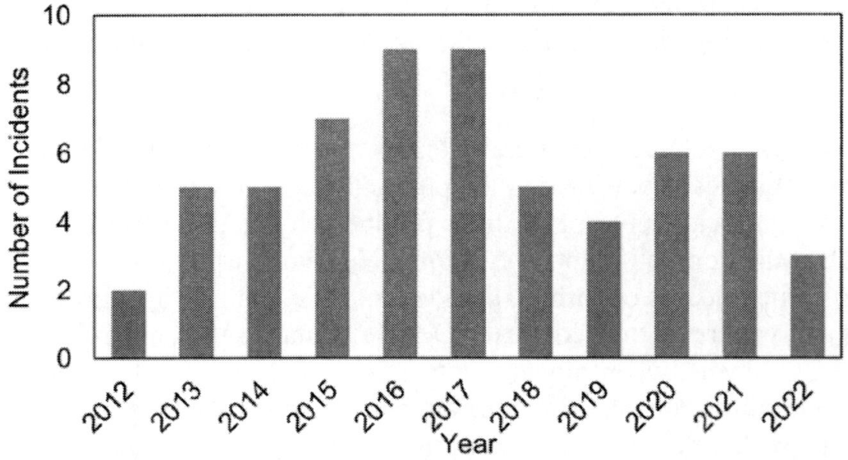

Figure 1. Number of CO_2 pipeline failure incidents between 2012 and 2022. Adapted from data provided by PHMSA [13].

Interestingly, a non-negligible fraction of CO_2 pipeline failures can be attributed to pipeline corrosion. In the timeframe between 2012 and 2022, 7 failures (11%) were reported to be caused by corrosion. When compared to corrosion-related failure rates for LNG, for example, only 3% of LNG pipeline failures over the same time are related to this cause. One possible cause for increased failure rates due to corrosion could be specifically due to carbonic acid corrosion. Carbonic acid corrosion of carbon steel is a recognized source of damage in oil-field equipment and gas pipelines and is commonly referred to as "sweet gas" corrosion [14]. The pipeline's material plays a critical role in its integrity, and numerous predictive tools, such as the de Waard/Milliams nomograph [15], exist to estimate corrosion under various operating temperatures and CO_2 partial pressures. In addition to the pipeline material itself, the amount of water present in the pipeline plays a critical role in carbonic acid formation, and technologies for removing water from CO_2 are well understood [16].

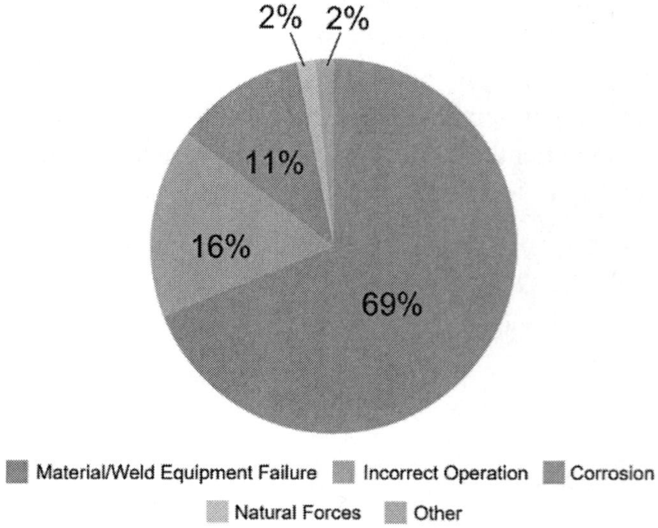

Figure 2. CO_2 pipeline incidents between 2012 and 2022 sorted by cause. Adapted from data provided by PHMSA [13].[1]

While this data from PHMSA provides a benchmark to compare to liquid natural gas (LNG) and natural gas (NG) pipeline failures, a lack of historical data makes failure frequency difficult to calculate for CO_2 pipelines [17]. According to Vitali *et al*, a total of 113 incidents have occurred between 1986 and 2021, and the only recognized database containing statistics on CO_2 pipeline incidents is PHMSA. Furthermore, CO_2 pipelines comprise only about 2% of the overall pipeline network in the United States; while CO_2 pipelines comprise only approximately 5,000 miles of onshore pipelines, hydrocarbon pipelines account for an estimated 340,000 miles.

While CO_2 pipeline mileage increased rapidly between 2004 and 2014, pipeline construction stagnated after 2014 [17]. Most existing CO_2 pipelines in the United States are related to enhanced oil recovery (EOR) activities. However, as carbon capture and storage technologies continue to grow, the need to develop pipelines to transport captured carbon dioxide to storage sites will grow. Preliminary estimates show that CO_2 pipeline failure frequencies are in the same range as those reported for hydrocarbon pipelines, but additional analyses of failure mechanisms and consequences of CO_2 releases from pipelines are necessary to develop risk assessments and hazard analyses for new CO_2 pipeline constructions.

Dense gas dispersion phenomena and topography

An additional complexity that must be considered for CO_2 pipelines is that CO_2 is denser than air. Upon rupture of a pipeline, this density difference may lead to CO_2 collection in depressions and low-lying topographical areas [14]. Unlike dispersion of other pipeline materials, such as conventional natural gas and sour gas, different dispersion characteristics and a lack of ignitability may also cause CO_2 leaks to remain undetected for longer periods of time. Taken together, the ability of CO_2 to settle in low-lying areas and go undetected for long periods of time poses a substantial exposure risk to the public upon pipeline failure [18]. Therefore, it is of critical importance to develop predictive tools that can accurately describe the CO_2 concentration distribution over time.

The development of models for dense gas dispersion is not dependent on the density difference between a dense gas and its surrounding environment alone. If we consider a density difference between a gas cloud and an ambient fluid, $g' = g[(\rho - \rho_a)/\rho_a]$, where g is gravitational acceleration and ρ and ρ_a are the density of the cloud and ambient fluid, respectively, the following criterion put forth by Britter and McQuaid describes the delineation between dispersion of passive (non-dense) and dense gases,

$$\frac{(g_0' q_0/D)^{1/3}}{U} \leq 0.15$$

[1] Note that "Other" combines PHMSA's separate categorizations "Other Outside Forces" and "All Other Causes." For CO_2, one incident is associated with "Other Outside Forces."

where g_0' is the source density difference, q_0 is the continuous source volume flowrate, D is the source dimension, and U is the ambient velocity [19]. When this criterion is less than or equal to 0.15, a continuous gas release effectively behaves as a passive gas.

This criterion clearly illustrates that density difference alone does not determine whether a release behaves as a dense gas. For example, a small release rate, release into a strong wind, or release over a large source area all decrease the propensity of a gas with a large g_0' to behave as a dense gas during release [20]. Furthermore, the temperature and pressure conditions play a significant role in determining g' because $\rho = \rho(T, P)$ and $\rho_a = \rho_a(T, P)$. In the case of CO_2 release from pipelines, these considerations are important because CO_2 is pressurized within pipelines and may be present in gas/liquid mixtures.

Using this criterion, we can also highlight the dispersion phenomena of CO_2 in the 2020 CO_2 pipeline failure in Satartia, MS as an example. With a known release quantity of approximately 31,000 barrels (4,900 m^3) of CO_2 over four hours, we estimate q_0 to be 7,750 barrels CO_2/hour, or approximately 1.2×10^3 m^3/hour, through a pipe with a diameter, D, equal to 61 cm and an ambient air velocity of approximately 2.2 m/s. With an air temperature of 7 °C on the day of the release, CO_2 exposed to atmospheric pressure had a density ρ_{CO_2} of approximately 1.9 kg/m^3 compared to the ambient air density ρ_a of 1.3 kg/m^3. Substituting these values into the above criterion yields a result of approximately 0.6, well above the threshold for behavior as a dense gas during this release.

In addition to these properties, the topography of the surrounding area plays a critical role in the impact of a hazardous gas release. Recent studies of topography on oil and gas pipelines reveal the importance of pipeline routing on accidental releases of natural gas, specifically focusing on the threat of landslides. In the case of LNG, landslide incidents often result in larger leaks, and greater threats to property and the environment [21]. Landslides can cause pipeline damage through a number of mechanisms, including lateral displacement, spanning, loading, and impact exposure. Schematic examples of lateral displacement and spanning are shown in Figure 3.

When constructing new pipelines, it is also essential to carry out field identification techniques to identify potential hazards, such as fresh cuts or ruptures, irregular topography, or sudden changes in slope to avoid construction in those areas [22]. While the best practice is to route along ridge crests and spurs, it is not always possible to completely avoid all geohazards. Minor adjustments to pipeline routes and construction of structures to isolate pipelines from hazards present just a few possible ways to minimize the risk of pipeline failure due to earth movement events. In the next section of this paper, we explore the 2020 failure of a CO_2 pipeline in Satartia, MS to assess the impacts of earth movement events on a pipeline failure involving dense gas.

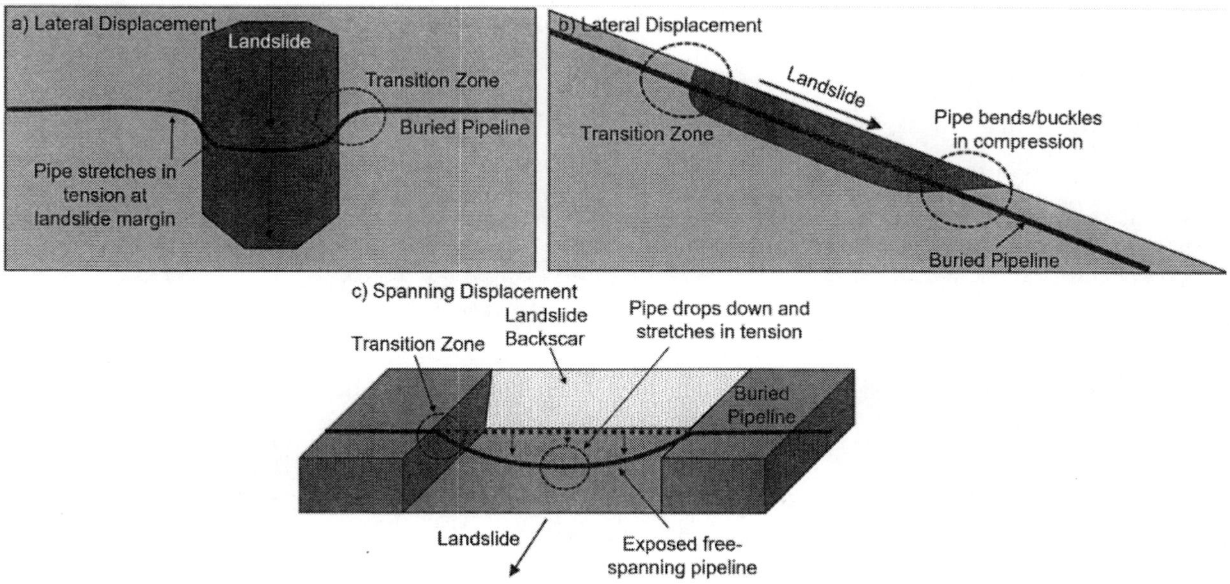

Figure 3. Schematic models for lateral displacement transverse (a) and parallel (b) to the pipeline and spanning (c). Adapted from Lee, Fookes, & Hart [21].

3 Failure experience with CO_2 pipelines

Pipeline failure modes in general (what causes pipes to break?)

According to the DOT, the most common causes of failure related to pipelines are corrosion, excavation damage, natural force damage, other outside force damage, material/weld failure, equipment failure and incorrect operation [23]. In Section 2.2, the causes were analyzed for CO_2 and LNG pipeline to highlight the frequency of occurrence for certain causes and determine which mechanism mainly causes failure of a given pipeline. As noted, both material/weld, equipment and incorrect operation appear to be the most dominant form of failure for both CO_2 and LNG pipelines. The main differences are between CO_2 and LNG causes of "other", and corrosion, which was noted in the previous section as being due to the presence of water and impurities in CO_2 forming corrosive compounds. The "other" causes, noted as being a combo of outside force and other causes not explicitly listed, is more common for LNG pipeline, which is most likely attributable to LNG pipeline networks being older and more extensive compared to CO_2 pipelines, leading them to being prone to additional failure mechanisms.

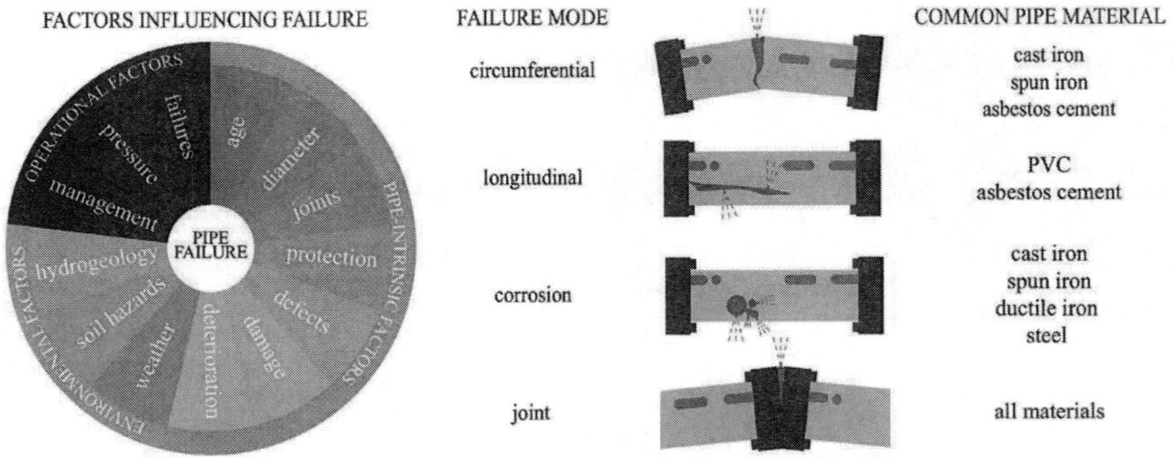

Figure 4. Failure modes of pipelines, factors influencing failure of pipelines and common pipe materials for failure modes [24].

Major causes of pipeline failures come from variety of sources, some of which originate when the pipelines are constructed, during transportation of the pipeline from when it was manufactured to where it will lay, or during its use while transporting material. When pipes themselves are being transported, they can be improperly handled or damaged inadvertently leading to dents, scratches, or local metal loss at the pipe [25]. When pipelines are being laid or when third party work is being done near the area where pipelines are located, mechanical equipment used to perform the work such as excavators can inadvertently cause damage to the pipelines [25]. Additionally, corrosion from internal sources such as impurities in the CO_2 mixture can cause unintentional reactions with the metal leading to erosion of the pipe wall. External sources of corrosion, such as galvanic corrosion due to weathering, are more likely to cause pipeline line failure, as unintended contaminants contact the surface of the pipe wall more frequently [26].

These causes typically lead to two common failure modes for pipelines, leak or a rupture [27]. The main difference between theses modes is if the crack/fracture stabilizes and does not lead to significant extension of the crack along the length/circumference of the pipe, then it is said to be a leak, otherwise it is a rupture. Leaks can be comparable and/or more severe than ruptures in certain cases, such as when they go undetected for long periods of time, allowing for sufficiently large quantities of material to be released [26]. When examining the type of cracks/fractures in pipelines, they exist as either a ductile or brittle fracture, which can divide them into four main categories: circumferential breaks, longitudinal splits, holes, and joint failures (Figure 4) [28].

CO_2 pipeline failure modes in particular (what specific issues does CO_2 introduce?)

CO_2 transportation in pipelines occurs mostly in North America, mainly for industrial process that occurs such as carbon capture and storage technologies and gas processing plants for enhanced oil recovery [29]. When transported in pipelines, CO_2 is typically pressurized and cooled to a supercritical fluid, either as pure compound (Figure 5) or containing impurities because of the process it originates from [30]. The code of regulations, title 49, defines CO_2 as a fluid consisting of more than 90% carbon dioxide molecules compressed to a supercritical state [31].

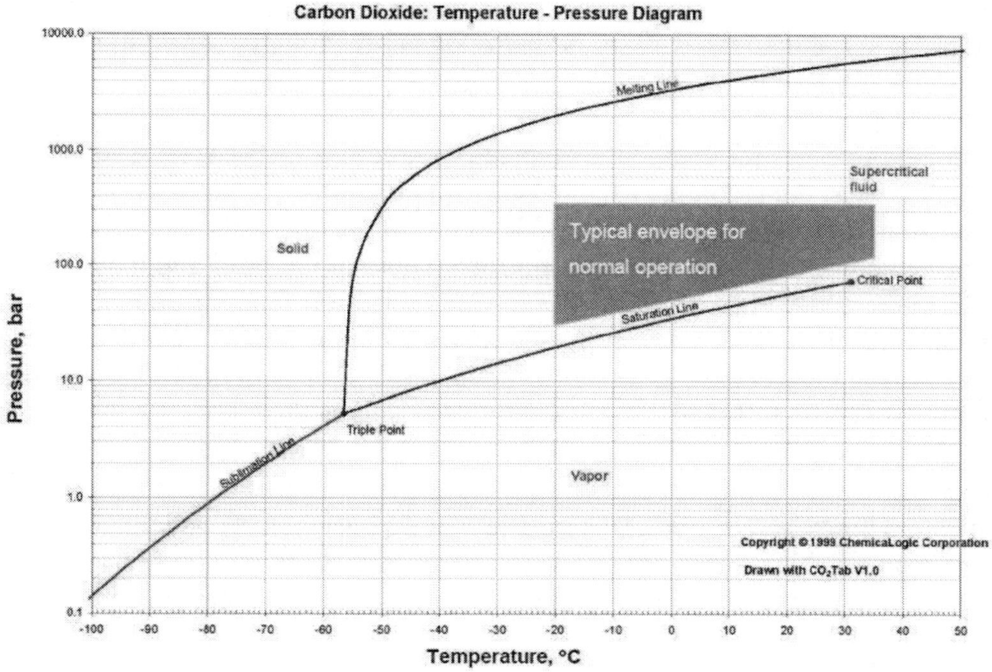

Figure 5. Phase diagram for pure CO_2. The gray trapezoid indicates the typical T&P conditions for supercritical CO_2 transport in pipelines [30].

Depending on the source and purpose for the CO_2, different constraints are imposed on the type and concentration of impurities that can exist in the CO_2 being transported. In the O&G industry, CO_2 captured pre or post combustion can generate different chemical components in the CO_2 flow stream such as Hydrocarbons, H_2O, H_2S, SO_x, NO_x, N_2, O_2, etc. [28]. Each of these contaminants changes the properties, risks, and hazards associated with transporting the CO_2 and the potential consequences in the event of a pipeline failure [32].

The presence of impurities in CO_2 can a have number of different effects on its transportation in pipelines and what risks are associated with it. Mixtures of CO_2 change the thermodynamic properties of CO_2 in pipelines, such as shifting the phase envelope and introducing the possibility of multi-phase flow within pipelines [30]. Additionally changes in the thermodynamic properties of the CO_2 effect the way it must be transported, to keep the fluid in the supercritical phase and apply sufficient re-pressurization along the distance of the pipeline [33]. With the possibility of multiphase flow of CO_2 comes the challenge of keeping impurities contained within the fluid, as each impurity has varying levels of solubility with CO_2 at different pressures and temperatures, which vary even more so, depending on the phase it is dispersed it [30]. The presence of such impurities as water, SO_x, NO_x, and H_2S can pose a significant challenge as the aqueous forms of these compounds can generate acids, and water itself can combine with the CO_2 to form carbonic acid, all of which can lead to internal corrosion of the pipeline [30]. Additionally, with water and other impurities, such as metal ion content that may be present from geological CO_2 sources, an additional form of galvanic corrosion can affect the integrity of pipeline walls. Corrosion mitigation strategies must be considered when designing CO_2 pipelines, and care consideration must be given to the impurity content with the fluid to understand the precautions that must be taken to mitigate the risks and reduce the likelihood of pipeline failure.

As mentioned earlier, due to the shifting in thermodynamic properties from impurities and challenges with multiphase flow developing because of shifting temperature and pressure within CO_2 pipelines, re-pressurization strategies must be integrated into the transportation of CO_2. As impurity content increases for compounds typically found in CO_2 because of sourcing from combustion processes, the recompression distance significantly decreases as the purity content increased, leading to further increased risk [33]. If a CO_2 pipeline fails and a leak or rupture occurs as a given point, because of the supercritical nature of the fluid, rapid depressurization will occur from the dense phase, leading to a throttling or Joule-Thompson like effect, rapidly cooling the pipe. As pipe is cooled, significant decreases in local toughness or resistance to failure, will occur rapidly and with the force of decompression from the CO_2, fracture propagation will occur rapidly [33]. Additionally, CO_2 will change in its decompression speed as it transitions from a supercritical fluid to a vapor, which will allow fracture propagation to continue and form long continuous cracks [30]. As such, a design consideration for CO_2 pipelines must be the integration of crack arrestors in the case of a propagating fracture, along with pipeline materials that resistance substantially drops in material properties such as toughness when exposed to extreme temperatures, such as the ones that occur with CO_2 decompression.

Case study: 2020 CO_2 pipeline failure in Satartia, MS

On February 22, 2020, at 7:06 p.m. (CST), a CO_2 pipeline operated by Denbury Gulf Coast Pipelines LCC ruptured in the proximity of the community in Satartia, Mississippi. The Denbury Delhi Pipeline, on which the rupture occurred, consists of 77 miles of 24-inch diameter pipeline, majority of which is located within Mississippi and flows east to west. The site of the rupture was on the northeast side of Highway 433 (HWY 433), approximately one mile southeast of Satartia. Prior to the rupture, heavy rains are believed to have led to a landslide, which created axial strain on the pipeline and results in circumferential girth weld failure. Sections of the failed section of the pipe were cut out and later to a laboratory for metallurgic analysis, conforming the onsite observation of a girth weld failure. Figure 6 shows a picture taken of the as found pipeline at the location of failure [34].

Figure 6. Failure location of the CO_2 pipeline in the Satartia incident [34].

As a results of the pipeline failure, authorities from multiple agencies and counties, quickly responded and participated in emergency response effort. Authorities were alert by individuals near HWY 433 of a "foul smell and green fog across the highway", leading them to initially believe the possibility of a chlorine gas release, which lead them to closing off HWY 433, directly west of the rupture site. Officials at Denbury were alerted nearly a minute after the rupture had occurred at the control room when its data acquisition system has altered them of a pressure drop. Denbury immediately began closing MLBVs downstream and upstream of the rupture site, isolating the rupture and minimizing CO_2 release, and confirming valve closure while locating the precise rupture location. Responders eventually got reports that the incident had been accompanied by a loud roar and that it sounded like a gas line had erupted, along with reports from community members, responders shifted their efforted towards a CO_2 release [34].

The volume of the release was approximately 5,000 m^3 with an estimated mass of 1,000 to 3,000 tonnes. Approximately 30 mins after the rupture of the pipeline, the national weather service (NWS) had generated a model for the plume/cloud of CO_2 (Figure 7) that was generated from the rupture and using meteorological data had predicted the spread of the cloud to the nearby town of Satartia, only a mile from the initial release site. Due to the weather conditions and the condition of the CO_2 as it has released from the pipeline, the dispersion of the released CO_2 gas behaved as a dense phase gas [34]. Section 2 examined this fact further by demonstrating the relative vapor density of CO_2 under the conditions when rupture occurred and the ambient weather, would predict CO_2 behaving as dense gas when dispersing. The topology/terrain near the point of the release from the pipeline created favorable conditions such that the dispersion of CO_2 naturally routed it way toward the town of Satartia [34].

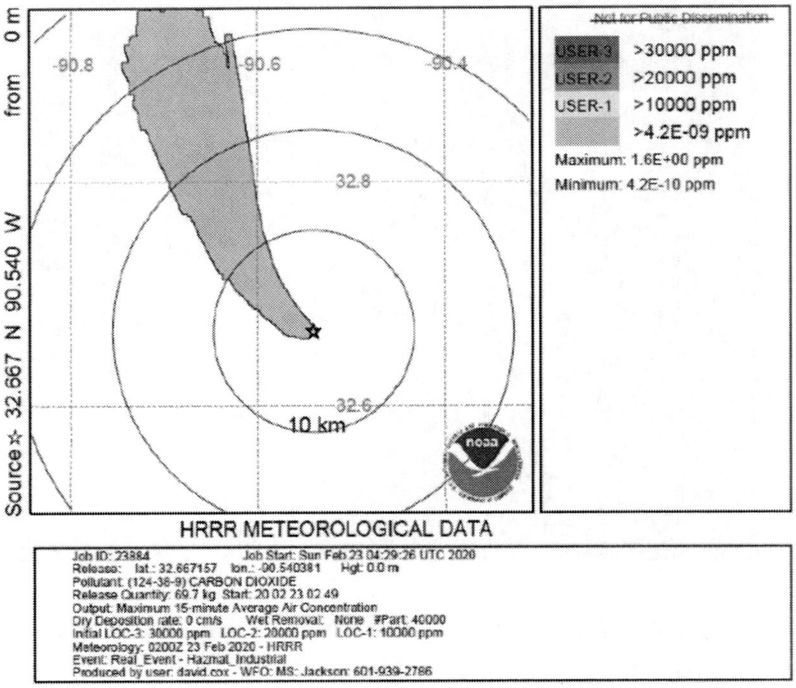

Figure 7. Plume model of CO_2 dispersion/spread generated by the National Weather Service/NOAA [34].

4 Pipeline integrity as the first line of defense

The first goal of pipeline safety is to keep the commodity in the pipeline and prevent accidental releases. Thus, pipeline integrity is the first line of defense in managing pipeline safety. In the US, CO_2 pipeline development is regulated at the federal level by the U.S. Department of Transportation's Pipeline and Hazardous Materials Safety Administration (PHSMA). In general, pipeline facilities used to transport hazardous liquids or liquefied carbon dioxide in the US are subject to 49 CFR 195. Prior to including CO_2, this regulation applied to the transportation of hazardous liquids, including petroleum, petroleum products, and anhydrous ammonia [35].[2] The regulation applies to CO_2 pipeline facilities and the transportation associated with those facilities in an interstate, international, or offshore context.

[2] The US regulations explicitly apply to pipeline transportation of CO_2 in a supercritical (or "dense vapor") phase, which was more desirable than transportation in the gaseous phase at the time the regulation was promulgated.

The federal regulations set forth in 49 CFR 195 are primarily focused on establishing and maintaining pipeline integrity by setting requirements for design, construction, pressure testing, operation and maintenance, training of personnel, and corrosion control. There are also reporting requirements associated with annual operations, safety-related conditions, and accidents, including loss of containment events.

PHSMA has four active projects to better determine impact areas for the safe operation of CO_2 pipelines, which include potential impact radii, leak detection methods, and material test and qualification for pipelines and underground storage facilities under consideration to be repurposed for CO_2 transport and storage [36]. These projects are anticipated to inform rulemaking at PHSMA, which PHSMA stated is a top priority for the agency and for which it anticipates issuing a notice of public rulemaking in early 2024.

However, at the time of this publication, facility siting and related permitting requirements are generally handled by local or state authorities [37]. The statutory requirements associated with building a CO_2 pipeline vary widely, and in many states do not exist. Thus, in the US, there lacks a single, consistent standard associated with pipeline routing and safety.

International guidelines for CO_2 pipelines have similar scope. For example, DNV (formerly Det Norske Veritas and subsequently DNVGL) has issued a guideline on the design and operation of carbon dioxide pipelines, which is focused on safety evaluation, concept development, and structural integrity of the pipelines [38]. Government regulations within Europe and Asia are in a state of flux, so specific design requirements are difficult to summarize at the time of this writing.

5 Risk-based approach to additional layers of protection

In the absence of specific prescriptive requirements codified by government regulations, risk based process safety provides a useful framework to address the safety issues associated with carbon dioxide pipelines. Risk based process safety is a process safety management system developed by the Center for Chemical Process Safety (CCPS), a technology affiliate of the American Institute of Chemical Engineers (AIChE) [39]. Risk based process safety is organized around twenty core elements ranging from process safety culture to management review and continuous improvement. Of these twenty elements, five are particularly relevant to carbon dioxide pipelines:

- Compliance with Standards: adopt pipeline design, construction, operation, and maintenance procedures that reduce a risk of CO_2 release to an acceptable level of residual risk.
- Stakeholder Outreach: establish a clear, honest, and convincing dialogue with the surrounding communities along the pipeline route so that they understand the safeguards in place to prevent or mitigate any risks posed by the CO_2 pipeline.

- Hazard Identification and Risk Analysis: engage in a comprehensive process to identify, evaluate, and control the risks of a CO_2 release.[3] Be especially vigilant for geophysical hazards to the pipeline such as seismic, flooding, or slope stability hazards.
- Asset Integrity and Reliability: ensure the integrity of the CO_2 pipeline system and its safeguards to prevent leakage and releases.
- Emergency Management: plan, communicate, train, and practice emergency response. Scale the level of effort so that it is commensurate with the risk to the target community.

Addressing these elements should enable future owners/operators of CO_2 pipelines to safely operate their facilities with confidence and to protect their social license to operate.

Facility siting considerations are especially important since this is perhaps the only opportunity to practice inherently safer design. In particular, consider the pipeline route in reference to the location of population centers, potentially vulnerable populations (hospitals, schools, care facilities), and topography. Topography is a major factor in the evaluation of geophysical hazards like slope stability (landslides and other ground motion) and hydrological hazards like flooding. Tectonic hazards like seismic activity are also of concern. In addition to pipeline routing, consideration should be given to hazard controls to prevent third party damage such as the depth of placement (deeper pipelines are more difficult to damage during normal construction activities by third parties) and pipeline location marking.

Safety management of the pipeline should evaluate the consequences of pipeline leakage or release. On flat terrain, this is strictly an exercise in dense gas dispersion modeling. However, dense gas dispersion is significantly affected by topography as illustrated by the Satartia Mississippi CO_2 release. One tradeoff to evaluate is the determination of a characteristic quantity of CO_2 release based on the spacing of emergency block valves, the response time of the control system, and the sensitivity of the leak detection algorithm employed by the pipeline control system.

Finally, the owner/operator should design an emergency management strategy that is commensurate with the risk posed to the surrounding communities. The potential consequences of an urban release are more severe than the same release in a rural area. Education should be offered to the host communities to raise awareness (direct mail, town hall meetings) and training should be potentially offered to emergency responders.

6 Conclusions

Carbon dioxide pipelines will become an essential component of the United States' carbon management strategy. Pipeline project proposals have been met with significant public resistance due to land use and safety concerns. Carbon dioxide is moderately toxic. In large releases it is denser than air and can displace oxygen (physical asphyxiant). A primary safety goal is to prevent leakage or rupture of a carbon dioxide pipeline. The historical record for carbon dioxide pipeline failure statistics is somewhat sparse, but design standards for material selection and

[3] See for example, DNV. CO2 RISKMAN. Guidance on CCS CO2 Safety and Environment Major Accident Hazard Risk Management, Levels 1-4, 2021.

pipeline design are available. Care must be exercised in the routing of a carbon dioxide pipeline to avoid, if possible, areas with geophysical hazards like slope instability, flooding, or seismic activity. The state of government regulation of carbon dioxide pipelines is in flux and uncertain. Thus, a risk-based process safety approach to carbon dioxide pipeline siting can help manage the safety concerns in a meaningful way.

7 References

[1] International Energy Agency Greenhouse Gas R&D Programme. Safety in carbon dioxide capture, transport and storage 2009.

[2] Congressional Research Service. Carbon Dioxide (CO_2) Pipeline Development: Federal Initiatives (IN12169) 2023.

[3] United States Department of Transportation Pipeline and Hazardous Materials Safety Administration. Background for Regulating the Transportation of Carbon Dioxide in a Gaseous State (PHMSA-2016-0049-0001) 2015.

[4] Gale J, Davison J. Transmission of CO_2 – safety and economic considerations. *Energy* 2004; **29**: 1319-1328.

[5] Scott JL, Kraemer DG, Keller RJ. Occupational hazards of carbon dioxide exposure. *J. Chem. Health & Safety* 2009; **16**: 18-22.

[6] Spitzenberger C, Flechas T. Carbon dioxide major accident hazards awareness. *Chem. Eng. Progress* 2023; 33-38.

[7] Gough C, O'Keefe L, Mander S. Public perceptions of CO_2 transportation in pipelines. *Energy Policy* 2014; **70**: 106-114.

[8] Martini M. CO_2 emission in volcanic areas: case histories and hazards, *Plant Responses to Elevated CO_2*. Cambridge University Press; 2010.

[9] United States Geological Survey Volcano Hazards Program. Volcanoes can affect climate. https://www.usgs.gov/programs/VHP/volcanoes-can-affect-climate#:~:text=Injected%20ash%20falls%20rapidly%20from,potential%20to%20promote%20global%20warming. Accessed 28 February 2024.

[10] Sigurdsson H, Devine JD, Tchua FM, Presser FM, Pringle MKW, Evans WC. Origin of the lethal gas burst from Lake Monoun, Cameroun. *J. Volcanology and Geothermal Res.* 1987; **31**: 1-16.

[11] Kling GW, Clark MA, Wagner GN, Compton HR, Humphrey AM, Devine JD, Evans WC, Lockwood JP, Tuttle ML, Koenigsberg EJ. The 1986 Lake Nyos Gas Disaster in Cameroon, West Africa. *Science* 1987; **236**: 4798.

[12] Harper P, Health and Safety Executive. Assessment of the major hazard potential of carbon dioxide (CO_2) 2011.

[13] United States Department of Transportation Pipeline and Hazardous Materials Safety Administration. PHMSA Pipeline Incidents for Hazardous Liquids, CO_2 2003-2022. https://www.phmsa.dot.gov/data-and-statistics/pipeline/pipeline-incident-20-year-trends. Accessed 23 October 2023.

[14] Barrie J, Brown K, Hatcher PR, Schellhase HU. Carbon dioxide pipelines: A preliminary review of design and risks. *Greenhouse Gas Control Tech.* 2005; **7**: 315-320.

[15] de Waard C, Lotz U. Prediction of CO_2 corrosion of carbon steel (51393-069), NACE International 1993.

[16] Ormiston RM, Luce MC. Surface processing of carbon dioxide in EOR projects. *J. Petroleum Tech.* 1986; **38**: 823-828.
[17] Vitali M. Zuliani C, Corvaro F, Marchetti B, Tallone F. Statistical analysis of incidents on onshore CO_2 pipelines based on PHMSA database. *J. Loss Prev. Process Ind.* 2022; **77**: 104799.
[18] Zammerilli A, Wallace R, United States Department of Energy National Energy Technology Laboratory Office of Fossil Energy. A review of the CO_2 pipeline infrastructure in the U.S. 2015.
[19] Britter RE, McQuaid J. Workbook on the dispersion of dense gases. Health & Safety Executive; 1988.
[20] Britter RE. Atmospheric dispersion of dense gases. *Annual Rev. Fluid Mech.* 1989; **21**: 317-344.
[21] Lee EM, Fookes PG, Hart AB. Landslide issues associated with oil and gas pipelines in mountainous terrain, *Quarterly Journal of Engineering Geology and Hydrogeology*. The Geological Society of London; 2016.
[22] Marinos V, Stoumpos G, Papathanassiou G, Grendas N, Papouli D, Papazachos C. Landslide geohazard for pipelines of natural gas transport. *Bulletin of the Geological Society of Greece* 2016; **L**: 845-853.
[23] United States Department of Transportation Pipeline and Hazardous Materials Safety Administration. Pipeline Failure Causes. https://www.phmsa.dot.gov/incident-reporting/accident-investigation-division/pipeline-failure-causes. Accessed 23 October 2023.
[24] Barton NA, Farewell TS, Hallett SH, Acland TF. Improving pipe failure predictions: factors affecting pipe failure in drinking water networks. *Water Res.* 2019; **164**: 114926.
[25] Vishnuvardhan S, Ramachanda Murthy A, Choudhury A. A review on pipeline failures, defects in pipelines and their assessment and fatigue life prediction methods. *Int. J. Pressure Vessels and Piping* 2023; **201**: 104853.
[26] United States Environmental Protection Agency. Chapter 11: Pipeline Failures, An Assessment of Potential Mining Impacts on Salmon Ecosystems of Bristol Bay, Alaska, Volume 1 – Main Report 2014.
[27] Sun H, Zhou W. Classification of failure modes of pipelines containing longitudinal surface cracks using mechanics-based and machine learning models. *J. Infrastruct. Preserv. Resil.* 2023; **4**: 5.
[28] Bilio M, Brown S, Fairweather M, Mahgrefteh H. CO_2 pipelines material and safety considerations. *Hazards XXI, IChemE* 2009; **155**: 423-429.
[29] Lu H, Ma X, Huang K, Fu L, Azimi M. Carbon dioxide transport via pipelines: a systematic review. *J. Cleaner Production* 2020; **266**: 121944.
[30] Det Norske Veritas. Design and operation of CO_2 pipelines (Recommended Practice DNV-RP-J202) 2010.
[31] National Archives and Records Administration. 49 CFR 191.2, Definitions. https://www.ecfr.gov/current/title-19/chapter-I/part-191/subpart-A/section-191.2. Accessed 23 October 2023.
[32] Holt H, Guidance on CCS CO_2 safety and environmental major accident hazard risk management, Rev. 3 (131JLJW-2). Det Norske Veritas 2021.

[33] Onyebuchi V, Kolios A, Hanak D, Biliyok C, Manovic V. A systematic review of key challenges of CO_2 transport via pipelines. *Renewable and Sustainable Energy Rev.* 2018; **81**: 2563-2583.
[34] United States Department of Transportation Pipeline and Hazardous Materials Safety Administration. Failure Investigation Report – Denbury Gulf Coast Pipelines LLC, Pipeline Rupture/Natural Force Damage 2022.
[35] Federal Register 26922 1991; **56**: 113.
[36] United States Department of Transportation Pipeline and Hazardous Materials Safety Administration. Written Statement of Tristan Brown Deputy Administrator, "Fueling America's Economy: Legislation to Improve Safety and Expand U.S. Pipeline Infrastructure." https://www.transportation.gov/fueling-americas-economy-legislation-improve-safety-and-expand-us-pipeline-infrastructure. Accessed 23 October 2023.
[37] National Association of Regulatory Utility Commissioners. Onshore U.S. carbon pipeline deployment: siting, safety, and regulation 2023.
[38] Det Norske Veritas. Design and operation of CO_2 pipelines (Recommended Practice DNV-RP-F104) 2021.
[39] American Institute of Chemical Engineers/Center for Chemical Process Safety. Guidelines for Risk Based Process Safety. John Wiley & Sons; 2007.

SPRING24
+20TH GCPS
A Joint AIChE and CCPS Meeting

Building Design Requirements - Utilizing Your Facility Siting Study Effectively

Karen R. Vilas
FORTRESS Protective Buildings®
757 N. Eldridge Pkwy #1000, Houston, TX 77079
kvilas@fortressprotectivebuildings.com

Thomas Mander
Baker Engineering and Risk Consultants, Inc. (BakerRisk®)
3330 Oakwell Ct #100, San Antonio, TX 78218
tmander@bakerrisk.com

Prepared for Presentation at
American Institute of Chemical Engineers
2024 Spring Meeting and 20th Global Congress on Process Safety
New Orleans, LA
March 24-28, 2024

Keywords: Facility Siting, Occupied Building Siting, Blast Analysis, Protective Building, Blast Resistant Building, Occupant Vulnerability, MRB

Abstract

Blast-resistant buildings have traditionally been quantified by the ability to resist blast loads for a prescribed level of structural damage. Focus has been placed on the structural performance (i.e., building response) without fully correlating this to occupant vulnerability. Fire and toxic material hazards and their impact on building occupants is often overlooked, resulting in buildings that do not comply with the full intent of API's Recommended Practices 752 and 753 (API RP 752/753). The definition of occupant vulnerability, per API RP 752/753, is the "portion of occupants that could potentially experience a life-threatening injury or fatality if a potential event were to occur." As summarized in the fourth edition of API RP 752, "owners/operators should understand and document the basis for the correlation [between building damage and occupant vulnerability] and assess its applicability." ***Why does this matter?***

This paper utilizes results from full-scale tests of modular building structures to demonstrate why occupant vulnerability matters and the repercussions of limiting the focus to building response to blast, fragments, fire, and toxic impacts. This paper also draws on occupied building siting results from a petrochemical facility case study discussion on locating buildings based on demonstrated occupant vulnerability to facility hazards and understanding the information to reduce your onsite risk profile.

1 Introduction

With the release of updates to the API's Recommended Practices 752 [1] and 753 [2] (API RP 752/753) in January of this year, it is the appropriate time to examine two guiding principles that facility owners and operators should consider for implementing facility siting recommendations for occupied buildings.

The most important guiding principle of the two is "The Golden Rule", which states: *whenever possible, locate your personnel in buildings away from hazardous areas.* However, this is not always possible. Facilities that have been in place for many years may not have the available land to expand beyond their existing footprint. Additionally, many hazardous processes require operators to be located near the units for ease of daily operations and quick response if an incident occurs. The remaining guiding principle dictates that occupied buildings should be designed to protect workers from hazards associated with the facility, which could include explosion, fire, and toxic material releases. When reviewing facility siting results and implementing recommendations, these two principles should drive decision making.

While this paper focuses on the siting of occupied buildings, it is important to note that it is equally important to protect critical equipment, which is often responsible for mitigating any potential hazardous event or preventing further injury or loss of life by providing a safe environment in which to bring operations under control. In fact, the Chemical Safety Board (CSB) report on the Philadelphia Energy Solutions (PES) incident (see Figure 1) resulted in five recommendations [3], one of which focused on safeguard reliability in critical buildings [4]:

Update API RP 751 Safe Operation of Hydrofluoric Acid Alkylation Units to require the following:

a. Protection of critical safeguards and associated control system components, including but not limited to wiring and cabling for control systems and primary and backup power supplies, from fire and explosion hazards, including radiant heat and flying projectiles; and

b. Installation of remotely-operated emergency isolation valves on the inlet(s) and outlet(s) of all hydrofluoric acid containing vessels, and hydrocarbon containing vessels meeting defined threshold quantities.

Figure 1. PES Fire and Explosions, Courtesy of WCAU Philadelphia [5]

This CSB recommendation acknowledges the need for buildings housing critical safeguards and associated control system components to be designed as robust multi-hazard resistant buildings (MRBs) –providing protection not only against blast events, but also from fire (radiant heat) and debris hazards. Whether this recommendation is adopted by the API, it is anticipated that industry will focus best practice on robust design for buildings housing critical equipment and controls. Therefore, best practice for facility siting would include addressing these critical buildings in addition to the occupied building situations discussed in this paper.

2 Protective Buildings

Industrial facilities are potentially vulnerable to hazards such as explosions, projectiles and debris, fires, unignited toxic gas leaks, and extreme weather. Hazards are increased for buildings in or near process units, which are typically where buildings housing essential operations personnel are sited out of necessity and/or convenience. When occupied buildings covered by API 752 and API 753 must defend against the full range of applicable hazards to ensure the safety of personnel and for conditions where a facility siting study shows exposure to multiple hazards, a MRB should be specified. A MRB is a holistic solution that provides personnel and equipment protection, peace of mind, and additional security even in high-risk environments containing numerous hazards [6].

Figure 2 shows a FORTRESS Protective Building designed as an Operator Shelter for a refinery. This building was designed to be a MRB to ensure field operators are protected and can quickly respond during an incident to mitigate the event (first line of response) in an area with high blast, debris, fire, and toxic hazards. For a properly designed MRB, there are two equally important aspects: the design must focus on the vulnerability of the people as much as the response of the building [6]. For more information on this important distinction, see the hazard specific sections below as well as the case study presented.

Figure 2. A FORTRESS Protective Building Utilized as an MRB for Operators [7]

2.1 API RP 752 and API RP 753 2024 Updates

Updated API RP 752 and API RP 753 recommendations were released on 16th January of 2024 [[1],[2]]. API Recommended Practices are not a collection of rules but rather a collection of industry's knowledge and experience rooted in sound engineering and operating practices. With the release of updated recommended practices with respect to occupied permanent and portable buildings, API is recognizing the need to adapt to new challenges and emerging technologies as well as demonstrating a desire to enhance safety and ensure compliance with the latest industry developments [8].

While the updates to API 752 and API 753 do not include substantial changes in the overall approach or definition of what to assess in facility siting, there is additional guidance provided in the form of tables, bulleted lists, and examples as to how to evaluate hazards and risk. One notable change is the emphasis on examining occupant vulnerability versus building response. While both are technically allowable per the recommended practices, because enough detail is provided on how to correlate building response to occupancy vulnerability, one should consider this best practice. In addition, the guidance on fire and toxic hazards have been enhanced to emphasize these hazards shall be addressed with the same rigor as blast hazards.

2.1.1 Applicable Hazards

In Section 5.2 of API 752, the consideration of applicable hazards states: "Owners/operators shall determine if the building intended for occupancy under consideration could be impacted by a credible explosion, flammable vapor release, thermal radiation, or toxic material release." It goes on to state that for each hazard that can impact the building, siting for that hazard shall be conducted per subsequent sections and documented [1]. While the method of assessment can vary from a spacing table approach to a consequence or risk-based approach, reviewing and addressing the full range of hazards, as applicable, is non-negotiable.

2.1.2 Occupant Vulnerability

As noted above, reviewing the hazard impacts on buildings (i.e., the building response) can no longer be considered best practice when discussing the safety of occupants. The updated recommended practices acknowledge that there are available published methodologies for determining occupant vulnerability based on different construction types and while there is not a requirement for siting to be based on vulnerability when taking a spacing table or consequence-based approach, API 752 does state, "…owners/operators should understand and document the basis for the correlation and asses its applicability." Furthermore, a risk-based approach should utilize robust vulnerability models in calculations.

Figure 3 shows the results of a BakerRisk testing program to examine the correlation between building response during an explosion event and internal debris hazards for different construction types [13]. Significant internal debris from an explosion is a better indicator of building occupant vulnerability than building response is. The example in Figure 3 shows an Iso Container that experienced a "low damage rating" (see Section 3 for response references) for the explosion test, and yet the interior non-structural debris would have resulted in severe injury or worse.

Figure 3. An Iso Container Before and After – Vulnerability Matters [13]

2.1.3 Risk Mitigation

Once an owner/operator understands the building response from applicable hazards as well as the vulnerability, a building mitigation plan can be established for existing buildings, or a protective building design specification can be established for new buildings based on the preliminary hazard and/or risk modeling. API 752 includes a hierarchy of mitigation measures in Table 1 of Section 5.5.3, which is a very helpful addition that includes ways to reduce risk through process changes, facility mitigations, or building improvements. The most prevalent issues that the authors of this paper see with respect to building design specifications for hazards are a clear definition of blast requirements for building response, with little to no specification for other hazards, and a requirement for occupant vulnerability and/or risk.

3 Blast Resistant Design

In addition to API RP 752 and API RP 753, other owner-enforced blast design guidelines often recommend that essential buildings are designed for a Low building response (i.e., limited to localized damage such that the building can be re-used) for the design-basis blast event(s) [8], [9], [10], [11], [12]. These documents provide analysis methodologies and acceptance criteria (i.e., damage limits) for blast loading, but no specific guidance on fire and toxic design considerations. Specifications and designs for new buildings tend to be focused on blast load requirements rather than occupant (personnel) vulnerability. The owner/operator should ensure that facility siting takes these analyses one step further to correlate building response to actual personnel impacts.

3.1 Structural Response

Blast design guidance documents define structural damage levels in terms of the peak deflection of a structural component. The deflection levels may be normalized to the deflection at which the component yields (ductility ratio) or the span of the component (support rotation). Most criteria were originally developed by the U.S. Department of Defense for the design of structures to resist accidental explosions [14]. It is these criteria that make up the basis of blast load and acceptable BDL, per Table 1.

Table 1. Building Damage Levels (BDL) for Blast Analysis and Design [10]

Damage Level	Description
Low	Localized component damage. Building can be used; however, repairs are required to restore integrity of structural envelope. Total cost of repairs is moderate.
Medium	Widespread component damage. Building should not be occupied until repaired. Total cost of repairs is significant.
High	Key components may have lost structural integrity and building collapse due to environmental conditions (i.e., wind, snow, rain) may occur. Total cost of repairs approaches replacement cost of building.

While a properly designed structure should protect building occupants from structural failures during a blast, there are other secondary hazards that are often overlooked when addressing building response that should not be overlooked when addressing facility siting. As discussed in a previous paper [6], non-structural items necessary for the function of a structure (cabinets, shelves, desks, electrical equipment, ducting, lighting, etc.) can become sources of hazardous debris to building occupants. This is especially true for building construction types where the wall and roof members are designed to deflect, which occurs with such speed that items near or attached to the deflected walls become projectiles. This has been observed in full-scale tests of metal buildings, both standard buildings and those designed to be blast resistant [13].

Conversely, buildings designed not to deflect such as those with robust precast reinforced concrete wall panels, have been tested in a shock tube with applied pressures exceeding 20 psi without generating hazardous non-structural debris [6]. In fact, these tests highlighted that the ASCE design criteria [10] does a poor job of characterizing building response for properly designed reinforced concrete structures. This was further demonstrated when BakerRisk® tested the FORTRESS Defender, a 6-inch precast reinforced concrete building, at their Box Canyon test facility in August of 2023. Per the ASCE response criteria, the Defender experienced Medium damage as shown in Figure 4 where the hairline cracks are marked; however, the structure is still in tact and usable. Furthermore, as shown in Figure 4, the post test results showed no internal debris or occupant vulnerability from the test.

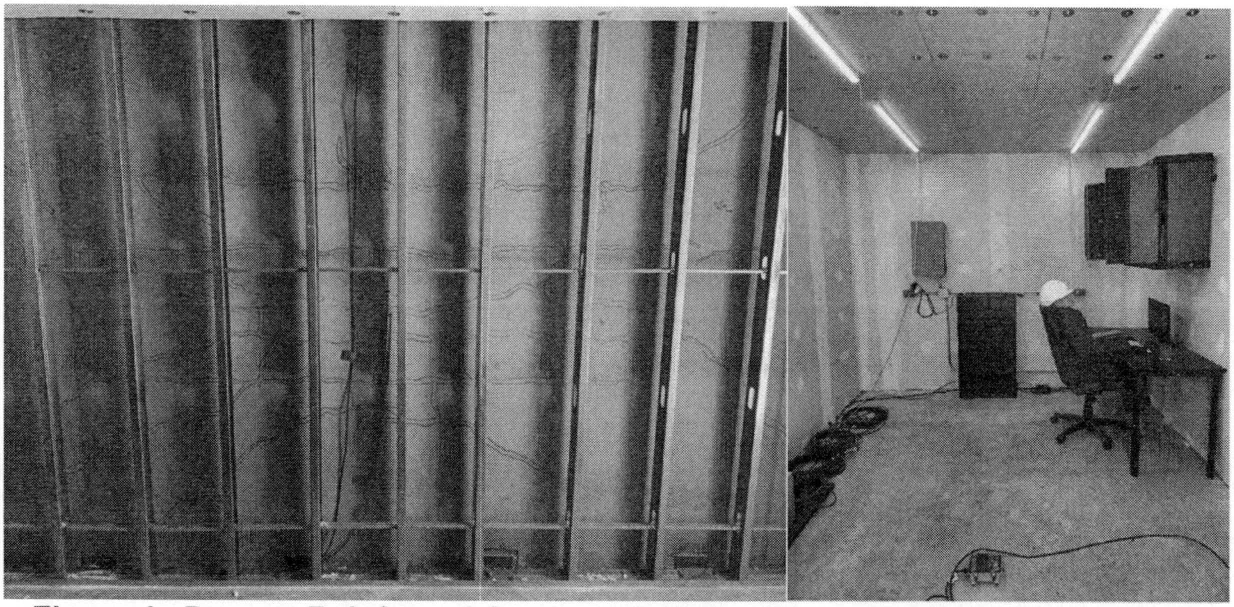

Figure 4. Precast Reinforced Concrete Building Test – Negligible Vulnerability

3.2 Other Occupant Vulnerability Considerations

Another way that non-structural debris can become an issue is from an unanchored building used in lightweight construction sliding. In the full-scale test program described above, an unanchored ISO container was subjected to deflagration blast loads [13]. A 4.9 psig test with a short duration of 22 ms resulted in the ISO container sliding 35 ft on its concrete foundation. Damage incurred by the ISO container walls and roof after this test is shown in Figure 5, and resulted in a significant amount of internal debris from both local wall and global sliding movements. However, as indicated in the photo, the resulting structural damage would be considered “Low” or “Medium”.

Figure 5. ISO Container After Testing [13]

Similar test results have been observed on full-scale steel Blast Resistant Modules (BRMs) when tested at approximately half the manufacturer's specified rating. The magnitude of sliding observed in these test programs produces velocities and accelerations that can cause injuries to personnel [6]. These testing programs are representative as to why facility siting decisions should not be made based on building response only.

4 Multi-hazard Resistant Design

As mentioned above, there are several guidelines that document how to analyze blast impacts on buildings; however, none of these address fragments, fires, toxics, or extreme weather. API RP 752 and API RP 753 are trying to close this gap with respect to occupied buildings by including information and references for not only how to calculate fire and toxic hazards, but also how to correlate that to occupant vulnerability. With these changes, a building asserting it is API RP 752/753 compliant should be capable of documenting its multi-hazard resistant design; however, it will be the responsibility of the owner/operator implementing facility siting results to confirm building compliance with hazards modeled.

4.1 Fragments (and Extreme Weather)

While API RP 752 and API RP 753 do not directly address fragment hazards, fragments are a known hazard on industrial facilities in the form of vessel ruptures, explosion debris, and extreme weather debris. Care should be given by the owner/operator in building construction material selection and building design when they are aware of hazards that may result in high-speed projectiles. This is especially true in locations that are subject to hurricanes and tornados since these events have a higher likelihood in many cases of catastrophic operational hazards. A true MRB will specify projectile ratings and/or have undergone some level of ICC-500 testing for natural hazards.

4.2 Fire

Industrial fires result in thermal radiation hazards, which shall be considered based on the updated API recommended practices. Fires to be considered include flash fires, jet fires (Figure 6), pool fires, and fireballs; however, it is noted that flash fires and fireballs typically only impact personnel in buildings during evacuation. Several methods are included as examples on analyzing fires, and Table 4 of API RP 752 highlights the primary fire effects on buildings, which include flame impingement; thermal radiation flux or dose; convection thermal flux or dose; pressure; and momentum [1].

Figure 6. Jet Fire Test from BakerRisk's Wilfred E. Baker Test Facility

The most impactful change in API RP 752 for fire hazards is where Occupant Vulnerability is addressed in Section 5, Table 5 – which is reproduced below in Table 2. This section highlights that not only are there concerns based on heat flux at the building location, but an owner/operator should also consider specific vulnerability impacts based on the scenario such as time temperature rise in a building and the impacts of smoke and other combustion products on people over time. In addition, details are also provided on how to evaluate a fire refuge building.

Table 2. Building Consequences Associated with Fire Type (per API RP 752)

	Hazard	Pool Fire	Jet Fire	Fireball	Flash Fire
Immediate effects to building occupants	Ignition of building envelope	YES	YES	YES	NO
	Failure of windows	NO*	YES	YES	NO
	Internal flash fire or VCE	NO	NO	N/A	YES

	Hazard	Pool Fire	Jet Fire	Fireball	Flash Fire
Time dependent effects to building occupants	Toxic fumes from building component /content degradation	YES	YES	NO	NO
	Temperature rise in building	YES	YES	NO	NO
	Ignition of building envelope/contents	YES	YES	NO	NO
	Structural failure due to thermal weakening	YES	YES	NO	NO
	Ingress of smoke and combustion products	YES	YES	NO	NO
Effects during evacuation to building occupants	Flame contact	YES	YES	YES	YES
	Thermal radiation / dose	YES	YES	YES	NO

The importance as to why an owner/operator should consider occupant vulnerability for fire impacts when locating an occupied building is highlighted in Figure 6. Due to their stated building response to blast loads, standard metal BRMs are often located within or very close to processing units. Unfortunately, BRMs have low fire resistance and can reach hazardous temperatures within minutes when exposed to a jet fire. The heating of the crimped steel plate and subsequent interior architectural items create toxic off-gassing at levels lethal to occupants. A ¼-inch saturated propane jet fire applied to a single wall of a BRM demonstrated these shortcomings; within a short duration, the fire destroyed the building interior as shown in Figure 7. The use of intumescent coatings can partially address jet fire hazards, but siting a BRM well away from jet fire hazards is generally required to fully address the hazard.

Figure 7. Interior of Steel BRM After External Jet Fire

There are some ways to upgrade existing buildings for fire impacts, which a protective building design consultant can assist with. This may include measures such as installing fire resistant panels around the building, erecting a concrete fire wall, or wrapping the building in concrete panels. For new building construction, fire is best mitigated against by the building construction type – with precast reinforced concrete providing the best thermal mass for fire protection. When implementing the facility siting study in building specifications, an owner/operator should ensure their design specifications request the right building construction for inherent protection.

4.3 Toxic

In the updated API RP 752 standard, it explicitly states that if a credible toxic material release can impact a building at acute levels such as ERPG-3, AEGL-3, or equivalent, a siting evaluation for toxic material release is required [1]. Figure 4 of API RP 752 Section 9.1 is a flowchart of the process for evaluating toxic impacts on buildings, which can include analysis using toxic concentration impacts or dose impacts for inside and outside the building (evacuation vs. shelter). Furthermore, Section 9.4 goes on to discuss the differences between a building analyzed for evacuation vs. sheltering (i.e., toxic refuge). For toxic refuge buildings, owners/operators should specify whether the building is a safe haven or shelter-in-place (SIP) location and buildings should be designed accordingly. For an example of a toxic refuge, see Figure 8 below.

Figure 8. Example Toxic Gas Dispersion and a Toxic Refuge Building

For details on performance requirements, see Section 9.4.3. It is important to understand that non-structural damage to buildings such as window glazing, door gaskets, HVAC connections, etc. may compromise a building's effectiveness as a toxic refuge. A summary of the requirements are as follows.

- Length of time that the toxic material impedes the refuge.
- Number of occupants in the refuge.
- Length of time that the toxic refuge meets the internal environmental quality criteria.
- Length of time occupants are required to remain in the refuge.
- Airtightness of the refuge.
- Length of time for an event to escalate so as to impair the refuge or evacuation.
- Ability of emergency response to evacuate the refuge if evacuation is needed.

Furthermore, Annex D in API 752 is provided as an informative appendix to provide typical building protection features for buildings used as refuges from flammable/toxic material exposure. Most importantly, Annex D includes Table D.1, which differentiates typical protections for safe havens and SIP locations and includes protection features such as facade, doors, and windows; ventilation; envelope tightness; gas detection; response to gas detection; PPE; and maintenance/inspection. Facility owners/operators should utilize this annex when implementing facility siting study results in new building design specifications.

5 Case Study

This case study is based on an analysis previously submitted in the 2020 AIChE GCPS paper *Leveraging Process Safety Techniques in Capital Projects* [15]. This paper introduced a stage gate review process to minimize personnel exposure, asset damage, and business interruption while also controlling costs and ensuring ease of project implementation. The implementation of facility siting results is one of the concepts presented in the Capital Projects paper and is worth reviewing in the context of this paper. Through the early application of facility siting, capital projects can improve designs in terms of risk reduction while minimizing project and facility lifetime cost impacts.

The case study is hypothetical and consists of a capital project associated with a new facility processing flammable and toxic materials. The facility has public receptors to the south and west, is land locked to the north, and has space for expansion to the east. The hypothetical facility has a Crude Unit, a Naphtha Unit, an Ammonia Unit, associated Utilities, and storage/truck loading facilities.

5.1 Layout and Spacing

Based on the preliminary hazard review for the three main units present, it was determined that there are explosion, flammable/fire, and toxic hazards present at this facility. The facility will have two Central Control Buildings (CCBs) – one for the Naphtha and Crude units, and one for the Ammonia and Utility units. Access to the site is from the north, the topography is flat, and the wind blows predominantly from the southwest. For the preliminary site layout, see Figure 9.

Based on the knowledge of Subject Matter Experts (SMEs), the capital project team qualitatively defined a preliminary estimate of hazardous impacts. For more information on this process, see the 2020 paper on this topic. Engineering judgement suggests the following [15]:

- Access and traffic must enter from the north.
- Toxic releases can reach outside the fence line.
- Hydrogen is present both as a pure supply and a major component of mixtures; therefore, deflagration-to-detonation (DDT) explosions are possible.
    - Likely pressures >10psig possible around areas of congestion/confinement.
    - Potential for full site to experience pressures >0.9psig, ruling out the ability to use light wood trailers per API RP 752/753 [[1],[2]].
- Concentrations above the lower flammable limit (LFL) are likely to reach beyond site boundaries.
- Thermal radiation at high intensity (>37 kw/m^2) is likely at or beyond the facility fence line.

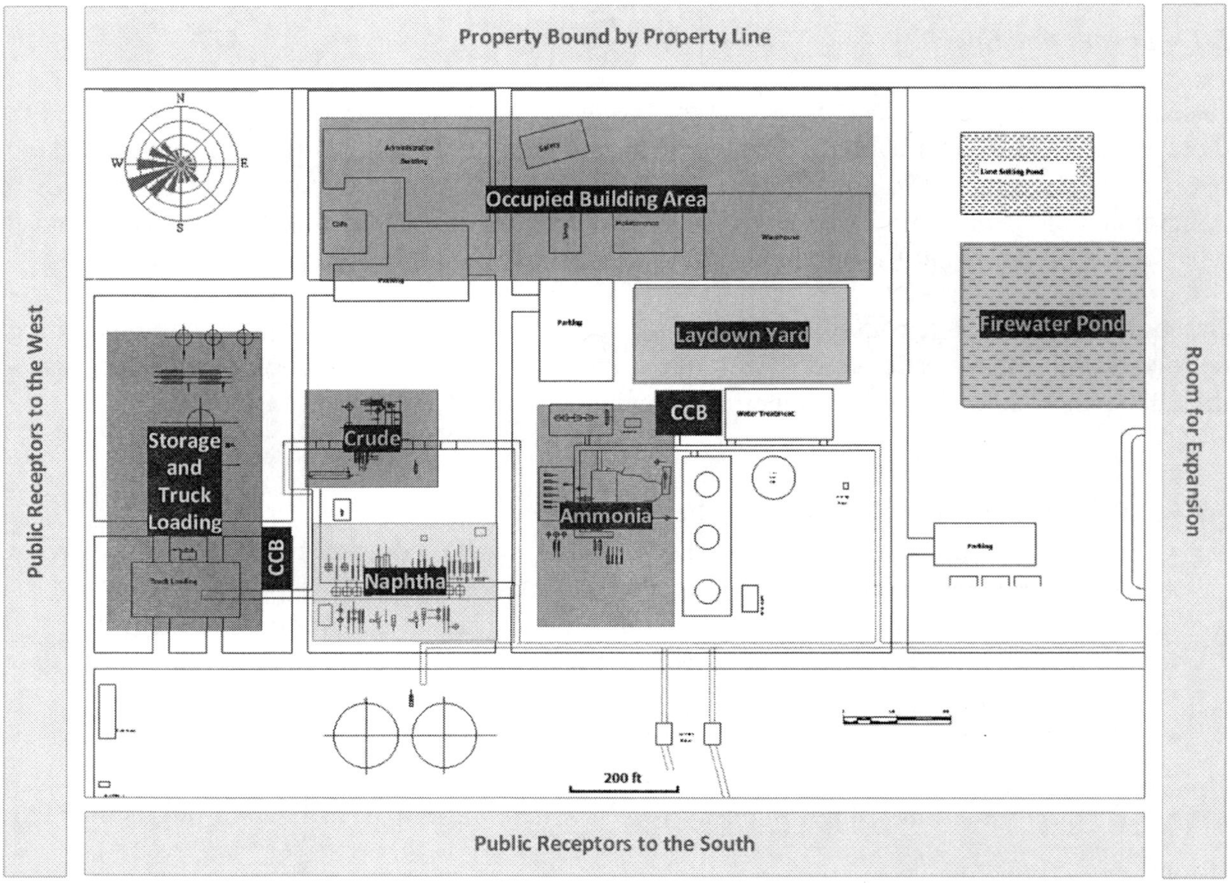

Figure 9. Site Layout and Spacing [15]

5.2 Facility Siting Study (Consequence and Risk Based)

For a greenfield facility like that described in this case study, the documentation to conduct a facility siting study to determine site hazards and risks is available in a preliminary format in the detailed engineering portion of the capital project (Define FEL-3, as described in the 2020 referenced paper). Therefore, it is at this phase that the preliminary facility siting study should be conducted with the goal to identify hazard and risk concerns; review risk mitigation options to determine cost effective mitigation options, where applicable; define occupied building design requirements; and confirm risk compliance based on the site layout and associated hazards.

At this stage of the project, one of the important facility siting-based decisions to be made is the design parameters of critical buildings such as Central Control Buildings (CCBs). The reason for the importance of this information is two-fold: first, in the event of an incident, the ability to maintain occupancy to safely shut down the facility in these locations is essential; and second, the design and construction of control buildings is often a significant investment for capital projects [15]. For the purposes of demonstrating how the results of a facility siting study can be successfully implemented, the CCBs will be reviewed. *Note that the same concepts apply for facility siting studies conducted for existing facilities. Control buildings are often critical buildings for both safe operations of the facility as well as protection of personnel; therefore, ensuring that multi-hazard occupant vulnerability is reviewed in a similar manner, albeit different process, is essential.*

The consequence-based facility siting study defines a range of hazards for facility operations based on a pre-defined methodology and determines the maximum hazards. The CCBs, which are occupied and critical to operations, are to be designed based on hazard exposure in anticipation of increased future staffing as well as anticipated future facility expansion, both of which will increase the risk profile for the CCBs.

The maximum hazards for the two control buildings are shown in Table 3. Due to the hazards present, API 752 indicates that the CCBs will need to be designed as Safe Haven locations with protection from blast, thermal, and gas ingress (flammable and toxic). Therefore, the specifications for the Detail Design phase of the project are to design the control buildings to maximize leak tightness, be designed for the blast loads shown in Table 3, and be constructed from materials that are thermal resistant.

Table 3. Maximum Hazards for Example Facility CCBs

Building	Average Occupancy	Maximum Hazard Exposure			
		Blast	Thermal	Flammable*	Toxic**
Crude/Naphtha CCB	2.78	4.7 psi, 54.2 psi-ms	>37.5 kW/m^2	>UFL	>90% OV
Ammonia/Utilities CCB	2.78	8.0 psi, 142.5 psi-ms	>37.5 kW/m^2	>UFL	>90% OV

*UFL – Upper Flammable Limit
** Occupant Vulnerability

Due to the multi-hazard protection requirement for the two CCBs, FORTRESS MRBs are specified as the building design requirements. Uniquely, FORTRESS has been subjected to a full-scale testing program to confirm occupants of the building are exposed to negligible vulnerability for design basis events. Based on the FORTRESS design specifications, the protection from these buildings includes the following:

- ***Blast:*** 8 psi overpressure at >> 200 ms (long duration), negligible occupant vulnerability
  - *ASCE Low Response (i.e., <1-degree of rotation) for the design load*
- ***Fragmentation:*** 13 lb projectile at 171 ft/s (116 mph) velocity
  - Very minor local spalling observed
- ***Thermal:*** 1-hour direct impingement for ¼-inch saturated propane jet fire
  - Local spalling observed on the building exterior, but internal air temperature < 110 °F and negligible smoke/toxic off-gassing
  - Tested at thermal radiation levels well beyond 37.5 kW/m2
- ***Toxic:*** < 0.1 ACH infiltration for main building and < 0.03 ACH infiltration for interior Shelter-In-Place (SIP) room
  - SIP Control Box, designed and engineered to provide system specification
- ***Extreme Weather:*** Resistant to high wind and debris from hurricane and tornado natural hazards

The risk-based facility siting aggregate (societal) risk results for the CCBs are shown in Table 4. This table indicates that the CCBs, as designed, are below risk tolerance criteria; however, it will be up to the project team to ensure due diligence is taken to ensure the buildings are constructed and operated to the specifications used in the facility siting modeling.

Table 4. Maximum Hazards for Example Facility CCBs – FORTRESS Construction

Building	Societal Risk (fatalities per year)				
	Blast	Fire	Flammable	Toxic	Total
Crude/Naphtha CCB	*Negligible*	4.7E-8	1.6E-9	2.6E-8	7.5E-8
Ammonia/Utilities CCB	*Negligible*	2.3E-7	2.3E-9	3.2E-7	5.5E-7

The full site risk profile is shown in Figure 10. This figure indicates that blast events, both Vapor Cloud Explosions (VCEs) as well as vessel ruptures and the AN Neutralizer runaway reaction drive the risk profile for the site. Therefore, any risk mitigation options investigation should address lowering the blast risk profile for indoor and outdoor populations.

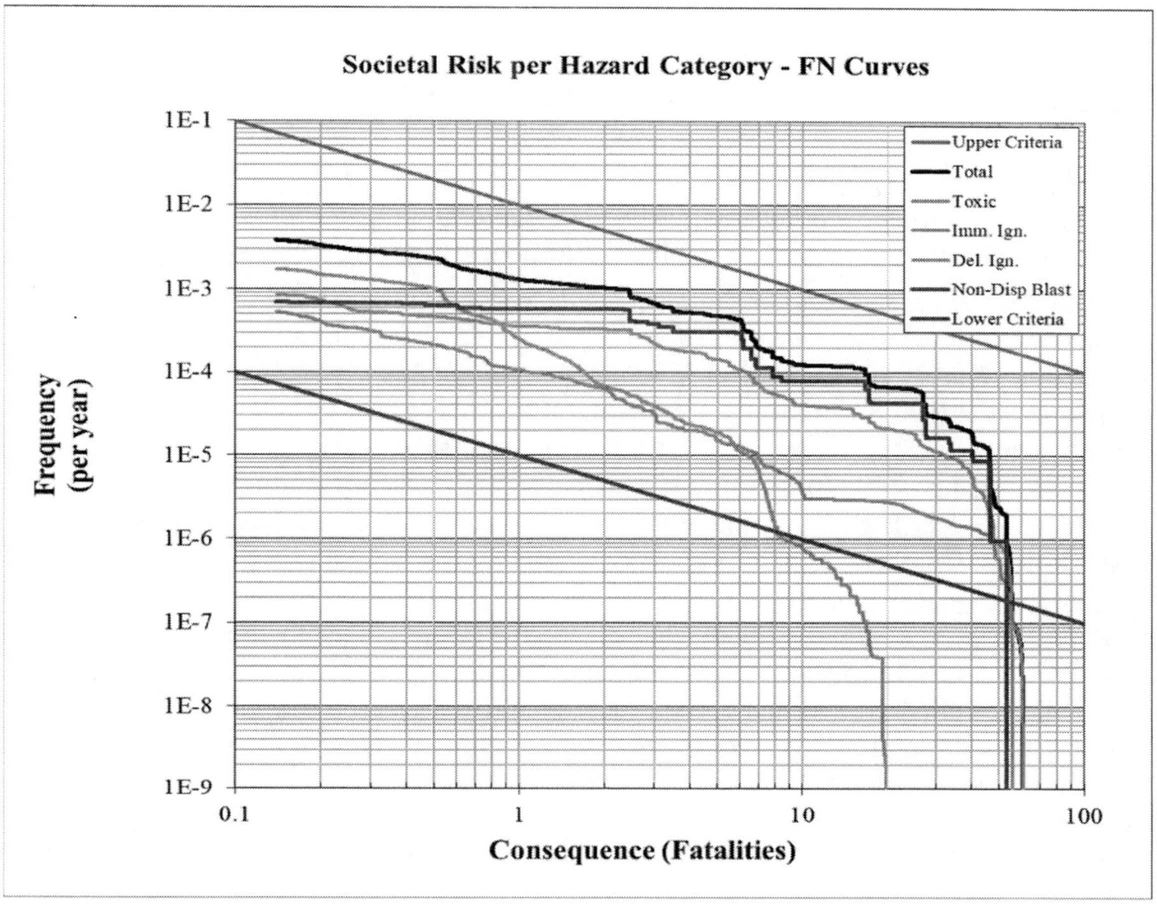

Figure 10. Facility QRA Results – FN Curve Total and by Hazard Type

6 Conclusion

Implementing the results of a facility siting study for building design is now easier than ever with the release of updated API RP 752/753 in January of this year. There is increased guidance as to how to analyze occupied buildings for not only blast, but also fire and toxic impacts. Furthermore, there is now a clear discussion on building response vs. occupant vulnerability and the requirement to document the correlation, depending on the way analysis is completed. To understand the impacts on vulnerability and not locate buildings based on impacts to people would be counter intuitive to properly applying facility siting to a given facility.

The number one takeaway from this paper should be that when implementing the results of a facility siting study to either design a new building or upgrade an existing building, the most important factor is to release a detailed building specification with precise requirements for blast, fire, toxic, and other hazards as part of the Request for Proposal process. If the building design requirements for facility siting are not properly stated, procurement may select an inferior building product that does not properly address all the facility siting results hazard types. Furthermore, if a certain vulnerability rating is required, then that must be explicitly stated. Resources are available for support with respect to constructing a building specification package for an MRB; for more information, please contact the authors of this paper.

7 References

[1] *American Petroleum Institute Recommended Practice 752 – API RP 752 Management of Hazards Associated with Location of Process Plant Permanent Buildings*. Washington, DC, American Petroleum Institute; 4th Edition, 16th January 2024.

[2] *American Petroleum Institute Recommended Practice 753 – API RP 753 Management of Hazards Associated with Location of Process Plant Portable Buildings*. Washington, DC, American Petroleum Institute; 3rd Edition, 16th January 2024.

[3] Wikipedia. 2019 Philadelphia Energy Solutions refinery explosion. https://en.wikipedia.org/wiki/2019_Philadelphia_Energy_Solutions_refinery_explosion. Accessed January 22, 2023.

[4] Chemical Safety Board. Philadelphia Energy Solutions (PES) Refinery Fire and Explosions. October 12, 2022. https://www.csb.gov/philadelphia-energy-solutions-pes-refinery-fire-and-explosions-/. Accessed January 2022, 2023.

[5] Stevens, Pippa. We finally know what caused the refinery blast that rocked Philadelphia. CNBC. October 16, 2019. https://www.cnbc.com/2019/10/16/we-finally-know-what-caused-the-refinery-blast-that-rocked-philadelphia.html. Accessed January 22, 2023.

[6] Vilas, K. and Mander, T. *Protective Building Design for Critical Assets*. American Institute of Chemical Engineers (AIChE) 2023 Spring Meeting and 19th Global Congress on Process Safety (GCPS). Houston, TX. 12th – 16th March 2023.

[7] FORTRESS Protective Buildings, LLC. FORTRESS Modular Shelters Make Excellent Security Guard Buildings. Pending. https://www.fortressprotectivebuildings.com/2023/02/15/fortress-modular-shelters-make-excellent-security-guard-buildings/. Accessed February 20, 2023.

[8] BakerRisk. *Changes in Facility Siting Standards: API 752, 753, & 756. September 2023.* https://www.bakerrisk.com/news/facility-siting-standards-api-752-753-756/. Accessed January 21, 2024.

[9] *Process Industry Practice (PIP) PIP STC01018 – Blast Resistant Building Design Criteria.* (2014). Austin, TX, Construction Industry Institute; 2014.

[10] ASCE Task Committee on Blast-Resistant Design, *Design of Blast-Resistant Buildings in Petrochemical Facilities*, Second Edition, Reston, VA, 2010.

[11] PDC TR-06-08, "Single Degree of Freedom Response Limits for Antiterrorism Design," USACE PDC *Technical Report 06-08*, Jan. 2008.

[12] U.S. Department of Defense (DoD). "Structures to Resist the Effects of Accidental Explosions," *Unified Facilities Criteria (UFC) 3-340-02*, Dec. 2008.

[13] Anderson, T.H., Horn, B.J., Thomas, J.K., Magenes, L. Deflagration load generator blast load testing of an ISO shipping container & blast resistant wood building, *ASCE Structures Congress*, 2017, https://doi.org/10.1061/9780784480397.012

[14] Departments of the Army, the Navy and the Air Force. *Structures to Resist the Effects of Accidental Explosions* Army TM 5-1300, Navy NAVFAC P-397, Air Force AFM 88-22, Washington, DC, June 1969.

[15] Vilas, K. and Broadribb. M. *Leveraging Process Safety Techniques in Early Phases of Capital Projects to Optimize Risk Reduction and Project Costs.* American Institute of

Chemical Engineers (AIChE) 2020 Spring Meeting and 16th Global Congress on Process Safety (GCPS). Houston, TX. March 30th – April 1st, 2020.

Implementing Facility Siting Study Recommendations—Lessons Learned

Larry Lepard
ERM (Environmental Resources Management)
8044 Montgomery Road | Suite 700-7336 | Cincinnati, Ohio | 45236
Larry.lepard@erm.com

Steve Hawkins
ERM (Environmental Resources Management)
8044 Montgomery Road | Suite 700-7336 | Cincinnati, Ohio | 45236
Steve.hawkins@erm.com

Keywords: Facility siting, recommendation, management of change, MOC, safety, risk management

Abstract

Facility siting studies (FSS) are performed for several reasons (i.e., it is a legal requirement for The Occupational Safety and Health Administration Process Safety Management (PSM) and US Environmental Protection Agency Risk Management Plan (RMP) facilities, they are often part of risk-based health and safety programs, and they are part of hazard review analysis associated with capital projects). Regardless of the reason for completing the study, it is not likely to add value or reduce risk if the recommendations are not effectively managed and/or if the study is not updated as site changes occur. This paper/presentation will focus on identifying when FSS should be done/updated and how to manage the subsequent recommendations.

In PSM facilities, recommendations and the subsequent actions and changes can come from a variety of sources (i.e., compliance audits, incident investigations, Process Hazard Analysis, FSS, Management of Change [MOC], and Mechanical Integrity inspections). These recommendations may require scope change for planned projects and/or extensive investment of capital and time for existing facilities. As such, managing all these recommendations and subsequent changes can be a struggle. Lack of action or taking improper action can have a negative impact on operations and employee health and safety.

This paper/presentation will include:

- An overview of when FSS should be done and/or updated;
- Use of MOC and other programs to drive effective management of studies and recommendations;
- Examples of successful and not so successful implementation of facility siting; and

- Discussion of lessons learned and best practices observed while conducting hundreds of compliance audits.

1 Introduction

In its simplest form, a FSS is a spatial analysis of the proximity of occupied structures and areas, both permanent and temporary, with respect to the potential negative impacts of fire, explosion and overpressure, and toxic release associated with hazardous processes. In many ways, the FSS forms the foundation to ensuring worker safety from potential impacts of catastrophic events and releases and is utilized to define "safe haven".

FSS are required to be conducted initially and revalidated every 5 years under the Process Hazard Analysis requirements of the PSM (29 CFR 1910.119(e)(5)(v)) and US Environmental Protection Agency Risk Management Plan (RMP) (40 CFR 68.67(c)(5)). The following are industry-specific guidelines and best practices related to conducting the studies:

- **American Petroleum Institute (API)**
    - *API RP 752, Management of Hazards Associated with Location of Process Plant Permanent Buildings, Third Edition, 2009*—provides guidance for managing the risk from explosions, fires, and toxic material releases to onsite personnel located in new and existing buildings intended for occupancy. [1]
    - *API RP 753, Management of Hazards Associated with Location of Process Plant Portable Buildings, First Edition, 2007*—provides guidance for reducing the risk to personnel located in portable buildings from potential explosion, fire, and toxic release hazards. [2]
    - *API RP 756, Management of Hazards Associated with Location of Process Plant Tents, Second Edition, 2024*—provides guidance for managing the risks from process related explosion, fire, and toxic material release hazards to personnel located in onsite tents. [3]
- **Center for Chemical Process Safety**
    - *Guidelines for Facility Siting and Layout, Second Edition*, 2018. [4]
    - *Guidelines for Evaluating Process Plant Buildings for External Explosions, Fires, and Toxic Releases, Second Edition*, 2012. [5]
    - *Guidelines for Chemical Process Quantitative Risk Assessment, Second Edition*, 1999. [6]

While there are a variety of technical approaches and tools available for modeling various hazard scenarios, including vapor cloud explosions, fire and thermal heat flux and radiation intensity computations, and toxic release dispersion modeling, it is important to remember that there are literally hundreds, if not thousands of technical parameters that are required to conduct complex facility siting analysis. As such, accounting for changes to the site or these parameters is of utmost importance in maintaining an "ever-green" FSS that accurately depicts current site conditions and existing risks.

2 Management of Change and the Facility Siting Lifecycle

2.1 Management of Change/Facility Siting Lifecycle

Manufacturing and processing sites are always changing, and it can be easy to fall into the trap of only considering an update or revalidation of the facility siting analysis every 5 years as is regulatorily required. However, it is imperative that all qualifying facility changes be evaluated and considered for the FSS revision as part of the Management of Change (MOC) process. In our experience of auditing and evaluating hundreds of PSM and RMP covered processing facilities, we often find that these changes are not considered as part of an FSS revision. These include:

- Changes in raw material or processing parameters;
- Changes in chemical inventory;
- Introduction of new chemicals or products;
- New construction or capital project activity;
- Changes in personnel location; and
- Additions of new occupied spaces, roadways, walkways, and/or means of egress, etc.

As a best practice, we have advised clients to embed facility siting change factor placeholders into the MOC review process to ensure that all potential change scenarios are evaluated like all other Process Safety Information and the study updated as needed. This level of rigor is often required to assure that the study accurately reflects the current site risks and that it remains valid throughout the lifecycle of the site.

3 Examples of Facility Siting Implementation

In this section, we will review some examples of facility siting implementation. These examples were observed by the author(s) while working in the process industry and as a consultant over the last 20-plus years. The goal of this section is to describe the nature of the change and/or what prompted the FSS, what actions were required/taken, and lessons that can be learned from these examples.

3.1 Installation of a Noise Barrier Wall

In this example, the facility was receiving noise complaints from surrounding neighbors. To reduce the noise levels offsite, the facility decided to install a noise barrier wall around a portion of the facility. The companies' standard procedures/practices for MOC, capital approval, and project implementation were followed, and the wall was installed. A few years after installation of the wall, during a compliance audit conduct by an external third party, it was observed that the wall was taller than the height of the facility emergency process blow down stack. The auditor asked if an FSS had been conducted to evaluate the potential impact of the wall on the emergency blow down system. A review of the MOC completed for installation of the wall indicated that an FSS had not been identified as applicable/required for this change by the MOC team and hence a study had not been conducted. The audit team made a recommendation to perform an FSS, including gas dispersion modeling.

Following the audit, an FSS, including gas dispersion modeling, was performed as part of the facilities audit corrective action plan. The study identified that the noise barrier wall could interfere with the emergency process gas blow down system. In fact, it identified that the wall could trap flammable process gas and funnel it to another part of the facility that was not in an electrical hazard classified area and could contain potential ignition sources. This resulted in the facility needing to develop/implement additional corrective measures including extending the height of the emergency process blow down stack.

The following are potential lessons that can be learned from this example:

- To address noise complaints, a non-process hazard, the facility had installed a noise barrier wall that created a potential process hazard and required additional corrective actions.
- The facility evaluated why the initial MOC had not identified the need to conduct an FSS prior to installing the noise barrier wall. In general, the evaluation concluded that the associated checklist and knowledge of the review team involved was not adequate to identify this hazard and the need to perform a formal siting study. The checklist focused on whether new process equipment was being installed, was existing process equipment being modified, was a new occupied structure being installed, etc. The MOC team at the time noted that none of these applied to the change and that the primary potential hazard/concern with installation of a wall at the perimeter of the property was the potential to strike underground utilities or process piping during construction. The team felt that a simple hazard screening was sufficient and did not require an FSS. As a result of this evaluation, the team made recommendations to revise the MOC process, including additional questions in the screening checklist and requiring additional employees/resource groups be involved in the review/approval process for these types of changes.

3.2 Installation of a Blast Deflection Wall

In this example, a client had purchased an existing facility and commissioned a PSM/RMP compliance audit of the facility. During the audit, it was observed that a ramped wall was installed between the process building and the operator control room. The audit team discovered that an FSS had been conducted at the request of the former owner that identified that the existing operator control room was within the potential blast zone and not of adequate design to withstand the expected blast wave of at least one explosion case modeled. The siting study included several recommendations such as relocating the control room outside of the blast zone or replacing the control room with a blast-resistant module (BRM). While no corrective action plan for this siting study could be found during the audit, it was stated that the wall had been installed following the study and was intended to deflect the blast up and away from the control room. During the audit, an MOC for installation of the deflection wall could not be found, no documentation of the design basis for the deflection wall could be found, and there was concern that the wall might not be adequate to protect the control room and operators. The audit team recommended that an additional FSS, including an engineering evaluation of the deflection wall, be conducted.

Following the audit, an FSS, including an engineering evaluation of the deflection wall, was performed as part of the audit corrective action plan. The study concluded that the deflection wall was not adequate to protect the control room or operators. In fact, it concluded that the wall was not likely to withstand the expected blast wave and debris from the wall could pose an additional hazard to the operator control room. It also identified that if the wall remained intact, the updraft created could create a vacuum behind the wall that could also cause the collapse of the existing operator control room. This new siting study recommended that the deflection wall be removed and that the control room either be relocated or replaced with a structure that could withstand the expected pressures.

The following are potential lessons that can be learned from this example:

- The prior owner either did not follow or did not maintain documentation of an MOC process to manage the installation of the deflection wall. As there was no MOC, the technical basis for choosing to install the wall, as opposed to relocating or replacing the control room, was not known. In addition, it did not appear that an appropriate evaluation of the effectiveness of and/or potential hazards of installing the wall was performed. Based on the new FSS, the hazard identified in the initial siting study had not been addressed and was potentially made worse by installation of the deflection wall. This meant that operators continued to be in harms way while feeling a false sense of security.
- Failure to follow an MOC process while taking the corrective action of installing the blast deflection wall resulted in additional corrective actions needing to be taken. The new facility owner had to redo the FSS based on installation of the wall. As the new study recommended the wall be removed, this was an additional cost on top of the original recommendations to either relocate or replace the existing control room, which still had to be done.

3.3 Installation of a Blast Resistant Module (BRM)

In this example, it was observed that the facility had installed a BRM and relocated the operator control room to the module. The facility had installed the module based on a recommendation from the FSS. The companies' standard procedures/practices for MOC, capital approval, and project implementation were followed for installation of the BRM. The former building/location of the operator control room had been designated as a non-occupied structure and the FSS revised to reflect the new occupied location. Documentation was available to demonstrate the BRM met all design requirements and had been installed properly. Initial indications were that the installation had successfully addressed the hazard identified in the FSS; however, additional employee interviews identified that some unintended consequences had been identified that had been corrected and/or still needed additional corrective actions.

The following are some of the unexpected/unintended consequences that resulted and the potential lessons that can be learned from this example:

- After relocating operations into the BRM, it was discovered that the BRM interfered with cell phone service and computer wifi connections. This resulted in unexpected/additional project costs to modify the BRM to have additional hardwire connection points for computers installed and the purchase/setup of a cell/wifi signal-boosting system.
- For various reasons, employees were propping the doors to the BRM open. The doors were required to remain closed for the BRM to provide the designed protection. The heating/cooling/air circulation system of the BRM was determined to be inadequate and required additional costs to correct/replace. When selecting/designing a BRM, it is important to consider the climate of where it will be installed, equipment that will be operated inside, and how many occupants are expected to adequately design the heating/cooling/air circulation system.
- Additional employee training was also needed to communicate why the doors of the BRM needed to be closed when not in use.
- When large groups of people not normally assigned to the facility were at the site, such as for compliance audits and engineering/construction projects, there was not enough room in the BRM for all the guests. Tables and chairs remained in the former control room area and this building was periodically occupied for up to 8 hours per day for several days/weeks at a time. Based on the FSS this building was not to be occupied. The facility had not identified/considered this issue and needed to develop a corrective action plan to address.

3.4 Temporary Control Room

In this example, a facility was impacted by a hurricane that caused significant damage to the site's Main Control Room. As an interim measure, a temporary control trailer, evaluated per API RP 753, was sited based on the results at an acceptable distance to continue process operations during the transition to recommissioning of the Main Control Room. However, after 6 months of operation, a decision was made by management to forego capital expenditure on the Main Control Room re-build and to use the temporary Control Trailer as a permanent solution. This included a decision to move the new Control Trailer adjacent to the former Main Control Room to help further provide cost savings associated with electrical, utility, and instrumentation tie-ins. However, in this instance, the move of the trailer was not evaluated utilizing the MOC program and there was an overarching assumption by those involved in the project that the control trailer was "designed to withstand a worst-case event" based on the previously completed FSS. The failure to evaluate the trailer move was identified in the site PSM compliance audit conducted by the author as a gap and corrective action was taken to update the facility siting analysis as appropriate based on the Control Trailer construction and location. When the modified FSS was completed, it identified the Control Trailer as NOT having a robust enough construction to withstand the predicted overpressure at its new location. Consequently, the site had to remove the Control Trailer and proceed with the Main Control Room modification project as originally planned with additional time delay and associated cost escalation.

The following are potential lessons that can be learned from this example:

- The facility did not use the MOC process to evaluate/manage the decision to relocate the temporary trailer and did not consider whether the trailer was safe for use at the new location as there was a FSS approving its use at a different location. This placed employees at an increased level of risk.
- The original recommendation to re-construct/modify the existing Main Control Room still had to be done and there was additional costs and time delay related to relocating and eventually removing the temporary trailer.

4 Conclusion

FSS are critical to ensure worker protection and safety from the adverse effects of potential catastrophic fires, explosions, and/or toxic releases. However, these studies and their conclusions are only as good as the information and assumptions used in the analysis. Failure to conduct an FSS, conducting the study based on incomplete/inaccurate information, and/or improper implementation of the recommended actions can leave risks unmanaged, create new risks, and lead to additional costs to resolve. To ensure that FSS remain accurate and correctly identify and mitigate risks, it is imperative that they be updated when there are changes to any of the parameters or variables that affect the study. Facilities should fully engage the MOC process to consider these types of changes and manage the implementation of recommended actions. The FSS should be treated as an "ever-green" document versus just one that is reviewed and revalidated every 5 years as required by regulation. It is only under these circumstances that there can be confidence that the study is truly providing the worker protection for which it was designed.

5 References

[1] API RP 752 (American Petroleum Institute Recommended Practice 752). 2009. *Management of Hazards Associated with Location of Process Plant Permanent Buildings, Third Edition 2009.* https://www.api.org/oil-and-natural-gas/health-and-safety/refinery-and-plant-safety/process-safety/process-safety-standards/rp-752 . Accessed 15 February 2024.

[2] API RP 753 (American Petroleum Institute Recommended Practice 753). 2007. *Management of Hazards Associated with Location of Process Plant Portable Buildings, First Edition, 2007.* https://www.api.org/oil-and-natural-gas/health-and-safety/refinery-and-plant-safety/process-safety/process-safety-standards/rp-753 . Accessed 15 February 2024.

[3] API RP 756 (American Petroleum Institute Recommended Practice 756). 2024. *Management of Hazards Associated with Location of Process Plant Tents, Second Edition, 2024.* https://www.api.org/news-policy-and-issues/news/2024/01/16/api-releases-new-updates-to-standards-for-hazard-management-in-process-plant-structures . Accessed 15 February 2024.

[4] CCPS (Center for Chemical Process Safety). 2018. *Guidelines for Facility Siting and Layout, Second Edition.* March 2018.

[5] CCPS (Center for Chemical Process Safety). 2012. *Guidelines for Evaluating Process Plant Buildings for External Explosions, Fires, and Toxic Releases, Second Edition*. April 2012.

[6] CCPS (Center for Chemical Process Safety). 1999. *Guidelines for Chemical Process Quantitative Risk Assessment, Second Edition*. October 1999.

A Detailed Strength, Weakness, and Opportunity Analysis of Existing Sustainability Metrics and the Case for Process Safety Inclusion

Mitchell Huffman, Qingsheng Wang, Faisal Khan
Mary Kay O'Connor Process Safety Center
Artie McFerrin Department of Chemical Engineering
Texas A&M University, College Station, TX 77843-3122, USA
mhuffman@tamu.edu, qwang@tamu.edu, fikhan@tamu.edu

Keywords: Process system sustainability; Sustainability metric; System safety; Safety assessment

Abstract

As the human population grows, the strain on resources and the environment grows exponentially. Sustainable practices are necessary to protect the planet and preserve the standard of living of those inhabiting it. The quantification of sustainability is vital to monitor and regulate industry practices, and to measure improvement in sustainability. This is the first phase of a research project towards developing a thorough sustainability metric for the process industry. Frameworks such as life cycle assessment, impact assessment, and the triple bottom line provide approaches to quantify, understand, and implement sustainability concepts. From extensive literature review, the development of the field of research is discussed, and it is demonstrated that sustainability metrics to this point have lacked an inclusion of safety, time dependence, and interrelations between sustainability pillars. To determine the validity of existing metrics, four different sustainability metrics from literature were utilized, and significant variation was found among the results. These inconsistencies imply the need for additional index considerations. This phase of research determines that the currently available sustainability metrics lack an interrelated, time-dependent quadruple bottom line methodology that would provide a thorough assessment of the sustainability of a process.

1 Introduction

Sustainability has been widely explored as a qualitative description of processes, with more recent studies exploring the quantitative nature of the topic.[1] While qualitative measurements have helped to increase discussion of sustainability issues, the definition of sustainability must allow for quantification to implement regulation and internal monitoring of a process. The lack of quantification of sustainability limits the efficacy of regulatory agencies when providing industrial guidelines.

As the world population continues to grow, industrial needs expand, increasing the environmental load of industry. Public pressure, political interest, and optimization efforts lead to the conclusion that more sustainable processes are essential. To accomplish this goal, there must be a method of measuring sustainability which can mark improvement or denote unsustainable practices. To develop a sustainability measurement that can be implemented in policy and action, there must be a metric which is easily interpretable and meaningful to the public.[2] There is a vital need for this in industry, as there has been a historic absence of measurement standards to be enforced.[3] This has improved in the last decade as various environmental regulations and standards such as emissions tracking have been implemented, however, there remains no standards for a multidimensional view on sustainability. This multidimensional sustainability is emphasized in the 17 Sustainable Development Goals put forward by the United Nations, which propose various goals of preserving and improving life economically, environmentally, and societally worldwide. It is often the case with the already existing goals of industry and the new goals of sustainable development that objectives now compete, and company decisions must prioritize between these various goals. Sustainability metrics provide an opportunity to evaluate the multiple priorities of a company or process in order to provide the optimal decision.

It is crucial to establish the importance of sustainability to the industrial sector, as the implementation of sustainability concepts requires full commitment to a proactive and forward-thinking shift to business operations.[4] As environmental concerns develop into new laws and guidance pertaining to sustainable business practices, companies certainly would favor a process that is prepared for future regulation rather than requiring costly retrofitting procedures. Sustainability can be an essential tool for cost reduction, risk management, new product creation, and fundamental changes to company culture and structure.[4] Because the impacts that are difficult to quantify and the different definitions of sustainability can make it difficult to decide how to evaluate a process, a concrete measure of sustainability for industry is certainly in high demand.[5,6]

2 Development of Sustainability Metrics

The measurement of sustainability is a critical research topic that has been explored in numerous ways in recent decades. However, much of this research has pertained to specific steps of the quantification process without creating the full measurement process. Many studies establish frameworks, indicators, and indices without exploring how to combine or utilize these different necessary components. However, it is promising that these different aspects have been explored for many different industries. These different explorations into the measurement of sustainability will be discussed below.

The criteria with which to establish a sustainability measurement is widely disputed, but perhaps the most widely emerging method of quantifying sustainability is through indicators and indices, as they allow for operations to have simple, quantitative objectives.[7] Indicators are a single quantity that is assigned to represent a more complex situation, in this case some aspect of sustainability.[8] This simplification may allow for a measurement that can be easily adopted in industry and at the design stage, however, indicators also result in loss of specificity and therefore a reduced quality of information.[9] In order to prevent this loss of information, multiple indicators can be utilized to account for the various aspects of a system. An accumulation of indicators into an index allows for a general prediction of trends pertaining to a process, and in order to accurately represent a complex process with simple indicators, the proper selection and verification of indicators is essential.[10] These sets of indicators must include information as to their inherent strengths, weaknesses, biases, and scale in order to properly inform decisions.[11–13]

Frameworks are different from indices in that rather than aggregating multiple indicators into representative values, they are methods of presenting the large amounts of indicators and thereby clarifying what to measure.[14] In many cases, these frameworks do not suffer from as much loss of information as indices, since the many indicators are individually displayed rather than being represented by an aggregate.[15] The disadvantage of using frameworks is that trend observation and prediction is much more difficult due to their more qualitative nature. These frameworks fall mainly into the following categories: objective-based, impact-based, influence-based, process-based, material flow accounting, and linkage-based.[16] Of these frameworks, impact-based and material flow analysis have been adopted in academia and industry. The impact-based framework focuses on the impact of a particular action, and has most notably developed the triple bottom line (TBL) principles, a proposition to expand the "bottom line" (solely economics) for businesses to include economic, environmental, and social aspects of their practices. This framework has in fact continued to shape most definitions of sustainability. Material flow analysis focuses on the flow and use of a material in a system, and is currently explored through the life cycle assessment (LCA), a thorough system analysis focusing on a cradle-to-grave approach toward a product.[17,18] Due to the extensive results provided from the analysis, it is widely used in system analysis for ongoing production.

Life Cycle Assessment

LCA has become perhaps the most utilized sustainability measurement for the environmental sustainability of a product or process. The scale of application can range from small-usage products to industrial scale processes, and in fact, Escobar-Hernandez et al. utilized the LCA to span this full range of scale, establishing the viability of metal-organic frameworks as an environmentally sustainable product at industrial production scale.[19] However, this is the limitation to LCA, which does not encompass the economic and social aspects of sustainability that are imperative to understanding sustainability.[20] Therefore, the LCA was expanded to a life cycle sustainability assessment (LCSA) for the purpose of calculating the integrated sustainability performance of chemical products. This process involved using the LCA for environmental sustainability, life cycle costing (LCC) for economic sustainability, and social life cycle assessment (SLCA) for societal sustainability.[21] Multiple studies have certainly demonstrated a promising, thorough assessment of sustainability from LCSA, developing excellent analysis tools that incorporate methods such as fuzzy inference, impact assessment, and multi-objective optimization.[22–24] However, the other weaknesses from the LCA are heightened in this framework. The data requirements also lead to the potential for high error within the assessment, as the raw amount of data means that highly accurate sources are necessary because any error will propagate from the various data sources.[25] In addition to this issue of uncertainty of data, the LCSA also struggles due to the multifaceted nature of sustainability leading to conflicting metrics. The amount of data necessary to perform a proper LCA presents a major difficulty in any design-stage application of the framework, and even more so for novel processes that have no previous large-scale operation data to utilize. Perhaps in the future, machine learning will provide a method to reduce the amount of necessary information for LCA.

Impact Assessment

Impact assessment is the estimation of the consequences, or impact, of a specific action.[26] Various impact assessment methodologies have been created such as environmental impact assessment (EIA), social impact assessment (SIA), health impact assessment (HIA), strategic environmental assessment (SEA), among others. Confusion within the field tends to revolve around the distinctiveness of these different methodologies. For example, EIA often considers impacts to nature and the planet, but also socio-economic and cultural impacts.[27] This would imply that SIA is actually a component of EIA rather than a separate analysis. Additionally, SIA has been very difficult to define, and due to the subjective nature of societal impact, quantitative valuation has been unsuccessful.[28] However, SIA provides a base for societal considerations in other societal sustainability frameworks and has been improved through the *International Principles for Social Impact Assessment*.[29] The field of impact assessment continues to evolve, with an increasing focus on integrative approaches, expanding EIA through environmental auditing and environmental management systems, and advancing to the policy level through the development of SEA. While impact assessment can be very intuitive in its original form of bottom-up analysis, the framework has not been successfully expanded for multi-dimensional sustainability analysis. Since sustainability measurement continues to include more aspects as the field advances, principles from impact assessment should certainly be referenced and utilized for single aspects of sustainability, but the multi-objective impact assessment frameworks require further development.

Triple Bottom Line

Perhaps the sustainability framework most readily adopted by industry is the triple bottom line, a framework to expand the "bottom line" of finances to a triple bottom line of economic, environmental, and social aspects.[30] The triple bottom line has presented sustainability in terms that are straightforward and easily understood by industry, leading to its widespread adoption. However, the framework has suffered from much criticism due to its simplification of sustainability and its adoption as a quantification methodology, when it was intended to be a decision-making mindset. Further, if the TBL is strictly adhered to as only its three aspects, the measure will disregard vital aspects of multi-dimensional sustainability that may be different for each industry. Research into the triple bottom line has mostly focused into each single aspect rather than interrelationships between the aspects.[31] For example, the environmental aspects of sustainability (such as air pollution) can have a health effect on the public (air quality), leading to an impact in the societal aspect, which would not be accounted for. Due to the lack of research in this area, some studies have attempted to address this interplay, but these studies have remained conceptual in nature.[32,33] Svensson et al. have provided the first quantitative exploration of the direct interrelationships between these elements, and the interactions of these interrelationships. Their conclusions provided proof that economic aspects have a significant impact on social aspects, and that social aspects have a significant aspect on environmental aspects.[31] This is very important as typical research has also focused most on the economic aspects of sustainability, with environmental sustainability being explored more recently due to climate change, but social sustainability continues to remain less explored due to being less concrete numerically. In fact, the societal aspect of the TBL is disregarded quite similarly to the SIA; for example, the majority of sustainable supply chain management papers fully disregarded the social aspect.[34] Therefore, the main limitations of the TBL are that it must be thorough to prevent loss of data, must be modified for its application, and should either be partnered with another framework or expanded for its specific application.

3 Analysis of Sustainability Metrics

On the most basic level, the sustainability of chemical elements themselves are of interest. The sustainability of raw materials have been classified by Ku and Hung in terms of supply risk.[35] This work led to the conclusion that long-term solutions will require cooperation between industries with regard to material supply chains. Graedel et al. established recycling rates for many different metals, including material criticality as a consideration, and helped to establish a quantification for recyclability of raw materials to be used in further sustainability assessments in the future.[36] The EU critical raw materials assessment also categorized critical materials, but included end of life recyclability.[37] Kolotzek et al. also attempted to calculate material sustainability using environmental and social aspects, but additionally included supply risk in their calculation.[38] The product sustainability index utilizes triple bottom line sub-indices to measure product life cycle sustainability, and attempts to account for recyclability. This index utilizes large amounts of data from a company for each product, requiring an abundance of information.[39] This is where the chemical element sustainability index is unique, as its use of sustainability indicators reduces the data collection necessary, and uses environmental, social, and economic indicators, as well as recyclability, in order to quantify the sustainability of chemical elements.[40] This index includes "hotspots" in its evaluation, providing a clear avenue to the aspect most requiring improvement, but also highlights a need for a global recycle rate data collection initiative.

Not all industries utilize the supply of raw materials as their input however, and therefore must include a measurement of the sustainability of their supplier. In order to determine the principles that are imperative for supplier selection, Lu et al. utilized multi-objective decision analysis.[41] In order to evaluate environmental performance of the supplier, Tuzkaya et al. presented a hybrid fuzzy multi-criteria decision method.[42] For a similar goal, Awasthi et al. established a fuzzy technique for order of preference by similarity to ideal solution (TOPSIS) approach.[43] In order to account for hazardous substance management, Hsu and Hu created a hazard substance-based analytic network process method.[44] In order to support sustainable supplier selection under incomplete information, Buyukozkan and Çifçi introduced a fuzzy multi-criteria decision framework.[45] Much of the difficulty in the area can be attributed to the multitude of activities that can be incorporated into a sustainability metric for supply chains.[46,47] This multitude of aspects can perhaps best be categorized using a triple bottom line approach, which has been utilized by Govindan et al.[48] This research utilizes the triple bottom line in combination with fuzzy TOPSIS to capture these many, and often qualitative, aspects of sustainable supplier selection. While these assessments of supply chain sustainability have been extensively explored, producing very in-depth analysis of the environmental and economic aspects, the social dimension has suffered from a lack of agreement on what considerations to make and how to measure the social dimension of sustainability, if it has not been neglected altogether.[49] Perhaps the most advanced study in this respect has been presented by Resat and Unsal, who proposed the use of a multi-objective optimization tool to assist decision making when establishing a supply chain.[50] This tool included optimization in all three aspects of the triple bottom line, and despite a low weighting of social aspects during usage, this presents an imperative first step toward more complete sustainability analyses of supply chains through optimization.

When exploring sustainability measurement in industry, it can be found that multiple frameworks exist for specific industry areas. One very common field for this quantification is agriculture. Many different indicator-based tools have been developed in the area[51–55], however this wide range of work has led to a multitude of indicators without a common consensus.[13,56–58] Part of the reason for this is that little work has been performed to validate the indicators that are suggested.[59] Therefore, as suggested by Talukder and Blay-Palmer, a multi-criteria decision analysis has been utilized in combination with a validation framework in order to create a new agricultural sustainability index.[59,60] This exploration both addressed the weaknesses in the area and integrated the use of economic, environmental, and social parameters into the sustainability metric, presenting the most thorough agricultural sustainability assessment to date. Additionally, an effort has been made to quantify the sustainability of the agriculture industry in a time-dependent manner. Despite calculating only the environmental sustainability, the sustainability index by Sands and Podmore is novel in its application of the index over different time scales.[61]

The steel industry has also been heavily involved in the formation of sustainability considerations, with forty-two companies that are members of the International Iron and Steel Institute contributing toward establishing sustainability indicators.[62] Other studies established a standard set of indicators for companies that attempts to address all aspects of sustainable development, as well as a model for integrated assessment of these indicators.[63,64] From these studies, Singh et al. attempted to develop composite indicators to include the economic, environmental, societal, governance, and technical aspects of sustainability in order to establish a composite sustainability performance index.[65] This study is the first in its area to expand on the base of economic, environmental, and societal focuses in order to fully and properly encompass the issues within the steel industry that should be quantified.

As the process industry has expanded and pushed for larger throughput, the demand for solvent-based processes has risen. In order to understand the sustainability issues of the process, many factors such as waste production, toxicity, recyclability, and life cycle impacts must be considered.[66–68] This has led to the proposition of a sustainability index specifically for solvent-based processes.[69] This proposed industrial environmental index attempts to quantify the sustainability of a process through material usage, equipment, energy requirements, environmental health and safety, and life cycle analysis. What is quite significant in this paper, however, is that it proposes the inclusion of process safety in a sustainability metric, a recent but steadily growing consideration in sustainability analysis.

As processes expand and the demand for more sustainable processes rises, process intensification arises as a method of improving environmental impact.[70] Therefore, the development of a sustainability evaluation through intensified alternatives was developed by Bravo-García et al.[71] This framework not only proposes the introduction of sustainability assessment to process intensification, but also presents a sustainability metric that can be performed in early process design and considers the inherent safety of a process.

It is widely understood that changes to a process are much simpler and more effective when applied in the design stage of a project, rather than approaching change during the operational stage.[72] Additionally, the sequential approach of developing a design then retrofitting often leads to suboptimal system designs when multiple objectives should be considered.[73] Specific sustainability aspects have been considered for conceptual design evaluation, such as integrating environmental impacts into process design for separation processes.[74] An index by Ordouei et al. was developed to continue this conceptual design sustainability measurement approach. This approach introduced a composite sustainability index that measures the environmental, energy, and risk reduction aspects of various process alternatives using the limited process data that would be available at the conceptual design stage.[75] By applying the index to a hydrogenation process, the authors demonstrated how the considerations of safety risk and impact of energy on the environment lead to differing results between their methodology and the waste reduction (WAR) algorithm. This case serves as an example that sustainability metrics can successfully determine the more sustainable design for real industrial processes with limited information.

Perhaps the most publicized industry for sustainability concerns is the energy industry, due to the environmental impact of energy generation from conventional methods. At the highest scale, the Energy Sustainability Universal Index proposed by Silva et al. focused on assisting the proper use of energy sources.[76] In a more narrow scope, Infante et al. propose a framework for the economic, environmental, and social aspects of the oil and gas industry.[77] Vital industries such as the food industry have also explored the sustainability of their own energy usage, with Hall and Howe developing an analysis of energy management and waste reduction, developing a triple bottom line analysis that could be furthered by a quantitative analysis.[78]

When analyzing potential renewable energy sources, the sustainability analyses are quite extensive, and biodiesel has certainly drawn attention due to its potential capability as an oil substitute, with many research endeavors exploring the economic, environmental, or social aspects of the process.[79–81] Three studies in the area presented especially unique measurements. The first was a conceptual framework which utilized system dynamics in order to provide a more thorough understanding of the interrelationships of the various aspects of sustainability within the process.[82] The second paper expands on the triple bottom line to analyze the biodiesel process using economic, environmental, social, political, and technological aspects.[83] The third was an energy integration study that aimed to assess the sustainability of an integrated biofuels system using indicators that go beyond the utility requirements and total annual cost to include environmental impacts and safety.[84] This study provided evidence that an optimal solution of sustainability and process safety improvement can be reached through partial integration, and provided an initial inclusion of process safety considerations for sustainability in this area.

Perhaps the most recent development in process sustainability measurement is the inclusion of process safety. While many safety indices have been developed, the inclusion of safety in sustainability indices is quite limited. However, in the process design stage, inherent safety is being increasingly considered, and therefore when establishing the sustainability aspects of a process at the design stage, it is optimal to consider inherent safety simultaneously to achieve the optimal process design. To this effect, the safety and sustainability weighted return on investment metric was developed as method of multi-objective optimization in process design. This expansion on the original sustainability weighted return on investment metric by El-Halwagi serves as an easily interpretable optimization framework for a process.[85] Specifically, this is a metric that uses return on investment as a familiar framework for process engineers, while also considering the safety and sustainability indicators of a project when determining if the return on investment calculation reaches a threshold value.[73] While there are few combined safety and sustainability analyses, they represent a shift and improvement in the inclusion of process safety in process sustainability.

Among the many methods of sustainability quantification, the aspect of time dependence remains heavily underrepresented in literature. As equipment ages, technology advances, processes advance, and the sustainability of the process does not remain a constant function. Additionally, the relationships between the main pillars of sustainability are rarely explored. Lee et al. noted that many assessments do not integrate the different aspects of sustainability, and developed a sustainability measurement which attempted to address the lack of connections within a multidimensional sustainability measurement. This was accomplished using system dynamics for the interrelationships between the sustainability pillars and the triple bottom line for the multidimensionality of product service systems.[86] Additionally, this metric was explored over time, in order to properly represent the dynamic nature and progression of sustainability. The work by Lee et al. sets a precedent for the proper way of addressing sustainability quantification, and establishes the precedent of using interrelated, multidimensional sustainability measurements.

4 Results and Discussion

The current state of research on sustainability measurement is analyzed using two sets of results. The first is a trend analysis of article publication over time. This consists of results from two scientific article databases, Science Direct and EBSCOhost, and provides the number of articles published each year that contain the keywords provided in the figures as a Boolean phrase within the article. The second result is a comparison of four existing sustainability indices, and the results they yielded when used for two different processes. These results provide a clear understanding of the development of sustainability quantification studies over time, as well as a comparison of the results from these studies, and implications of the necessary next steps in the field.

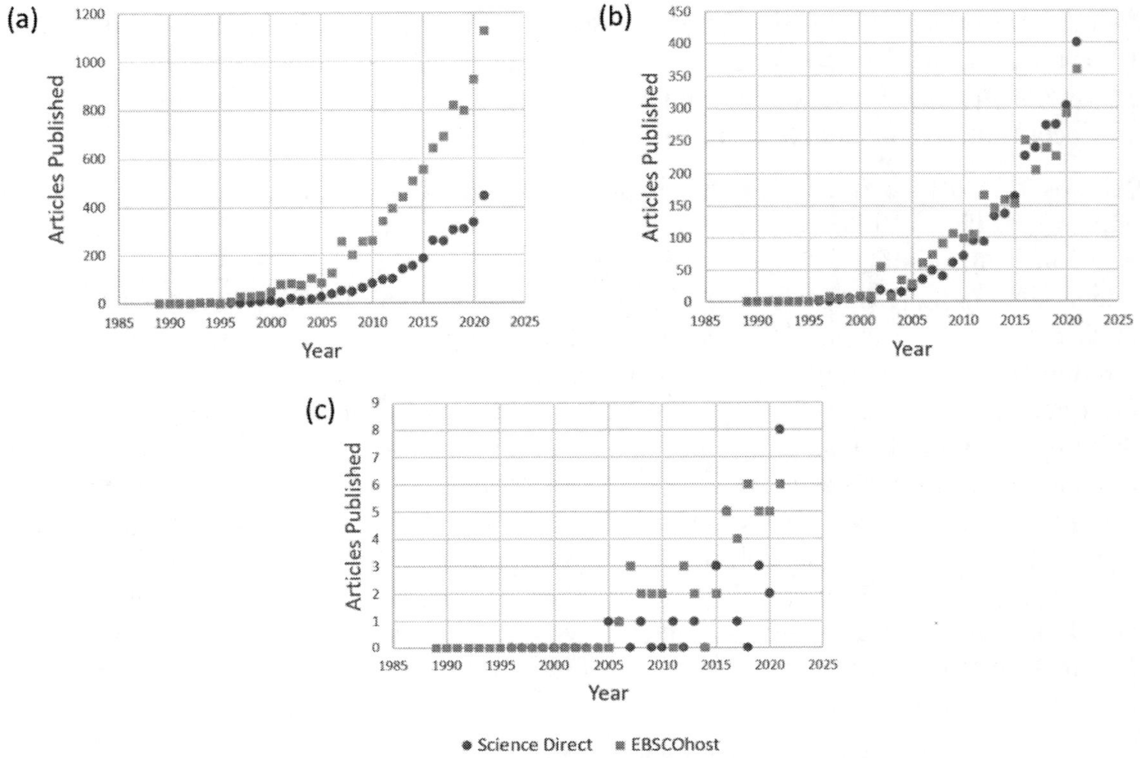

Figure 1: **(a)** Articles published for specified databases using search terms "sustainability index" **(b)** Articles published for specified databases using search terms "sustainability index" and "environment" **(c)** Articles published for specified databases using search terms "sustainability index" and "process safety"

Figure 1(a) depicts the development over time of research into sustainability quantification. As the most common measure of sustainability, sustainability indicators and indices adequately reflect the trend of sustainability quantification efforts. Research in the area mostly began in 1989, in which the Brundtland Commission established their definition of sustainability and began the worldwide discussion of sustainability as a field. Research in the area remained quite limited until the late 1990s, with the research in the area consistently increasing beginning around 2002. This is significant as the World Summit on Sustainable Development occurred in 2002, and the United Nations adopted their current definition of sustainability, leading to a central definition within research efforts as well. However, most of these sustainability indices focused solely on the economic aspects of sustainability or disregarded the environmental concerns.

Figure 1(b) depicts the inclusion of environmental considerations into the research that quantifies sustainability. The first research in this area was developed in 1994, at the introduction of the triple bottom line. Just as the TBL concept introduced to industry the importance of environmental considerations rather than solely economic aspects, the focus in research also developed to consider both economic and environmental effects. It can be seen that research remained rather limited again until 2002, in which there is a spike in research articles. The rise of research in the area reflects the importance of developing an all-encompassing sustainability metric that considers the TBL aspects.

Figure 1(c) depicts the inclusion of process safety into the research that quantifies sustainability. The first research in this area was published in 2005, but the research remains at an early stage, with no discernible trend in publications. It can be stated that there is comparatively less research including process safety in sustainability measurement despite the increase in process safety research in the past decades.

In order to evaluate the gaps within the field of sustainability measurement, it is necessary to observe the performance of existing metrics. Specifically, four sustainability indices have been utilized to compare their performance and evaluate the commonalities and discrepancies among the results ([Index paper 1][87]; [Index paper 2][88]; [Index paper 3][89]; [Index paper 4][90]). This analysis was performed for two processes, ethylene production and hydrogen production via steam-methane reforming. Ethylene production was selected as it presents a historical process that is imperative for many industries and will continue to be utilized long-term. Hydrogen was selected as it is a rapidly advancing process that is the subject of tremendous research due to its potential applicability as an energy solution, and therefore its potential usefulness for renewable energy in the future. These indices were separated into their economic, environmental, and societal indices, normalized to a 100-point scale, and compared.

Many existing sustainability indices refrain from including a scale or categorization for the results, rather preferring to present the raw numerical data to be interpreted by the user. While this allows for different criteria to assess different systems, it may allow the numerical results to be misinterpreted or misunderstood. However, for those that do provide scales for their measurement, the range between different categories tends to be between 5-20 on a 100-point scale. Therefore, values with anywhere greater than a 5-point difference can result in a different categorical result, such as the difference between low and medium sustainability ratings. This analysis will therefore consider a difference of 10 points to be significant.

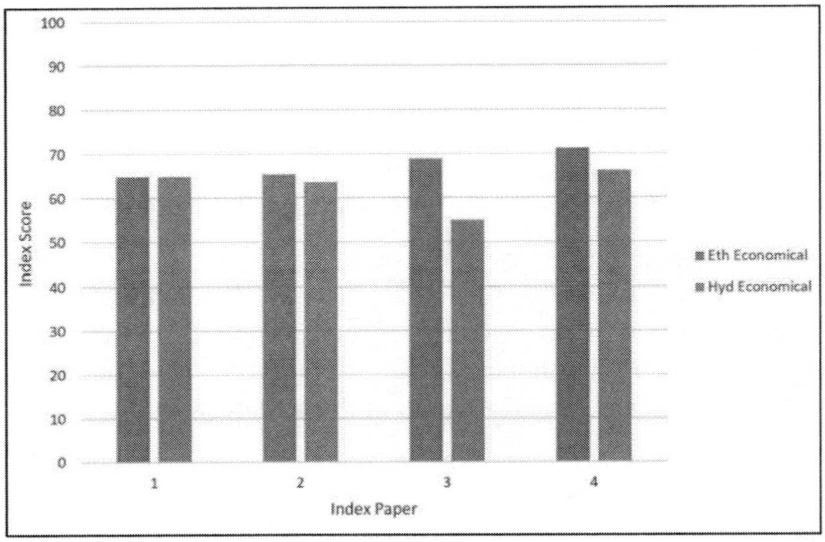

Figure 2: Economic Pillar Comparison

For ethylene production, the results of the economic pillar of the sustainability indices was rather consistent, remaining within a range of 6.25, representing each index establishing a relatively similar value for the economic sustainability of the process. Hydrogen production resulted in significant variability in ratings, however, with a range of 11.25. The final result of determining which process is more economically sustainable is mostly consistent, with three indices rating ethylene production as more economically sustainable and only one index resulting in the same rating for both processes. Therefore, the economic pillar demonstrates significant variability in the sustainability rating of hydrogen production, but consistent determination of the more economically sustainable process.

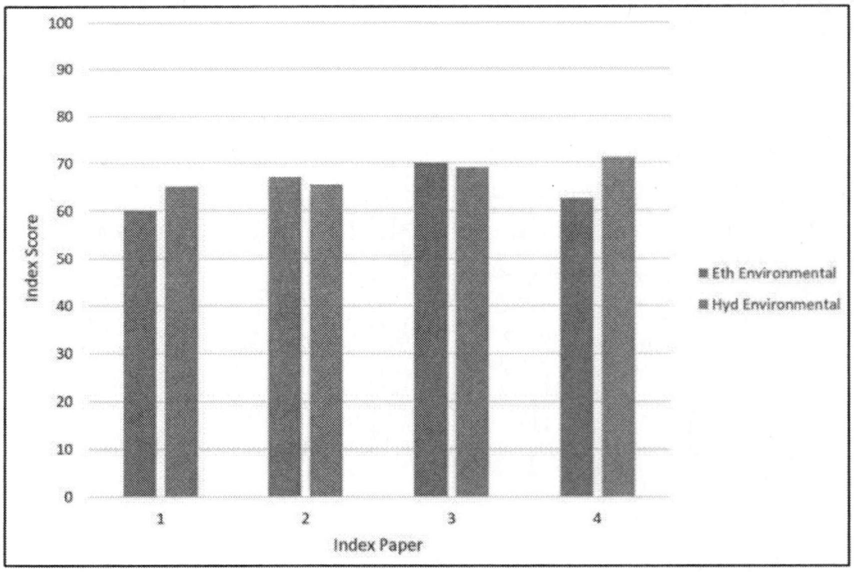

Figure 3: Environmental Pillar Comparison

For ethylene, the range of results between these indices varies by an index score of up to 10. This reaches the levels at which the results would be considered inconsistent. The hydrogen results are more consistent, remaining within a range of 6.25, therefore the indices similarly rank the environmental sustainability of hydrogen production. When comparing the results of both columns to judge the consistency of which process would be determined "more sustainable", there is significant variability. Indices 1 and 4 both depict hydrogen production as being more environmentally sustainable, while indices 2 and 3 depict ethylene production as being more environmentally sustainable. Therefore, the environmental pillar demonstrates significant inconsistency in the determination of a more sustainable process, and in the environmental sustainability rating of ethylene production.

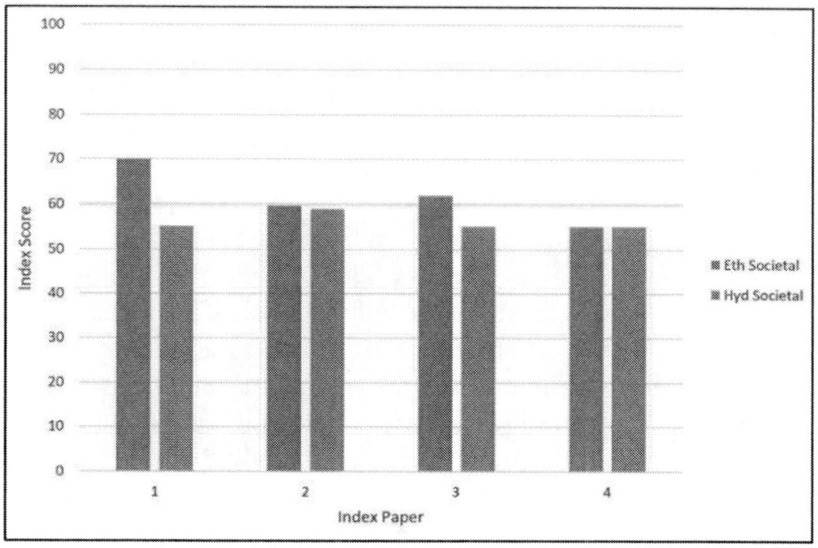

Figure 4: Social Pillar Comparison

The ethylene ratings for the social pillar vary greatly, with a range of 15 in the index score. This is a very significant difference that would certainly imply disagreement between the results. The hydrogen ratings are much more consistent, remaining within a range of 3.93, and demonstrating very consistent societal sustainability ratings. The comparison of the two processes also yields consistent results, with three indices determining that ethylene is more socially sustainable and one index resulting in the same rating for the two processes. Therefore, the social pillar demonstrates significant variability in the sustainability rating of ethylene production, but consistent determination of the more socially sustainable process.

The variability among results of the various indices presents an underlying challenge. The utilized sustainability indices present significant variation in results, as the ratings of one process range by 10 or more in their index score in each figure. It is also not simply due to variations with a single process, as this significant range of values exists twice for ethylene production and once for hydrogen production. The environmental pillar also presents inconsistencies in the determination of which process is in fact more sustainable. Therefore, the tested sustainability metrics demonstrate significant inconsistency in their results, despite the expansive field of research into sustainability metrics.

5 Discussion

When analyzing the main pillars of sustainability, the major explored categories include the TBL economic, environmental, and social pillars as well as technology, region, and politics expansions. However, what is lacking in this consideration is safety. As seen in Figure 1(c), there is comparatively limited research including process safety within a sustainability index, despite the increase in process safety research in the past decades. At the most fundamental argument for its inclusion, an unsafe process will lead to major incidents that have tremendous economic, environmental, and societal costs. Therefore, safety affects the other main pillars of sustainability, leading to the argument for its inclusion as a pillar of sustainability. Process safety has been evolving, with many companies now committing full teams to process safety considerations. This has led to a plethora of information to utilize when considering how to factor the concept of process safety into sustainability. Like sustainability indices, safety indices have been developed to simply display the safety of a given process. The Dow Fire & Explosion Index considers material and process factors to establish a safety level. The hazard identification and ranking methodology was developed to improve upon this measure by providing additional focus on the surrounding environment and the operating conditions.[91] This established a basis for future work into the ranking of other process-related risks such as the thermal risk from reaction hazards, which provides greater understanding of the risks being undertaken by processes that utilize reactions.[92] As noted, the most efficient and successful process decisions are made at the design stage, which has also already been explored within this field as inherent safety. As inherent safety indices are developed and implemented, the multi-criteria decision analysis models that have been utilized for sustainability can quite logically be expanded to include safety decisions.[93]

With the various inconsistencies between the existing indices that have been presented, it is certainly plausible that there is a missing category that impacts the various pillars of sustainability. Process safety is a fourth measure that highly influences each of the other pillars of sustainability and could provide a more complete and consistent measure of sustainability. Moreover, accounting for the interrelations between the different aspects of sustainability may address the inconsistencies presented by providing connections between indicators from different sustainability pillars, adding interactions that may not have been considered.

6 Conclusions

As the world population grows and the environmental burden increases, sustainability measurement is vital to properly assess, benchmark, and regulate industrial processes. Various approaches have been taken toward representing sustainability, with frameworks, indicators, and indices being prevalent methods. Common frameworks such as life cycle assessment, impact assessment, and triple bottom line serve as the basis for quantification techniques. The triple bottom line has provided perhaps the most readily understandable and adopted framework by industry due to its simplicity as a focus on the economic, environmental, and social aspects of a process. An ideal sustainability metric would analyze all of the economic, environmental, and social aspects of sustainability as well as considering the interrelations between these pillars.

Many sustainability measures have been developed for specific industries such as the agricultural or steel industries, or specific processes such as solvent-based processes. To determine the validity of existing metrics, the consistency of their results must be compared. When utilizing existing sustainability indices, significant variation was found both in the final quantification of sustainability for a single process, and in the determination of a more sustainable process when comparing two processes. These inconsistencies imply the need for additional considerations within the various indices.

7 Future Research

Some of the most recent studies into sustainability have included time dependence, process safety, or triple bottom line expansion. The optimal sustainability metric is one that can be incorporated in the design phase of a project, as this results in the most effective integration into a process rather than inefficient retrofitting. The emerging prioritization of safety within industry as well as the importance of process safety to all pillars of sustainability justifies its inclusion as another pillar in an expanded triple bottom line for the process industry. As a process advances, the sustainability of a system changes, and therefore sustainability is not a constant value. Therefore, dynamic measurement is essential to properly track sustainability. This work proposes that a paradigm shift is necessary, in which sustainability is considered by a time-dependent quadruple bottom line of economics, environment, safety, and society with interrelations between each of these pillars of sustainability.

The next phase of research will consist of developing a new sustainability index that accounts for this paradigm shift, with the expansion of the triple bottom line to include process safety providing a basis for measuring process sustainability. Indicators can be focused to use information available at the process design stage. To properly compare these indicators, they must be accounted for in the same unitless scale. Therefore, the indicators can be normalized by establishing boundary conditions for how each indicator affects system sustainability, then developing a curve to demonstrate the behavior between these boundary conditions. System dynamics concepts can provide a methodology to develop the interrelations between these four pillars of sustainability. This would involve establishing positive, negative, or zero relationships between each pair of indicators, then setting a multiplicative effect to account for those indicators with an interrelation. This procedure provides a way forward towards shifting the paradigm of sustainability measurement.

Acknowledgments

The authors thankfully acknowledge the financial support provided by the Mary Kay O'Connor Process Safety Centre at Texas A&M University.

8 References

1. Mayer AL. Strengths and weaknesses of common sustainability indices for multidimensional systems. *Environ Int*. 2008;34(2):277-291. doi:10.1016/j.envint.2007.09.004

2. Becker J. Making sustainable development evaluations work. *J Sustain Dev*. 2004;12:200-211.

3. Fiksel J, McDaniel J, Spitzley D. Measuring product sustainability. *J Sustain Prod Des*. 1998;6:7-19.

4. Azapagic A. SYSTEMS APPROACH TO CORPORATE SUSTAINABILITY A General Management Framework. *Process Saf Environ Prot*. 2003;81(September).

5. Glavic P, Lukman R. Review of sustainability terms and their definitions. *J Clean Prod*. 2007;15:1875-1885.

6. Lee KH, Saen RF. Measuring corporate sustainability management: A data envelopment analysis approach. *Int J Prod Econ*. 2012;140:219-226.

7. Feil AA, Schreiber D, Haetinger C, Strasburg VJ, Barkert CL. Sustainability indicators for industrial organizations: Systematic review of literature. *Sustain*. 2019;11(3):1-15. doi:10.3390/su11030854

8. Linke BS, Corman GJ, Dornfeld DA, Toenissen S. Sustainability indicators for discrete manufacturing processes applied to grinding technology. *J Manuf Syst*. 2013;32:556-563.

9. Lodhia S, Martin N. Corporate Sustainability Indicators: An Australian mining case study. *J Clean Prod*. 2014;84:107-115.

10. Schuch Bork CA, de Souza JF, de Oliveira Gomes J, Venancio Pappetti Canhete V, De Barba DJ. Methodological tools for assessing the sustainability index (SI) of industrial production processes. *Int J Adv Manuf Technol*. 2016;87(5-8):1313-1325. doi:10.1007/s00170-014-6684-8

11. Morse S, Fraser EDG. Making 'dirty' nations look clean? The nation state and the problem of selecting and weighting indices as tools for measuring progress towards sustainability. *Geoforum*. Published online 2005.

12. Ness B, Urbel-Piirsalu E, Anderberg S, Olsson L. Categorising tools for sustainability assessment. *Ecol Econ*. 2007;60:498-508.

13. Parris TM, Kates RW. Characterizing and measuring sustainable development. *Annu Rev Environ Resour*. 2003;28:559-586.

14. Olalla-Tárraga MÁ. A conceptual framework to assess sustainability in urban

ecological systems. *Int J Sustain Dev World Ecol*. 2006;13:1-15.

15. Anand S, Sen AK. Human Development Index: methodology and measurement. *UNDP Hum Dev Off*. Published online 1994.

16. Waheed B, Khan F, Veitch B. Linkage-based frameworks for sustainability assessment: Making a case for driving force-pressure-state-exposure-effect-action (DPSEEA) frameworks. *Sustainability*. 2009;1(3):441-463. doi:10.3390/su1030441

17. SETAC. A Technological Framework for Life Cycle Assessment. *Soc Environ Toxicol Chem Washington, DC, USA*. Published online 1991.

18. Wernick IK, Irwin FH. Material Flow Accounts: A Tool for Making Environmental Policy. *WRI Mater Resour Proj Washington, DC, USA*. Published online 2005.

19. Escobar-Hernandez HU, Quan Y, Papadaki MI, Wang Q. Life cycle assessment of metal-organic frameworks: Sustainability study of zeolitic imidazolate framework-67. *ACS Sustain Chem Eng*. Published online 2023.

20. Ren J. Life cycle aggregated sustainability index for the prioritization of industrial systems under data uncertainties. *Comput Chem Eng*. 2018;113:253-263. doi:10.1016/j.compchemeng.2018.03.015

21. Klöpffer W. Life-cycle based methods for sustainable product development. *Int J Life Cycle Assess*. 2003;8:157.

22. Kouloumpis V, Azapagic A. Integrated life cycle sustainability assessment using fuzzy inference: A novel FELICITA model. *Sustain Prod Consum*. 2018;15:25-34. doi:10.1016/j.spc.2018.03.002

23. Salem A, Hegab H, Kishawy HA. An integrated approach for sustainable machining processes: Assessment, performance analysis, and optimization. *Sustain Prod Consum*. 2021;25:450-470. doi:10.1016/j.spc.2020.11.021

24. Zortea RB, Maciel VG, Passuello A. Sustainability assessment of soybean production in Southern Brazil: A life cycle approach. *Sustain Prod Consum*. 2018;13(November):102-112. doi:10.1016/j.spc.2017.11.002

25. Sheikholeslami Z, Ehteshami M, Nazif S, Semiarian A. The environmental assessment of tertiary treatment technologies for wastewater reuse by considering LCA uncertainty. *Process Saf Environ Prot*. 2022;168(October):928-941. doi:10.1016/j.psep.2022.10.074

26. Vanclay F, Bronstein D. The state of the art of Impact Assessment. In Environmental and Social Impact Assessment. *Chichester: Wiley*. Published online 1995:xi-xiii.

27. Harvey N. Environmental Impact Assessment: Procedures, Practice and Prospects in Australia. *Melb Oxford Univ Press*. Published online 1998.

28. Vanclay F. The triple bottom line and impact assessment: How do TBL, EIA, SIA, SEA and EMS relate to each other. *Sci World Co, Publ Policy, Environ Assess*. 2004;6(3).

29. Vanclay F. International principles for social impact assessment. *Impact Assess Proj Apprais*. 2003;21:5-11.

30. Elkington J. Cannibals with Forks: The Triple Bottom Line of Twenty-first Century Business. *Capstone, Oxford, UK*. Published online 1997.

31. Svensson G, Ferro C, Høgevold N, Padin C, Carlos Sosa Varela J, Sarstedt M. Framing the triple bottom line approach: Direct and mediation effects between economic, social and environmental elements. *J Clean Prod*. 2018;197:972-991. doi:10.1016/j.jclepro.2018.06.226

32. Padin C, Svensson G. Intertwining ecological, social and economic efforts and achievements of business sustainability in business networks. *Int J Logist Econ Glob*. 2015;6:195-204.

33. Svensson G, Wagner B. Implementing and managing economic, social and environmental efforts of business sustainability propositions for measurement and structural models. *Manag Environ Qual An Int J*. 2015;26(2):195-213. doi:10.1108/MEQ-09-2013-0099

34. Walker PH, Seuring PS, Sarkis PJ, Klassen PR. Sustainable operations management: recent trends and future directions. *Int J Oper Prod Manag*. 2014;34.

35. Ku A, Hung S. Manage raw material supply risks. *Chem Eng Prog*. 2014;110:28-35.

36. Graedel TE, Allwood J, Birat JP, et al. What do we know about metal recycling rates? *J Ind Ecol*. 2011;15:355-366.

37. Report of the Ad-hoc Working Group on defining critical raw materials. Report on Critical Raw Materials for the EU. *Eur Comm*. Published online 2014.

38. Kolotzek C, Helbig C, Thorenz A, Reller A, Tuma A. A company-oriented model for the assessment of raw material supply risks, environmental impact and social implications. *J Clean Prod*. 2018;176:566-580.

39. Shuaib M, Seevers D, Zhang X, Badurdeen F, Rouch KE, Jawahir IS. Product Sustainability Index (ProdSI). *J Ind Ecol*. 2014;18:491-507.

40. Smith L, Ibn-Mohammed T, Reaney IM, Koh SCL. A Chemical Element Sustainability Index. *Resour Conserv Recycl*. 2021;166(November 2020):105317.

doi:10.1016/j.resconrec.2020.105317

41. Lu LYY, Wu CH, Kuo TC. Environmental principles applicable to green supplier evaluation by using multi objective decision analysis. *Int J Prod Res*. 2007;45:4317-4331.

42. Tuzkaya G, Oagen A, Ozgen D, Tuzkaya UR. Environmental performance evaluation of suppliers: a hybrid fuzzy multi-criteria decision approach. *Int J Environ Sci Technol*. 2009;6:477-490.

43. Awasthi A, Chauhan SS, Goyal SK. A fuzzy multi criteria approach for evaluating environmental performance of suppliers. *Int J Prod Econ*. 2010;126:370-378.

44. Hsu CW, Hu AH. Applying hazardous substance management to supplier selection using analytic network process. *J Clean Prod*. 2009;17:255-264.

45. Buyukozkan G, Çifçi G. A novel fuzzy multi-criteria decision framework for sustainable supplier selection with incomplete information. *Comput Ind*. 2011;62:164-174.

46. Eltayeb TK, Zailani S, Ramayah T. Green Supply Chain Initiatives among Certified Companies in Malaysia and Environmental Sustainability: Investigating the Outcomes. *Resour Conserv Recycl*. 2011;55:495-506.

47. Sarkis J. Greening the supply chain. *Springer, Berlin*. Published online 2006.

48. Govindan K, Khodaverdi R, Jafarian A. A fuzzy multi criteria approach for measuring sustainability performance of a supplier based on triple bottom line approach. *J Clean Prod*. 2013;47:345-354. doi:10.1016/j.jclepro.2012.04.014

49. Desiderio E, García-Herrero L, Hall D, Segrè A, Vittuari M. Social sustainability tools and indicators for the food supply chain: A systematic literature review. *Sustain Prod Consum*. 2022;30:527-540. doi:10.1016/j.spc.2021.12.015

50. Resat HG, Unsal B. A novel multi-objective optimization approach for sustainable supply chain: A case study in packaging industry. *Sustain Prod Consum*. 2019;20:29-39. doi:10.1016/j.spc.2019.04.008

51. de Olde EM, Moller H, Marchand F, et al. When experts disagree: the need to rethink indicator selection for assessing sustainability of agriculture. *Environ Dev Sustain*. 2016;19:1327-1342.

52. FAO. Biofuels and the sustainability challenge: a global assessment of policies for biofuels and related feedstocks. Published online 2013.

53. Keulen HV, van Ittersum M, Leffelaar P. Multiscale methodological framework to derive criteria and indicators for sustainability evaluation of peasant natural resource management systems. *Environ Dev Sustain*. 2005;7:51-69.

54. Marchand F, Debruyne L, Triste L, Gerrard C, Padel S, Lauwers L. Key characteristics for tool choice in indicator-based sustainability assessment at farm level. *Ecol Soc*. Published online 2014.

55. Martín-Gamboa M, Iribarren D, García-Gusano D, Dufour J. A review of life-cycle approaches coupled with data envelopment analysis within multicriteria decision analysis for sustainability assessment of energy systems. *J Clean Prod*. 2017;150:164-174.

56. Bell S, Morse S. Sustainability Indicators: Measuring the Immeasurable? *Earthscan, London*. Published online 2008.

57. Bockstaller C, Guichard L, Keichinger O, Girardin P, Galan MB, Gaillard G. Comparison of methods to assess the sustainability of agricultural systems. *J Sustain Dev*. 2009;29:223-235.

58. Schoggl JP, Fritz MM, Baumgartner RJ. Toward supply chain-wide sustainability assessment: a conceptual framework and an aggregation method to assess supply chain performance. *J Clean Prod*. 2016;131:822-835.

59. Valizadeh N, Hayati D. Development and validation of an index to measure agricultural sustainability. *J Clean Prod*. 2021;280:123797. doi:10.1016/j.jclepro.2020.123797

60. Talukder B, Blay-Palmer A. Comparison of methods to assess agricultural sustainability. *Sustain Agric Rev*. Published online 2017:149-168.

61. Sands, G. R. & Podmore TH. A generalized environmental sustainability index for agricultural systems. *Agric Ecosyst Environ*. 2000;79:29-41.

62. IISI. The Measure of Our Sustainability—Report of the World Steel Industry. *Int Iron Steel Inst*. Published online 2004.

63. Krajnc D, Glavič P. How to compare companies on relevant dimensions of sustainability. *Ecol Econ*. 2005;4:551-563.

64. Krajnc D, Glavič P. A model for integrated assessment of sustainable development. *Resour Conserv Recycl*. 2005;43:189-208.

65. Singh RK, Murty HR, Gupta SK, Dikshit AK. Development of composite sustainability performance index for steel industry. *Ecol Indic*. 2007;7(3):565-588. doi:10.1016/j.ecolind.2006.06.004

66. Jiménez-González C, Constable DJC, Ponder CS. Evaluating the "greenness" of chemical processes and products in the pharmaceutical industry - a green metrics primer. *Chem Soc Rev*. 2012;41:1485-1498.

67. Sheldon RA, Sanders JPM. Toward concise metrics for the production of

chemicals from renewable biomass. *Catal Today*. 2015;239:3-6.

68. Velasco MI, Kinen CO, de Rossi RH, Rossi LI. A green alternative to synthetize azo compounds. *Dye Pigment*. 2011;90:259-264.

69. Fadel C, Tarabieh K. Development of an industrial environmental index to assess the sustainability of industrial solvent-based processes. *Resources*. 2019;8(2):1-13. doi:10.3390/resources8020115

70. Hofstetter TB, Capello C, Hungerbühler K. Environmentally preferable treatment options for industrial waste solvent management. A case study of a toluene containing waste solvent. *Process Saf Environ Prot*. 2003;81:189-202.

71. Bravo-García J, Huerta-Rosas B, Sánchez-Ramírez E, Segovia-Hernández JG. Sustainability evaluation of intensified alternatives applied to the recovery of nylon industry effluents. *Process Saf Environ Prot*. 2021;147:505-517. doi:10.1016/j.psep.2020.11.040

72. Chen H, Shonnard DR. Systematic framework for environmentally conscious chemical process design: early and detailed design stages. *Ind Eng Chem Res*. 2004;43:535-552.

73. Guillen-Cuevas K, Ortiz-Espinoza AP, Ozinan E, Jiménez-Gutiérrez A, Kazantzis NK, El-Halwagi MM. Incorporation of Safety and Sustainability in Conceptual Design via a Return on Investment Metric. *ACS Sustain Chem Eng*. 2018;6(1):1411-1416. doi:10.1021/acssuschemeng.7b03802

74. Li C, Zhang X, Zhang S. Environmentally conscious design of chemical processes and products: multi-optimization method. *Chem Eng Res Des*. 2009;87:233-243.

75. Ordouei MH, Elsholkami M, Elkamel A, Croiset E. New composite sustainability indices for the assessment of a chemical process in the conceptual design stage: Case study on hydrogenation plant. *J Clean Prod*. 2016;124:132-141. doi:10.1016/j.jclepro.2016.02.107

76. Silva RM, Silva ARC, Lima TM, Charrua-Santos F, Osório GJ. Energy Sustainability Universal Index (ESUI): A proposed framework applied to the decision-making evaluation in power system generation. *J Clean Prod*. 2020;275. doi:10.1016/j.jclepro.2020.124167

77. Infante C, Mendonça F, Purcidonio P, Valle R. Triple bottom line analysis of oil and gas industry with multicriteria decision making. *J Clean Prod*. 2013;52:289-300.

78. Hall GM, Howe J. Energy from waste and the food processing industry. *Process Saf Environ Prot*. 2011;90(3):203-212. doi:10.1016/j.psep.2011.09.005

79. Buchholz T, Luzadis V, Volk T. Sustainability criteria for bioenergy systems:

results from an expert survey. *J Clean Prod.* 2009;17:S86-S98.

80. Cramer JM. Testing Framework for Sustainable Biomass; Final report from the project group "Sustainable production of biomass." Published online 2007.

81. Kurka T, Blackwood D. Participatory selection of sustainability criteria and indicators for bioenergy developments. *Renew Sustain Energy Rev.* 2013;24:92-102.

82. Musango JK, Brent AC, Amigun B, Pretorius L, Müller H. Technology sustainability assessment of biodiesel development in South Africa: a system dynamics approach. *Energy.* 2011;36:6922-6940.

83. Bautista S, Enjolras M, Narvaez P, Camargo M, Morel L. Biodiesel-triple bottom line (TBL): A new hierarchical sustainability assessment framework of principles criteria & indicators (PC&I) for biodiesel production. Part II-validation. *Ecol Indic.* 2016;69:803-817. doi:10.1016/j.ecolind.2016.04.046

84. Gabriel J, Hernández S, Gómez-castro FI, Romero-izquierdo AG, Conde-mejía C, López-molina A. Partial energy integration between biofuels production processes: Effect on costs, CO2 emissions and process safety. *Process Saf Environ Prot.* 2022;159:918-930. doi:10.1016/j.psep.2022.01.069

85. El-Halwagi M. A Return on Investment Metric for Incorporating Sustainability in Process Integration and Improvement Projects. *Clean Technol Environ Policy.* Published online 2016.

86. Lee S, Geum Y, Lee H, Park Y. Dynamic and multidimensional measurement of product-service system (PSS) sustainability: A triple bottom line (TBL)-based system dynamics approach. *J Clean Prod.* 2012;32:173-182. doi:10.1016/j.jclepro.2012.03.032

87. Jawahir IS, Dillon Jr. OW, Rouch KE, Joshi KJ, Venkatachalam A, Jaafar IH. Total life-cycle considerations in product design for sustainability: A framework for comprehensive evaluation. *Proc TMT 2006, Keynote Pap.* Published online 2006:1-10.

88. Harik R, El Hachem W, Medini K, Bernard A. Towards a holistic sustainability index for measuring sustainability of manufacturing companies. *Int J Prod Res.* 2015;53:4117-4139.

89. Hubbard G. Measuring Organizational Performance: Beyond the Triple Bottom Line. *Bus Strateg Environ.* 2006;19:177-191.

90. Nicoletti Jr A, de Oliveira MC, Helleno AL. Sustainability evaluation model for manufacturing systems based on the correlation between triple bottom line dimensions and balanced scorecard perspective. *J Clean Prod.* 2018;190:84-93.

91. Khan FI, Abbasi SA. Multivariate hazard identification and ranking system. *Process Saf Prog*. 1998;17:157-170.

92. Wang Q, Rogers WJ, Mannan MS. Thermal risk assessment and rankings for reaction hazards in process safety. *J Therm Anal Calorim*. 2009;98:225-233.

93. Janošovský J, Rosa I, Vincent G, et al. Methodology for selection of inherently safer process design alternatives based on safety indices. *Process Saf Environ Prot*. 2022;160:513-526. doi:10.1016/j.psep.2022.02.043

Application of Inherent Safety Principles to Projects

Tim Hoff
ExxonMobil Technology and Engineering Company
22777 Springwoods Village Parkway
Spring, TX 77389
Timothy.P.Hoff@exxonmobil.com

Keywords: Projects, "Inherently Safe", "Inherent Safety"

Abstract

Projects, especially large capital projects, often increase the residual risk profile of an operating plant through the addition or modification of equipment and processes. Management of those new risks is normally achieved through hazard identification and subsequent application of administrative or engineered safeguards. Maintaining those risks at an acceptable level over the life of a facility requires maintenance of those safeguards through mechanical integrity programs, training and procedures, and appropriate management of change.

The best way to manage risk in a new operation is to never create it, instead applying inherent safety principles during the project phase, which serve to reduce risk by the removal or reduction of the hazard instead of controlling the hazard. By properly applying these principles in the early stages of a project's timeline, operational risks can be eliminated or greatly reduced and, as an additional benefit, total project costs can likely be reduced through elimination of project scope. Application of inherent safety principles can also reduce operating costs through elimination of installed safeguards and their associated life cycle maintenance costs.

1 Introduction

This paper provides guidance on the approach to be taken to complete inherent safety, health, and environmental Reviews (ISHERs) during the development and design of new facilities.

One basic principle underlying the incorporation of safety into plant design is the recognition and elimination of hazards, including the consideration of fire, explosion, and other accident potentials. Hazards associated with a process should be identified and thoroughly evaluated during the conceptual stage. Efforts should then be made to eliminate or reduce the hazard through the application of inherent safety, health, and environmental principles. The focus of inherent safety is to avoid creating a hazard or to minimize one if its inclusion is required for the design, rather than creating a hazard and relying on the use of one or more safeguards to control it. The recognition of hazards and inclusion of inherently safer technology can be particularly difficult in sustainability projects, as hazard awareness may not be as robust as with more mature technologies.

To minimize the impact on project execution, inherent safety should be incorporated as early as possible in the design. As a project progresses through the project management system, the ability to cost-effectively develop an inherently safer design diminishes and additional safeguards (e.g., SHE critical devices, SIFs) might be required to mitigate risks identified later in the project timeline. Properly applied, inherent safety can result in cost reduction because vessels can be eliminated or their volumes reduced, operating conditions can become less severe, and less safety equipment may subsequently be required. However, even after inherent safety principles are applied, some extrinsic safety devices will still likely be required to further reduce the risk of a loss of containment, but properly applied inherent safety principles will minimize the quantity and/or extent of such safeguards.

2 ISHE Principles

There are four inherent safety, health, and environmental (ISHE) principles which can reduce the risk associated with hazardous materials:

1. Substitution

   "Substitute a more hazardous material with a less hazardous one, where feasible"

   The substitution can involve process raw materials or utilities. For instance, in selecting water treatment chemicals, chlorine can be replaced by sodium hypochlorite. Process raw materials need to be considered early in the development stage, but utilities (e.g., low-pressure steam vs high-pressure steam) may be substituted even in later design stages.

2. Reduction

   "Reduce the inventory in storage and process vessels to a practical minimum"

   Inventory can be reduced by reducing hold-up or surge capacity or by adding a drumless condenser. Reduction can also be applied by decreasing flanged connections and increasing the use of welded pipe. Less obvious ways include increasing the reaction rate, possibly through better mixing or increasing yield to reduce recycle rate; both of which reduce reactor volume.

3. Attenuation

   "Reduce the severity of the conditions under which hazardous material is used"

   For exothermic or decomposition reactions, heat input might be limited by selection of heating medium with an appropriate temperature limit. Also, the reactant might be diluted to limit the exotherm. Liquid phase exothermic reactors limit runaway reacting temperature rise to the liquid boiling point, whereas vapor phase exothermic reactors are more likely to rise in temperature until a runaway reaction occurs. Avoiding the generation of static electricity is another form of attenuation.

4. Simplification

   "Simplify the process to reduce the potential for operator error"

   Operational flexibility should be weighed against the increased complexity which it creates and the increased associated risk of operating error. In tank farms, jumpovers connecting multiple tanks increase the potential for incorrect lineups that could result in loss of containment from adverse chemical reactions or creation of an unexpected flammable atmosphere.

The ISHE focus is on the avoidance of hazards rather than on their control by the addition of protective equipment. To strive for inherently safer designs, efforts should be made to eliminate or reduce the hazards through applications of these four ISHE principles.

3 ISHER Methodology

The preferred methodology for a large project ISHER is a hybrid combination of checklist review and what-if analysis, although for simple, smaller projects, a single method can be used. The ISHER Leader should determine the best methodology for the project, based upon their understanding of the scope and familiarity with the hazards. A summary of the various ISHER methods is below.

The checklist review and what-if analysis are both beneficial methods for early design stage ISHERs, as they promote "out-of-the-box" thinking and the generation of ideas that might be suitable in an existing facility or future plant concept.

3.1 Checklist Review

A checklist review is an experience-based approach and is versatile. The type of evaluation performed with a checklist can vary; it can be used quickly for simple evaluations or for more extensive in-depth evaluations. The success of this technique depends on the experience of the checklist developer as well as those who use the checklist.

The ISHER leader must use their own experience to generate a project-specific checklist centered on ISHE principles. Example checklist questions include:

ISHE Principle	Question
Substitution	Can a flammable solvent be replaced with water?
	Is it possible to completely eliminate hazardous raw materials, process intermediates, or by-products by using an alternative process or chemistry?
Minimization	Can hazardous finished product inventory be reduced?
	Can alternative equipment with reduced hazardous material inventory be used?
Moderation	Is it possible to limit the supply pressure of raw materials to less than the maximum allowable working pressure of the vessels to which they are delivered?
	Is it possible to make the reaction conditions (temperature, pressure) less severe by using a catalyst or a different catalyst (e.g., structured or monolithic vs. packed bed)?
Simplification	Can equipment be designed such that it is difficult or impossible to create a potentially hazardous situation due to an operating or maintenance error?

There is no single checklist that can or should be followed in successfully completing an ISHER. The checklist that works well for the circumstances of a particular project might not be appropriate for another project. For example, utilizing a process hazard analysis checklist for a project focused on control systems or analyzers might not successfully accomplish the ISHER objectives.

3.2 What-If Analysis

The what-if analysis technique is a brainstorming approach in which a group of experienced people familiar with the subject process ask questions or voice concerns about possible undesired events. The what-if analysis is not as structured as some other techniques. Instead, it is a fit-for-purpose technique that allows the leader to adapt the base concept to the project's particular process or installation.

The what-if analysis concept encourages the hazard evaluation team to think of questions that begin with "What-if…" and should include all modes of operations such as steady state, start-up, and shut-down. However, any process safety concern can be voiced, even if it is not phrased as a question. For example:

- "I'm concerned about the supplier delivering the wrong reactive material."
- "What if Pump A stops running during start-up?"
- "What if the operator opens valve B instead of valve A?"

For a what-if analysis to be effective as an ISHER, team members must be well-versed in the principles of ISHE. The team would brainstorm process deviation scenarios and identify inherent safety improvements for reducing or eliminating the potential for the scenario to develop. The inherent safety principles can be used to determine the safety improvements.

For example, a team might ask "What if the chlorine container at the cooling tower developed a leak in the tubing?" The analysis might determine that the vaporized chlorine could affect a nearby public facility. After discussing existing safeguards, a recommendation might be developed to consider replacing the chlorine cylinder with a solid biocide (substitution). The potential for chemical reactivity hazards associated with the use of the biocide would need to be evaluated as part of the determination of residual risk.

Another analysis might evaluate the question of "What if the sulfuric acid was inadvertently loaded into the 50% caustic tank?" The team may conclude, after evaluation of the chemical reactivity hazard and the severity of consequences, as well as existing safeguards (e.g., procedural verifications, unique connections, and emergency relief protection), that no action is warranted. Discussion might nevertheless proceed to whether opportunities exist to substitute, minimize, moderate, or simplify.

The what-if methodology can be limiting in sustainability projects where the ISHER team is not familiar with the technology or hazards. It is imperative that the ISHER facilitator researches industry knowledge or utilizes external technology specialists to identify the major hazards of concern.

3.3 What-If / Checklist Analysis

This hybrid method capitalizes on the strengths and compensates for the individual shortcomings of the separate approaches.

The checklist analysis is an experience-based technique, and the quality of a hazard evaluation performed using it is highly dependent on the experience of the checklist's authors. If the checklist is incomplete, then the analysis may not effectively address a hazardous situation. The what-if analysis encourages the hazard evaluation team to consider potential abnormal situations and consequences that are beyond the experience of the authors of a good checklist, and thus are not covered on the checklist. Conversely, the checklist portion of this combined technique provides a more systematic balance to the what-if analysis.

Like most other hazard evaluation techniques, the method works best when performed by a team experienced in the subject process. This technique is generally used to analyze the most common hazards that exist in a process. Although it can evaluate the significance of incidents at almost any level of detail, the what-if/checklist analysis method usually focuses on a less-detailed level of resolution than other techniques. Often, a what-if/checklist analysis is the first hazard evaluation performed on a process and is a precursor for subsequent, more detailed studies.

For a what-if/checklist analysis, the hazard team leader obtains or develops an appropriate checklist for the team to use in conjunction with the what-if analysis prior to the team session. The team will first complete a what-if analysis consistent with the guidance found above in the What-If section. Once the team has identified all the questions and issues it can in a particular area or step of the process or activity, the hazard team leader will then use the checklist to fill any gaps. It is preferred to complete the what-if analysis prior to the checklist analysis to avoid letting the checklist restrict the creativity and imagination of the team.

4 Timing and Scope of ISHER

The benefit of a focused ISHER decreases as a project timeline progresses. The ideal time for hazard identification and mitigation through inherent safe design is early in the project, before scope definition is finalized and equipment development begins, when it will provide maximum benefits.

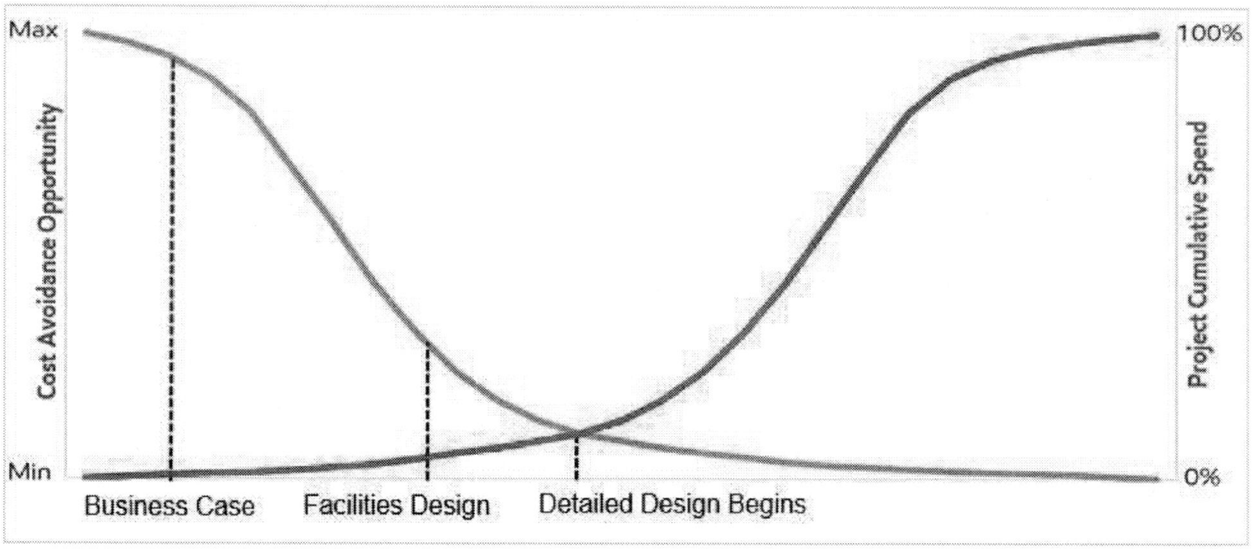

4.1 Business Planning Stage

During this stage, an ISHER should cover all viable alternatives and scopes. The ISHER completed at this time will be beneficial for selection of the lead case, minimizing the potential for cost growth at later stages. As scope and details are limited at this stage, the focus should be on identifying hazards that can be showstoppers, or high-risk items that would drive looking for an alternative technology or project as a whole.

For example, is the project planning on installing a large light ends pressurized vessel close to the site boundary? This installation may not meet facility siting guidelines and could lead to BLEVE concerns to public receptors. The recommendation at this stage would be for the project team to consider relocation of the vessel where BLEVE would not have public impacts. If identified later in the project when the plot space is considered final, relocation of the vessel could be cost-prohibitive, and a project would be required to install designed safeguards to mitigate the risk. But in this instance, the hazard to public would still remain if the safeguards were to fail.

Another example, what is the routing of a proposed CO2 pipeline? If a project has several potential route options, the ISHER should give direct commentary on the inherent safety concerns of each option and which of the routes would result in the inherently safest option (e.g., lowest exposure to the public or environment), regardless of expected cost. Late project stage discovery that the pipeline routing has major public impacts can lead to socioeconomic consequences (such as buying land or relocating communities) or installation of costly safeguards, such as gas detection and automated response.

4.2 Concept Selection Stage

During this stage, an ISHER should review the lead project design case. The ISHER at this time will be beneficial to potentially reduce the number of safeguards that might be necessary to manage the project's risks. As estimated design conditions and plot plans should be available, the focus should be on equipment and spacing modifications that can reduce overall project risk profile.

For example, can some vessel design contingencies be made "non-credible" if the maximum allowable working pressure is increased slightly and therefore greatly reduce the size and quantity of the required safety valves?

Further, is equipment spaced such that fire-fighting and suppression is difficult? In dealing with flammable or highly volatile chemicals, a successful inherent safety approach is to adequately space equipment such that a singular loss of containment or fire will not escalate through inclusion of proximate equipment. Through this approach, risk reduction is not heavily dependent on automated or human response to the fire.

4.3 Definition/FEED Stage

After concept selection stage, the benefit of a formal and focused ISHER decreases. ISHER can still be completed at the request of the project team at this stage, but the focus should be on proper resolutions and inclusion of inherently safer design principles from prior ISHERs instead of identifying new opportunities.

While inherent safety is still expected to be applied within the project, applying inherent safety principles becomes costly upon discovery at this stage, and preventing or mitigating safeguards are generally applied instead. Hazard and risk identification is generally completed through the HAZOP process at this and later stages.

5 ISHER Team Composition

Ideally, an ISHER team should consist of four to seven members, but membership will be dependent on the hazards and scope of the project. Smaller complexity or scope project ISHERs can generally be completed with a smaller team (individuals can serve in multiple roles), while ISHERs for larger projects will require a more complex analysis and involve a team from various disciplines.

5.1 Team Membership by ISHER Timing

Timing of the ISHER is also key in determining membership. While a smaller project won't require a large membership due to limited scope and hazards, a large project should require membership as reflected in the table below:

	ISHER Timing		
Role	Business Planning	Concept Selection	Definition/FEED
Process Safety Engineer	Required	Required	Required
Environmental Engineer	Required	Required	Required
Technology Expert	Required	Required	Required
Industrial Hygienist	Required	Required	Required
Project Development Lead	Required	Optional	Optional
Process Designer	Optional	Required	Required
Operations Representative	Optional	Required	Required
Maintenance Representative	Optional	Required	Required
Vendor Representative	Optional	Optional	Optional

5.2 Team Member Roles and Responsibilities

While all ISHER team members should be well-versed in the principles of inherent safety, each ISHER team member provides a specialized set of skills to the review.

Process Safety Engineer

- Facilitator for safety component of ISHER
- Provides expertise in inherent process safety considerations
- Responsible for ensuring ISHER team is aware of the project's particular hazards
- Develops checklist for ISHER and facilitates What-If conversations
- Leads team through process safety checklists
- Facilitates conversations and ensures documentation correctly captures ISHER concerns

Environmental Engineer

- Facilitator for environmental component of ISHER
- Provides expertise in inherent environmental considerations
- Leads ISHER team through environmental discussion

Process Technology Expert

- Provides knowledge of technical process and how hazards are controlled at similar installations
- Deeper understanding of process to bridge gap of knowledge among other disciplines

Industrial Hygienist

- Provides expertise in understanding of hazards inherent to personnel exposure to materials upon opening of equipment (e.g., delayed coking, catalyst change-outs) or loss of containment.

Project Development Lead

- Provides background and understanding of project scope and basis
- Brings explanation of project alternatives and can discuss rationales of each

Process Designer

- Provides technical scope and explanation of design and operating considerations

Operations Representative

- Provides critical information on operation and maintainability of equipment and process units and associated hazards

Maintenance Representative

- Provides critical information on maintainability of process equipment and associated hazards

Vendor Representative

- Provides specific knowledge to the vendor scope and details that might not readily be available otherwise
- In new technologies, provides deeper hazard recognition that might not be readily available to the ISHER team

6 Documentation and Resolution of Findings

No matter the methodology chosen for the ISHER, documentation must be completed that captures the discussions around inherent safety in a project and associated recommendations. This documentation is critical in reflecting that a systematic process and review was completed and whether the inherently safest technology was considered and incorporated.

The ISHER documentation is provided to the project team for actioning of the findings. It is important to note that while the application of inherent safety principles will result in a lower risk profile, they are not always cost-beneficial to apply (e.g., larger plots and greater equipment spacing will decrease risk due to fire escalation, but cost of piping and plot space will increase). It is the responsibility of the project team to align with the risk owner to determine the ideal course of action between application of inherent safety, designed safeguards, and risk acceptance.

7 Sustainability ISHER Case Study

A grassroots hydrogen plant is being developed and has progressed into the Concept Selection Stage. The technology being used is familiar to the business and ISHER team, but a majority of the scope is vendor design. While inherent safety principles have been applied to the vendor and non-vendor scope, an ISHER is being completed to review if the inherent process safety of the project scope can be further improved, resulting in a decrease in the number of designed safeguards.

The ISHER facilitator determines that a What-If approach would be the most beneficial approach in this installation, as the project is a combination of known technology and vendor scope. This approach also allows for the Operations personnel who may not be familiar with the technology to be more open to question all parts of the process.

The ISHER facilitator worked with the Project Management representative to determine the proper attendees for an ISHER at this stage. This particular project incorporates technologies on gas treating, converters, utility generation, and general fractionation and ancillary equipment. As the project is considered complex in terms of amount of equipment, existing unit interfaces, and the hazards associated with hydrogen production, it was determined a more diverse and sizeable team was required. Because of the inclusion of the vendor scope, the vendors were also invited to participate to discuss the details and known hazards of their design, and how inherent safety principles have been applied. The final ISHER membership was captured as such:

Title
Process Safety Engineer (Facilitator)
Technology Specialist
Development Planning Engineer
Process Design Engineer
Technology Integration Manager
Offsites & Utilities Engineer
Industrial Hygienist
Operations Lead
Operations Supervisor
Lead Facilities Engineer
Environment & Regulatory Advisor
Process Engineer
Vendor Representatives

The ISHER facilitator leads the team through a systematic line-by-line review of the process flow diagrams that were provided by the project team and vendor. The facilitator allows for open dialogue and receives 'What-If' questions from attendees as they arise, considering all expected operating modes (steady state, start-up, shutdown).

During discussion on the shutdown of the autothermal reformer, the team identifies a potential inherent safety gap associated with oxygen ingress upon shutdown. The project has designed a shutdown system with oxygen flow, but is there an inherently safer way to mitigate this hazard? The facilitator documents the discussion thusly:

ITEM / LOCATION	CONCERN	HAZARD / CONSEQUENCE	SAFEGUARDS	RECOMMENDATION
Autothermal Reformer	Oxygen ingress into other systems after a shutdown if oxygen flow continues for a short period of time.	Potential for formation of a flammable mixture, ignition, overpressure.	Shutdown system, purging.	Consider evaluating the potential for formation of flammable mixtures due to oxygen ingress as systems cool/depressure after a shutdown.

While the facilitator could not formally identify a better way to apply the inherent safety principles, the project is expected to further analyze the process and see if those principles can resolve the base risk.

A closer look at the high-pressure steam feed preheater and downstream piping reflects that the materials of selection may be inadequate, as there is the potential for an infrequent temperature excursion that could exceed the limits of the Nelson Curve. A loss of containment could occur in between downtime inspections, and in this service could result in severe consequences. Applying the substitution principle on materials of construction can fully resolve this risk, and so the finding is documented thusly:

ITEM / LOCATION	CONCERN	HAZARD / CONSEQUENCE	SAFEGUARDS	RECOMMENDATION
HP Steam Preheater	The preheater metallurgy and downstream piping is deficient per Nelson Curve.	Potential for high temperature hydrogen attack at elevated temperature and hydrogen concentrations, resulting in mechanical failure and loss of containment.	None	Upgrade metallurgy to be adequate for operating conditions per Nelson Curve

Applying the inherent safety principle at this time results in risk resolution, instead of having to consider the design and installation of high temperature cut-outs if the existing metallurgy is utilized.

As the conversation shifted to facility siting, the ISHER team identified that the location of the unit plot near a boundary fence line could lead to public exposures in the event of a loss of containment and fire scenario. As the flammability of the material is inherent to the process, the hazard cannot be removed and must be worked around. The project team has not completed any radiation or overpressure estimates yet, so the risk is unknown. Since the design is not proven inherently safe in this aspect, the team documents the concern and recommendation, as inherent safety principles (plot relocation eliminating public exposure) can be applied at this point of the project.

ITEM / LOCATION	CONCERN	HAZARD / CONSEQUENCE	SAFEGUARDS	RECOMMENDATION
Reaction Section	Overpressure contours can be in exceedance of nearby occupied buildings	Potential for fatalities in event of VCE from H2 plant	None	Determine expected overpressure contours associated with new plot, as well as radiant heat levels from a fire; if deficient, consider relocating unit plot to resolve risk to public

During the ISHER, the vendor representatives were key to all of the inherent safety discussions, particularly in discussing the processing requirements of the equipment, reflecting that their design had met the ISHE principles (most notably "attenuation" through reduction of process intensity).

The team documented all findings and discussions, and then formally handed over the documentation to the project team for resolutions. At the conclusion of the ISHER, the ISHER team ensured that the project team understood the concerns as documented and how to develop and progress the recommendations.

8 Conclusion

Application of inherent safety principles to a project has a more substantial and permanent reduction in risk profile, as the focus of inherent safety is the reduction or elimination of hazards. Awareness and understanding of the hazards are critical to the consideration of the inherently safest technology, but that recognition can be difficult in sustainability projects that involve technology or processes new to an organization.

The successful application of inherent safety is dependent on a timely, structured, and focused review of the proposed facility by a multi-disciplined team that is knowledgeable in the technology, associated hazards, and inherent safety principles. Recommendations brought forth by the ISHER team will aid a project in developing a facility that will require fewer designed and installed safeguards.

However, while application of the inherent safety principles into a project will lead to a lower risk, full application might be cost-prohibitive. It is the responsibility of the project team to align with the risk owner to determine the ideal course of action between application of inherent safety, designed safeguards, and risk acceptance.

9 References

Book:
[1] Wiley. *Guidelines for Hazard Evaluation Procedures, Third Edition*. CCPS; 2008.

[2] Wiley. *Guidelines for Inherently Safer Chemical Processes: A Life Cycle Approach, Third Edition.* CCPS; 2019.

Facility Siting Study for CCUS Onshore Facility

Soraya Tamara A Malik
PETRONAS Malaysia
soraya.malik@petronas.com

Shaiful Khalid[1], M Shaik Fareed[1], Chong Shin Lee[1]
[1]PETRONAS Malaysia

Keywords: CCUS, Safety Distance, Carbon Dioxide, QRA, Toxic

Abstract

The development of onshore facility for carbon capture, utilization, and storage (CCUS) involves a centralized facility where captured carbon dioxide (CO_2) emissions from various sources are collected, processed, and stored. While CCUS technology has the potential to mitigate climate change, CCUS onshore facility pose certain hazards and challenges. Performing facility siting studies for CCUS onshore facility is crucial during the early stages of a project to determine the safe location of the facility and minimize potential risks to nearby local communities.

However, the lack of availability of process parameters at the early project stage presents a challenge in providing reliable consequence modeling results, which may lead to delays in project execution. Hence, typical hazard distances for CCUS streams were established based on various operating conditions through consequence modeling. These typical hazard distances are utilized as references for CCUS projects for facility siting requirements to provide a preliminary proposal for the safe location of CCUS onshore facility. This paper presents the methodology used for facility siting and the resultant consequence modeling contours based on various key parameters.

1 Introduction

After the pivotal Paris Agreement of 2016, carbon capture, utilization, and storage (CCUS) has emerged as the coveted green technology implemented in numerous global facilities. This signifies the industry's resolute dedication to curbing greenhouse gas emissions. Within PETRONAS, the swift adoption of CCUS technology in both offshore and onshore facilities emphasizes the company's unwavering commitment to achieving net zero emissions by 2025.

CCUS technology functions to capture CO_2 from substantial point sources, such as power generators or industrial facilities, and subsequently transports the CO_2 for utilization or permanent storage. The rise of CCUS technology results in a significant increase in the handling of CO_2 inventories. Within the field of hazard management, dealing with large volumes of CO_2 represents uncharted territory for most specialists in the field. Here are some potential hazards that may arise from the implementation of CCUS technology:

TOXIC EFFECT:

CO_2 may cause asphyxiation. Inhalation of elevated concentration of CO_2 can cause adverse effects on respiratory system.

INTERNAL CORROSION:

CO_2 in combination with free water will form carbonic acid which is higly corrosive to carbon steels which potentially result in loss of containment (LOC).

LOW TEMPERATURE EMBRITTLEMENT:

Rapid depressurization of CO_2 could lead to low temperature condition 'as low as -56°C' and if the temperature is below the brittle transition temperature, a failure could occur.

THERMAL EXPANSION:

Increase of temperature of an isolated inventory of CO_2 can lead to overpressurization as CO_2 density is sensitive to temperature changes.

A significant CO_2 leak from CCUS has the potential to produce severe consequences, impacting a large area and posing risk to a large number of people. Identifying the CO_2 stream at an early stage of a CCUS project is crucial for effective risk management, especially concerning the substantial risks associated with storing and handling large CO_2 inventories at onshore hubs.

2 Case Study

Projected to store around 10 million tonnes of CO_2 annually by 2030, CCUS hubs are potentially located near industrial clusters and residential areas. Performing a facility siting study in the early project phases is essential to identify secure locations and mitigate potential risks to neighboring communities. However, obtaining accurate process parameters at this stage is challenging, impeding reliable consequence modeling results and potentially causing project execution delays.

Thus, an initiative was established to develop typical hazard distances based on various operating conditions and parameters. The concept is replicated from 'Hazard Radii Look Up Table' outlined in *EI Part 15: Area classification for installations handling flammable fluids*, which is derived from dispersion modeling. The hazard distances are then utilized as references for CCUS projects for facility siting requirements to provide a preliminary proposal for the safe location of CCUS onshore facilities.

Below are the steps in determining the typical hazard distances for CO_2 releases. The resultant hazard distances were then compared with an actual facility siting study performed for an onshore CCUS facilities in an Eastern Malaysia project.

3 Methodology

The methodology for establishing the hazard distances for CCUS stream is in accordance with the generic methodology of consequence modeling as used for offshore installations in the oil and gas industries. The steps involved in the analysis are depicted in Figure 3.1.

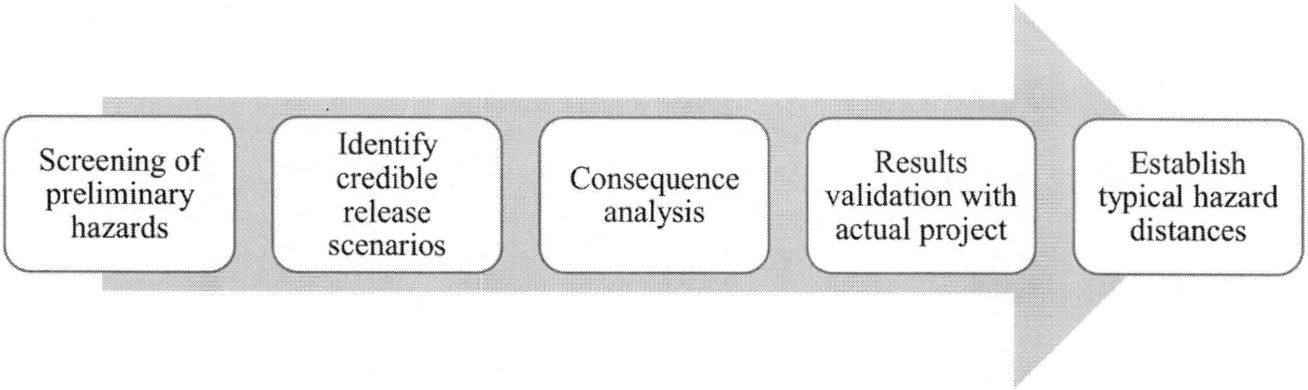

Figure 3-1 Methodology of Determination of Typical Hazard Distance for CCUS Stream

The detailed methodology and the basis are described in Table 3-1 below.

Table 3-1 Detailed Methodology of Determination of Typical Hazard Distances

Step	Description
Screening of preliminary hazards	Identify potential exposures from new facility to the surrounding area, or to a new facility from an adjacent hazard. For risk evaluation, the following information should be made available but not limited to, for risk evaluation: ▪ List of raw materials, intermediates, products, and by-products. ▪ Description of how the feedstocks and products are transported into, within, and from the facility and storage facilities. ▪ Preliminary process operating conditions and operating modes (batch/continuous). ▪ Environmental discharges such as potential air plumes from stack and vents.
Identify credible release scenarios	The credible release scenario is based on a credible event which has reasonable probability of occurrence considering the operating conditions and inventory in the facility. The credible release scenario should consider the following: ▪ Industry accident history which shows that many releases often occur from equipment failures and/or process upsets. ▪ Inherently safer process designs to prevent the potential of incident escalation or 'domino' effects. ▪ Multiple, reliable, and independent safeguards are in place and will not fail at the same time, thus mitigating the most severe.

Step	Description
Consequence analysis	Consequence analysis utilizing recognized software to simulate the potential releases and assess the extent of impact. The following basis is formed before conducting the analysis: **1) <u>Release Hole Size:</u>** The selection of hole sizes for determining hazard distances is critical because they directly influence the extent of the hazard resulting from a release. Larger hole sizes will lead to more severe consequences. **2) <u>Impairment Acceptance Criteria:</u>** Acceptance criteria for assessing the impact of accidental CO_2 releases are defined to evaluate their consequences on sensitive targets. **3) <u>Key Parameters for Dispersion Modeling</u>** Worst case key parameters should be used to model worst case dispersion distances (e.g., lower wind speed, stable weather stability, lower surface roughness length, etc).
Results validation with actual project	Results obtained from the consequence analysis are validated with the actual facility siting study performed for the onshore CCUS hub project. The comparison will guide additional parameters / margins to be considered in establishing the typical hazard distances.
Establish typical hazard distances	Typical hazard distances are input into **"Hazard Distance Look Up Table"**. The hazard extent of the credible worst-case scenario will be used as basis of safety distance between the proposed facility location and nearby sensitive receptors.

4 Results

4.1 Key Parameters

Following key parameters are used for consequence modeling in developing the **Hazard Distance Look Up Table.** This is to evaluate the potential impacted radius during emergency (i.e., leak scenarios).

Table 4.1 Key Parameters Used in Consequence Modeling

Parameter	Data Input	Basis / References
Release scenario	Pure CO_2 leak from Vessel / Piping / Storages	Worst case credible event from CCUS onshore streams. There is no hydrocarbon anticipated in onshore streams as only captured CO_2 from offshore will be transported to onshore facility.
Ambient temperature	30°C	Based on Malaysia's average ambient temperature.
Relative humidity	80%	Based on Malaysia's average humidity level.
Operating temperature	20°C	Fixed operating temperature. The likely operating temperature for CCUS onshore facility.

Parameter	Data Input	Basis / References
Operating pressure	5 barg, 10 barg, 20 barg, 30 barg, 50 barg, 80 barg, 100 barg	Dispersion distances are mainly affected by operating pressure. Therefore, a range of operating pressures is included in the dispersion modeling.
Hole size	5 mm, 25 mm, 50 mm, 100 mm	Based on more likely release scenarios.
Weather condition	Wind speed: 1.5 m/s Stability Class F: Stable weather condition at night with moderate clouds and moderate wind	The dispersion of toxic or flammable gas will be more severe during periods of low wind speed and stable weather conditions.
Surface roughness length	0.2 mm open water	Worst case surface roughness. A shorter length with fewer obstructions will result in greater dispersion distances.
OSH industrial acceptance criteria	▪ Immediately dangerous to life or health (IDLH) of 40,000 ppm. ▪ Short-term exposure limit (STEL) of 30,000 ppm.	Based on regulatory OSH industrial acceptance criteria.

4.2 Hazard Distance Look Up Table

Hazard distances are calculated via dispersion modeling in PHAST DNV software version 8.71. Below are the results, extracted from the modeling and input into **Hazard Distance Look Up Table.**

Table 4.2 Hazard Distance Look Up Table

Composition (molar %)	Release Pressure (barg)	Hazard Distance to 30,000 ppm (m) *STEL of CO_2*				Hazard Distance to 40,000 ppm (m) *IDLH of CO_2*			
		Leak Hole Size (mm)							
		5	25	50	100	5	25	50	100
100% CO_2	5	1.4	4.9	10.0	21.7	1.0	3.7	7.4	15.4
	10	1.4	6.6	13.6	31.0	1.4	5.0	10.1	21.3
	20	1.9	9.3	20.1	47.6	1.6	7.0	14.5	33.4
	30	2.3	11.9	26.9	63.3	1.8	8.9	18.8	44.7
	50	3.2	17.0	40.2	93.5	2.4	12.5	28.0	65.5
	80	11.8	84.9	174.9	342.2	8.9	60.7	137.7	263.4
	100	12.1	84.8	175.7	345.0	9.1	60.8	131.1	265.7

The **Hazard Distance Look Up Table** was established for operating pressures ranging from 5 barg to 100 barg, as the operating pressure is the dominant operating parameter in determining the hazard distance of a toxic release with same composition. In this table, operating temperature was set at 20°C, close to ambient temperature, for all simulation scenarios, since temperature is less dominant, as compared to operating pressure.

The hazard distances of various leak hole sizes were simulated at the STEL (30,000 ppm) and IDLH (40,000 ppm) of CO_2. Hazard distance to the STEL can be referred to for emergency response and evacuation planning while hazard distance to the IDLH concentration should be referred to for potential immediate impairment to public or workers at nearby facilities.

Smaller leak hole sizes (i.e., 5 mm, 25 mm and 50 mm) were considered for potential leakage from small bore piping such as instrument connections, drain connections, sampling connections, or vent connections. Rupture cases are not considered for hazard distance determination as these cases have lower likelihood as compared to other release sizes. It is deemed reasonable to focus on more immediate and likely scenarios for safety distance determination.

4.3 Validation with Actual Project Facility Siting Study

Consequence modeling results from an actual project's (Eastern CO_2 Hub) facility siting study, on process streams with similar operating conditions were extracted into **Table 4.3**, to validate and compare with the above **Hazard Distance Look Up Table** figures.

Table 4.3 Consequence Modeling Results from Eastern CO2 Hub Facility Siting Study

Parameter	Value			
Composition	100% mol. CO_2			
Operating Temperature	0°C			
Operating Pressure	100 barg			
Release Height	1 metre			
Leak Hole Size	50 mm		100 mm	
Wind Condition	1.5 F	5 D	1.5 F	5 D
Hazard Distance (m) to 30,000 ppm *STEL of CO_2*	180	212	346	430
Hazard Distance (m) to 40,000 ppm *IDLH of CO_2*	138	152	271	320

Table 4.3 presented the simulated hazard distances of a pure CO_2 process stream with operating conditions at 100 barg and 0°C. Consequence modeling of toxic hazard distances of CO_2 STEL (30,000ppm) and IDLH (40,000ppm) were performed for leak hole sizes of 50mm and 100mm, at wind conditions of 1.5F and 5D. The key parameters and assumptions were aligned with the above established **Hazard Distance Look Up Table**, except for operating temperature. The difference between the results is presented in **Table 4.4**.

Table 4.4 Variance of Consequence Modeling Results

Parameter	Value					
Leak Hole Size	50 mm			100 mm		
Sources	**Look Up Table**	**Actual Project**	**Variance (%)**	**Look Up Table**	**Actual Project**	**Variance (%)**
Hazard Distance (m) to 30,000 ppm *STEL of CO_2*	176	180	**2.5**	345	346	**0.3**
Hazard Distance (m) to 40,000 ppm *IDLH of CO_2*	131	138	**4.7**	266	271	**2.0**

The variance between **Hazard Distance Look Up Table** and actual case study is 0.3 – 4.7%, despite the difference in operating temperature. Furthermore, based on the facility siting study results presented in Table 4.3, it was noticed that at wind condition of 5D, the hazard distances increase up to 25%, as compared to expected more severe dispersion results at weather condition of 1.5F.

Hazard distance is recommended to be calculated with a 20 - 30% margin during the conceptual stage to account for uncertainties or changes in operating conditions or wind conditions that may arise in subsequent project stages. Alternatively, the project team may verify the safety distance by performing a single consequence modeling scenario for the process stream with the highest operating pressure, i.e. the worst-case scenario.

In the case where actual compositions of fluids or dispersion parameters vary widely from the **Hazard Distance Look Up Table**, specific dispersion modeling should be carried out to verify the hazard distances.

5 Conclusion and Future Works

The establishment of the **Hazard Distance Look Up Table** will benefit project teams in the early stages of projects when there is a limitation of resources and information prior to the actual consequence modeling. To avoid underestimation of safety distance for locating CCUS onshore hubs, the project should consider 20 – 30% margins to accommodate for any uncertainties effectively.

The conditions of both the process and the site, such as process temperature (0°C) and wind speed or weather stability class, may differ from project to project. It is therefore crucial to rigorously validate the outcomes under the most extreme fluid conditions possible.

6 References

1) CCPS, *Siting and Layout of Facilities, 2nd Edition*. John Wiley & Sons; 2003.
2) CO2RISKMAN JIP, *Guidance on CCS CO_2 Safety and Environment Major Accident Hazard Risk Management, Level 1 – Executive Summary*, DNV GL, 2013
3) CO2RISKMAN JIP, *Guidance on CCS CO_2 Safety and Environment Major Accident Hazard Risk Management, Level 2 – Overview*, DNV GL, 2013
4) CO2RISKMAN JIP, *Guidance on CCS CO_2 Safety and Environment Major Accident Hazard Risk Management, Level 3 – Generic Guidance*, DNV GL, 2013
5) CO2RISKMAN JIP, Guidance on CCS CO2 Safety and Environment Major Accident Hazard Risk Management, Level 4 – Specific CCS Chain Guidance, DNV GL, 2013
6) EI, *Model Code of Safe Practice, Part 15, Area Classification for Installation Handling Flammable Fluids*, Energy Institute London, Fourth Edition, 2015
7) ISO 17349, *Petroleum and natural gas industries - Offshore platforms handling streams with high content of CO_2 at high pressures*, ISO, First Edition, 2016
8) EI, *Good Plant Design & Operation for Offshore and Onshore Carbon Capture Installations and Onshore Pipelines,* Energy Institute London, First Edition, 2010.
9) EI, *Technical Guidance on Hazard Analysis for Onshore Carbon Capture Installations and Onshore Pipelines*, Energy Institute London, First Edition, 2010.
10) IOGP, *Gap Analysis of Standards, and Guide for Carbon Capture, Transport & Storage*; IOGP, First Revision, 2022
11) IOGP, *Risk Assessment Data Directory, Process Release Frequencies, Report 434-01*, IOGP, 2019.

Exploring the ALARP Zone

Nicole B. Loontjens
Americas Styrenics, LLC
24 Waterway Ave., Suite 1200
The Woodlands, TX 77380
nbloontjens@amsty.com

Keywords: ALARP, F-N curve, CFD, facility siting, quantitative risk analysis

Abstract

You've done a facility siting study and know that your building or site is "safe" - in the ALARP zone - great! But ALARP means "as low as reasonably practicable." How do you know if you're actually "as low as reasonably practicable?" And how do you know when you should dig deeper? At Americas Styrenics (AmSty), we have a situation that is common amongst several of our plants, where one certain hazard has an outsize impact on the overall risk level. With this information, we sought to better understand the nuances of this hazard and learn if there were reasonably practicable changes we could implement to help protect our employees and assets from the consequences. With the help of third-party experts, we found that there were some simple and relatively inexpensive modifications we could make to mitigate the impacts of these scenarios.

1 Introduction

At several of AmSty's polystyrene facilities, the process areas are completely outdoors until the last step of the process. The final processing step occurs indoors, where molten polystyrene is extruded through a die like endless strands of spaghetti, pulled through a waterbath to cool, then dried and cut into tiny cylindrical pellets. The room where this happens is called the finishing room. Commonly (particularly in cooler climates), the finishing room is adjacent to the control room, usually with a glass window between the two because this area requires frequent monitoring and interaction by operators. Some sites also have offices adjacent to or above the finishing room.

AmSty conducted facility siting studies at these plants and compared the results to the corporate risk tolerance criteria. The good news was that the overall risk levels associated with these plants were in the tolerable region. But the risk levels were not in the "broadly acceptable region"; they fell in the ALARP zone.

ALARP stands for *as low as reasonably practicable*. The Center for Chemical Process Safety's (*Guidelines for Developing Quantitative Safety Risk Criteria*, 2009) definition of ALARP is:

> The concept that efforts to reduce risk should be continued until the incremental sacrifice (in terms of time, effort, cost, or other expenditure of resources) is grossly disproportionate to the value of the incremental risk reduction achieved.

For a while, at AmSty we were happy knowing that the results of our facility siting study showed that our plants were in the ALARP region. Eventually, being a Responsible Care® company, and in line with the elements of risk-based process safety, we wanted to know if there were ways to reduce risk – move farther down the ALARP zone towards the "generally acceptable" region. Were there improvements we could make that were simple and cost-effective?

2 Risk Tolerance Curves

When evaluating risk, the United Kingdom's Health and Safety Executive (UK HSE) framework for the tolerability of risk (Figure 1) provides a good visual:

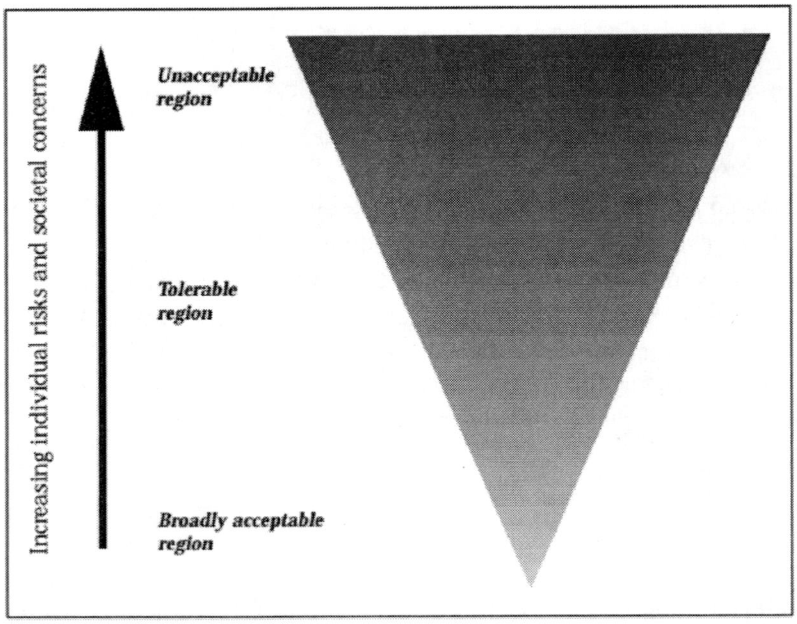

Figure 1: HSE Framework for the tolerability of risk. "The triangle represents an increasing level of risk for a particular hazardous activity." (HSE, 2001)

The top of the triangle – representing a very high-risk activity – is regarded as universally unacceptable (or intolerable) – the benefits will never outweigh the risks. At the bottom point of the triangle, the risk is considered broadly acceptable – the risks are insignificant or adequately controlled compared to the benefits. The middle region is what represents the ALARP zone. Understanding the risk associated with the polystyrene plant falls in the "tolerable region," we wanted to understand if there was anything we could do to drive the risk *as low as reasonably practicable* – towards the "broadly acceptable region."

To put this in perspective, let's take the example of swimming. Most people probably find it generally acceptable or tolerable to stand in a children's pool or sit on the edge of the shallow end of a swimming pool. A somewhat smaller number of people may find it acceptable to swim in a swimming pool with a deep end. A still smaller number of people may find it acceptable to swim in a lake or the ocean. Almost no one would find it acceptable to jump into waters with shark fins circling around.

The middle area of the swimming continuum – where you are technically swimming as long as there aren't sharks around – is similar to the ALARP zone. Some people may only be comfortable where they can see the bottom of a pool, while others may be comfortable jumping off a boat in the middle of the ocean. Either way, there are incremental, *reasonably practicable*, steps you could take to make the swimming experience more tolerable (i.e., safer) for most people to be comfortable – such as wearing a life jacket and/or ensuring a lifeguard is on duty.

Coming back to process safety, societal risk measures associated with facility siting may be represented on a graph called a frequency-number (F-N) curve. According to the Center for Chemical Process Safety (CCPS), "An F-N curve is a plot of the frequency distribution of multiple-casualty events, where F is the cumulative frequency of all events leading to N or more casualties (typically expressed as the number of fatalities)" (2009, p. 35). Here is an example of the F-N curve representing Hong Kong's societal risk criteria:

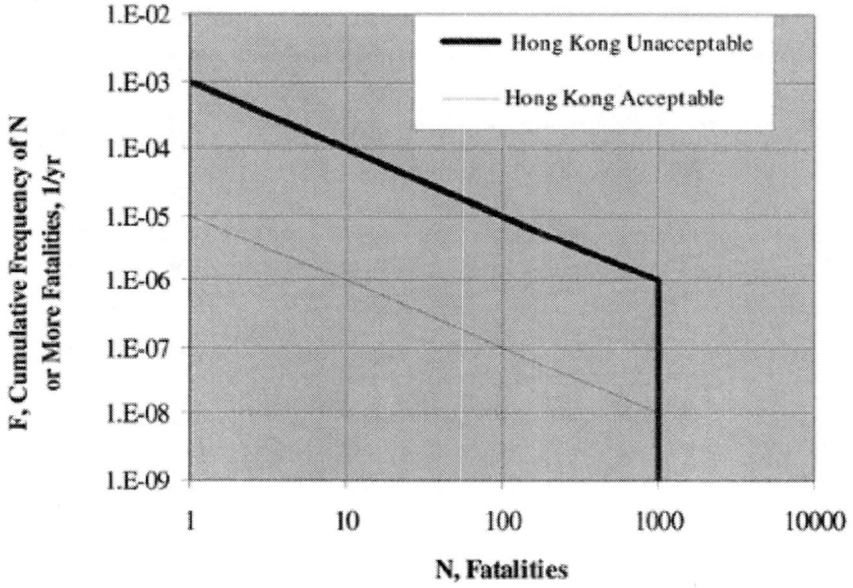

Figure 2: F-N Curve representing Hong Kong societal risk criteria (Center for Chemical Process Safety, 2009)

It may help to think of the F-N curve as turning the HSE triangle 90-degrees clockwise, where the area above and to the right of the "Hong Kong Unacceptable" line is the top of the triangle (i.e., the "unacceptable region") and the area below and to the left of the "Hong Kong Acceptable" line is the bottom of the triangle (i.e., the "broadly acceptable region"). The area in between the two lines is the ALARP zone. When it comes to high-consequence, catastrophic events, we want to drive the frequency (or likelihood) down and the consequence to the left.

3 Finishing Room Analysis

From the facility siting study of the polystyrene plant, we knew that a major contributor of risk came from the scenarios related to flammable vapors entering the finishing room. Usually the polystyrene is 99+% polymer when it is extruded from the dies, but certain failures – such as vacuum system failure or level control failure – could result in lower percentages of polymer and higher levels of hot volatiles being pumped into the finishing room. The finishing rooms are unclassified electrical areas, so in the event the concentration of vapors reaches flammable levels, there is a high likelihood that the vapor cloud would ignite, causing an explosion that our facility siting study indicated could impact the control room, lab, and offices (i.e., result in fatalities).

As stated previously, the facility siting study showed this plant is in the ALARP zone and these scenarios have several independent protection layers. The layers of protection analysis (LOPA) scenarios are closed and the independent protection layers (IPLs) have been validated and verified. This has been a known risk for a long time and AmSty takes it seriously. In addition, there are uncredited, mitigative safeguards in the finishing rooms such as oxygen analyzers and continuous ventilation.

Even given all the protective measures, once we knew the quantitative consequences, it was unsettling to know that the only thing separating the control room from the finishing room is a wall of windows. We considered replacing the windows with a steel blast wall. However, as mentioned in the introduction, the windows provide a practical means for the operators to monitor the finishing room. In addition, the windows provide a lot of visual interest and information – it's the only place in the polystyrene process where you can see what is being produced – all upstream processing is inside closed equipment without sight glasses. We wondered if there was anything else we could do to prevent an explosion or mitigate the consequences besides removing the windows and installing a blast wall.

Bids were solicited from a few consulting companies, and one was selected to conduct a computational fluid dynamics analysis of these scenarios at one of our facilities. We hoped the results from this study could be applied to the similarly designed sister plants.

The consultants visited the plant and developed a three-dimensional model of the finishing room. We specified various release rate scenarios for flammables entering the finishing room, ranging from "most likely" to "maximum probable" based on current production rates. We provided ventilation rates and building design information. The consultant created a computational fluid dynamics (CFD) model of the explosion for each of the identified release rates based on this information.

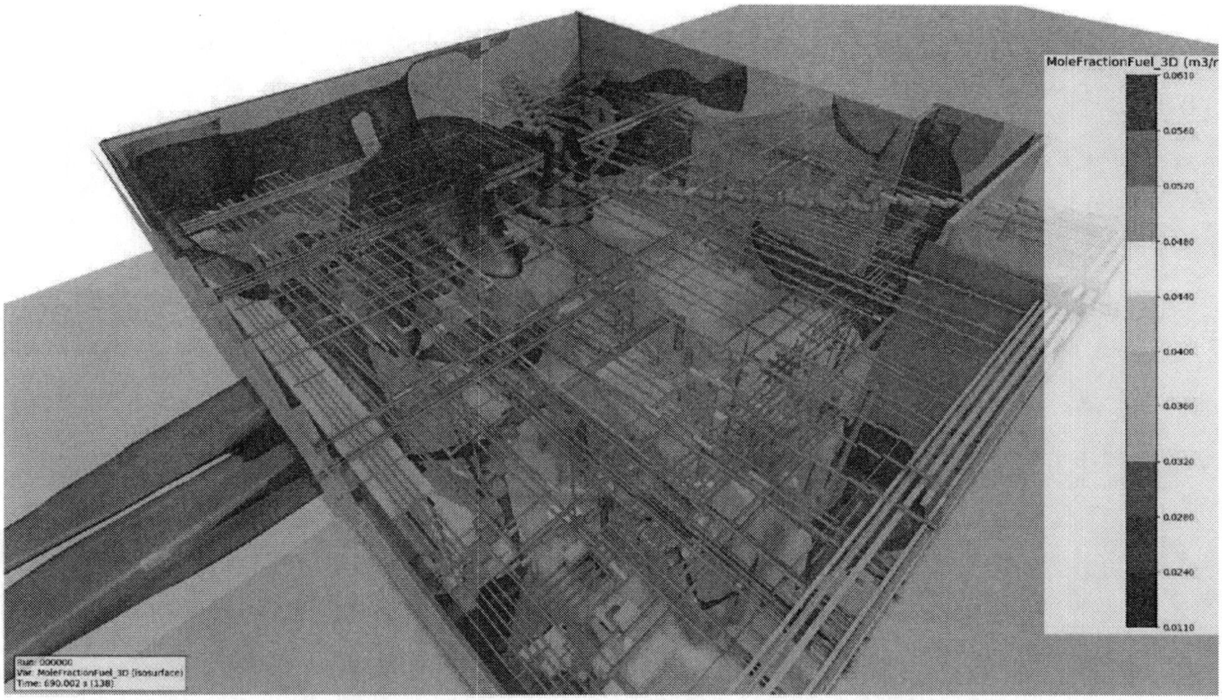

Figure 3: CFD model showing how the vapors would disperse through the finishing room.

The results were interesting and a little surprising!

The model assumed an additional louver was in place to cover the makeup air required for the ventilation fans. Such louvers didn't exist, so installing equivalent-sized louvers became one of the action items.

It turns out that the control room modeled was well-designed. While most of the walls are CMU, one wall is essentially a frangible wall, made of corrugated metal. That wall faces a typically unoccupied, non-hazardous area - a bank of polystyrene storage silos. Also, this is a one-story building - the ceiling of the finishing room is also the roof - there aren't any offices above it.

The surprising part is that the window wall is quite strong! The only scenario where the model predicted the window wall to fail was the worst-case scenario. With the lower release rate scenarios, the corrugated metal, frangible wall and the roof failed before the window wall, thereby directing the force of the explosion out the frangible wall and sparing the control room and offices from impact.

Figure 4: The scenario one step down from the worst-case scenario, showing that the corrugated steel wall fails, but the walls between the occupied areas remain intact.

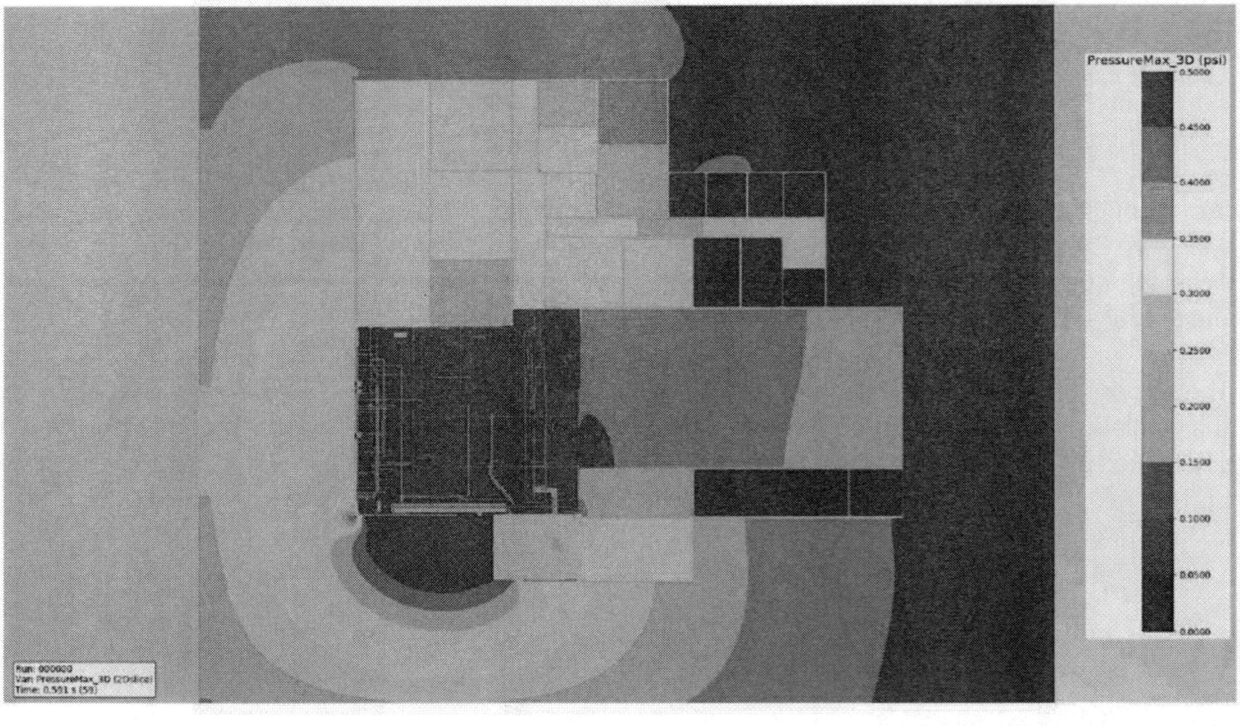

Figure 5: The worst-case scenario - the orange, yellow, and light green areas show the control room, lab, and offices being impacted.

4 Conclusions

Following this study, we completed a formal internal risk review. One of the recommendations from the study was to reverse the way two sets of doors close. One set of doors opened from the finishing room into the lab and the other set opened from the finishing room into the maintenance shop. We decided to reverse the jambs so that the doors open into the finishing room. This will make them less likely to blow open in the event of a vapor cloud explosion because the door will seal against the jamb. We also had the action item to add the louvers to the ceiling.

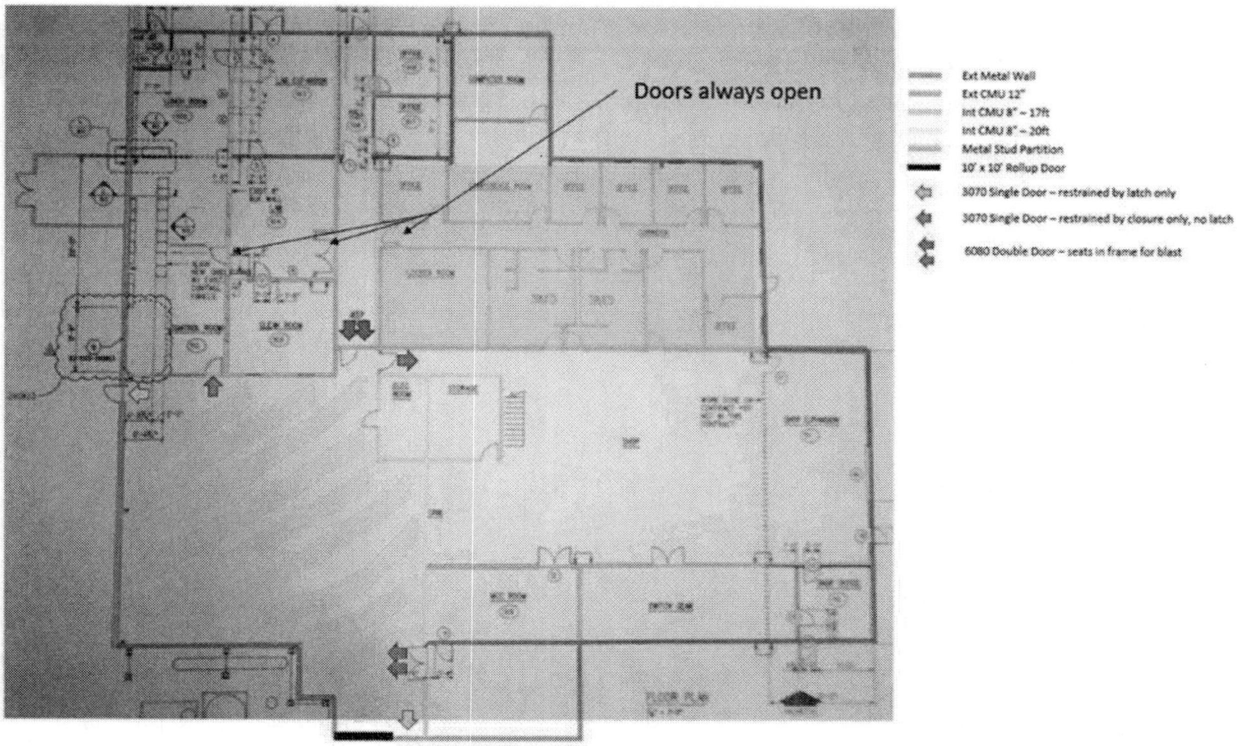

Figure 6: The yellow hashed area is the finishing room. The green arrows show the doors that opened from the finishing room into the control room and lab.

We considered other safeguards, such as interlocking a combustible gas detector to the ventilation fans and polymer gear pumps, for example, but because the process is highly automated and there are numerous protection layers in place, we expect to see other things going wrong, causing the pumps to shut off and alarms to sound. Also, because the wall between the finishing room and the control room is windows, operators would likely see the problem before the analyzers alarmed.

For about $30,000 worth of improvements, we have high confidence that we are well-protected against these worst-case scenarios and have moved down the ALARP zone.

As far as the applicability of this study to the two sister plants, unfortunately, neither plant has such an inherently smart design. All the walls are concrete block and have second floor offices above them. In fact, the plant managers' offices are on top of the finishing rooms in both cases. After discussions with plant managers regarding the study and its results, the plant managers engaged in offering solutions, such as converting windows to the outside into frangible panels.

Fortunately, we recently created an executive leadership-level Risk Review Board that is tasked with understanding and resolving significant risks in the company. When this study was reviewed by them, they fully supported conducting the studies at the other two plants, and those studies are now underway.

In conclusion, to apply AmSty's experience with this study to your facility:

- If you don't have risk tolerance criteria, develop some!
- If you're in the ALARP zone, understand where the risk is coming from. Is a significant portion coming from one place?
- If it is, consider doing a more detailed analysis to see if there is anything "reasonably practicable" you can do to mitigate that risk. It probably won't change your risk profile, but you may be able to add more layers of cheese to your Swiss cheese model.

5 References

[1] Center for Chemical Process Safety. (2009). *Guidelines for Developing Quantitative Safety Risk Criteria.* Hoboken, NJ: John Wiley & Sons, Inc.

[2] HSE. (2001). *Reducing Risks, Protecting People - HSE's Decision-making Process.* London: Her Majesty's Stationery Office.

Application and Challenges of Layer of Protection Analysis (LOPA) in Mining Processes: Insights into Benefits and Limitations

Laura Anato,
Principal Advisor Process Safety
Rio Tinto
1190 Avenue des Canadiens de Montreal,
Montreal, QC, H3C 1B3, Canada
Laura.anato@riotinto.com@riotinto.com

Luis Carrero[1], Guy Brouillard[1], Catherine Morar[2]
Principal Advisors Process Safety
Rio Tinto

Keywords: LOPA, Risk-Based Decision-Making, Hazard Identification, Mining and Metals Industry Safety

Abstract

Layer of Protection Analysis (LOPA) is a semiquantitative technique widely used in process industries for assessing hazardous scenarios and supporting risk-informed decision-making. It provides a balance between the simplicity of qualitative and the detail of quantitative analyses. This paper discusses the authors' experiences with the application of LOPA in the mining and metals industry, combined with traditional methods like HAZOP and HAZID to identify risks. Several LOPA's limitations became evident, and scenarios involving human factors, natural events, and asset integrity were excluded from analysis. Certain mining and metals processes, often complex and heavily reliant on manual operations and pose unique challenges to LOPA's effectiveness due to difficulties in isolating independent protection layers. For example, those involving induction furnaces where overheat scenarios can lead to explosive phase transitions upon contact with water and molten metal. Despite these challenges, the advantages of LOPA, such as enhanced understanding of protection layers and fostering effective safety improvements, are significant. The paper anticipates continued use of LOPA within the company, complemented by Safety Critical Task Analysis to manage human errors and enhance safety controls in critical situations.

1 Introduction

Layer of Protection Analysis (LOPA) presents unique challenges when applied to the mining and metals industry, mainly due to its operational nature and diverse risk profiles. The facilities range from mines to smelters, each with specialized technologies and risk types, which further complicates LOPA's application and the gathering of the required failure data, essential for an accurate and comprehensive risk assessment. This is particularly challenging as the databases for initiating events, probability of failure on demand of layers of protection and common consequences such as ignition probability, are not available in the mining and metals industry.

In mining and metals, scenarios involving unusual hazards, for example, those related to molten materials, test the limits of traditional LOPA methodologies. The industry's heavy reliance on specific operational procedures also makes it challenging to apply LOPA in its conventional form. This is further complicated by the complexity of the operations and equipment, especially when it comes to monitoring all the required process variables to verify if the controls in place are effective enough to be classified as Independent Protection Layers (IPL). Regardless of the challenges of applying LOPA, Rio Tinto sees this as a valuable tool in our Hazard Identification and Risk Assessment approach, particularly as continued improvement and As Low As Reasonably Practicable (ALARP) demonstration are becoming more expected within the industry. Highlighting both the adaptability and limitations of LOPA within mining and metals operations, the paper offers valuable insights into how the company navigates these challenges. Recognizing areas where LOPA falls short, especially in scenarios with significant human involvement, Rio Tinto is exploring Safety Critical Task Analysis (SCTA). This approach aims to better manage human-related scenarios, aligning with the company's ongoing efforts to refine risk analysis methods.

2 Layer of Protection Analysis process

To effectively implement LOPA within Rio Tinto's operations and ensure a standardized approach, a detailed guidance note has been developed. This section introduces an overview of the structured process outlined in the guidance note (Figure 3), providing additional details only on those sections that are particularly relevant or have been specifically adapted for mining and metals operations.

2.1 LOPA approaches used within Rio Tinto

There are two ways that a LOPA workshop can be performed at Rio Tinto:

1. *A standalone activity* that is completed after a Hazard Identification (HAZID), Hazard and Operability (HAZOP) or the application of other Process Safety Hazard Identification and Risk Assessment Methodology; or
2. *An integrated process (*refer to section 2.4) that is executed as part of the process hazard analysis (PHA). Methods primarily used by Rio Tinto are Hazard Identification (HAZID) and or Hazard and Operability (HAZOP) study.

2.2 *Scenario review*

As a minimum, the scenarios from the PHA that satisfy the following criteria shall be considered for LOPA:

- Health & Safety (including onsite and offsite fatalities) or Environment consequence; and
- The current risk has a health and safety consequence of one or more fatalities, and an environmental impact greater than one year to remediate.

Rio Tinto guidance on risk targets is derived from the industry standard guidance particularly the United Kingdom (UK) Health and Safety Executive (HSE) Guidance on ALARP demonstration.

2.3 *Human response to an abnormal condition*

While technological advancements and automation have made substantial advances in mining and metals industries, many tasks cannot yet be fully automated. Humans are also essential for the maintenance, operation, and supervision of mining and metals equipment, ensuring that machinery is functioning correctly and safely, and intervening in case of malfunctions.

The Rio Tinto LOPA guidance note requires compliance with the following criteria for the human responses to an alarm to be allowed as an IPL:

- The alarm system is effective - alarm management has been implemented and steady state, alarm floods and standard alarms are maintained within the benchmarks of [1] or [2].

- The alarm is auditable, and there is an operator present to receive the alarm.

- There is a clear action to take when the alarm is activated (supported by an operating procedure).

- Operator training is in place that specifically addresses the alarm response.

- The process safety time exceeds 10 minutes for a control room response or 20 minutes for a field response (to resolve the alarm condition), and the response time does not take more than 50% of the time for the event to occur. Drills are regularly performed to test that operator response in the real world matches the estimated time. Any tool or equipment that is required by the operator to successfully perform their action is available and in good condition.

2.4 *Rio Tinto Integrated LOPA Process*

Rio Tinto methodology for HAZID and HAZOP uses risk reduction factors (RRF) that can be considered as a preliminary and simplified LOPA without considering enabling conditions and/or conditional modifiers (refer to Figures 1 and 2 below). The next step to complete the

LOPA is the addition of these enabling conditions and conditional modifiers as needed and comparison against the company's risk targets.

If the LOPA activity is done at the same time as the HAZID or HAZOP, there are process steps (from Section 6) that can be removed as they will already be addressed through conformance to Rio Tinto Group Procedures; Hazard Identification (HAZID) or Hazard & Operability (HAZOP) Review.

Node: Boiler 5 Outlet

Drawings / References: 06072-M-0104, Sheet 2

Design Conditions/Parameters: Outlet Temperature - 350 F
Outlet Pressure - 300 psi
Outlet Piping Size - 10 inch

Deviations	Causes	Consequences	CAT	Inherent Risk			Effective Controls	CAT	RRF	Residual Risk			Recommendations		Incidents	Remarks
				S	L	RR				S	L	RR	Recommendations	RRF		
1. High Pressure	1. Closed or partially closed valves on 10" Boiler 607-BLR-5 outlet	1. Restricted or no flow leading to high pressure in Boiler 607-BLR-5, resulting in overpressure and equipment damage.	PER	4	C	Critical	1. Pressure relief system composed of 2 x PSVs	Preventive	100	4	E	High	1. Implement periodic testing of safety functions identified in the HAZOP to ensure safeties will be available as required and provide functionality status to the operator.	5		
							2. Boiler Trip: i) boiler high pressure; ii) boiler high temperature; iii) boiler outlet low flow	Preventive	5				2. Install a boiler high water pressure alarm to alert operators and provide adequate reaction time for operators to take action to avoid a high high pressure condition and potential rupture at electrical boilers (Boilers 607-BLR-1 and 607-BLR-5).	5		
							3. Operator intervention to regulate boiler outlet flow;	Preventive	10							
2. No / Low Pressure	1. No credible causes identified															

Figure 1 Example of Rio Tinto HAZOP review template

								Inherent risk				Controls				Current risk						Residual risk					
Hazard categ	Hazard	Hazardous scenario	Central event	Causes	Consequences	Operational sta	CF	S	L		RR	Controls	Type	CAT	RF	S	L		RR	Recommendations	RF	S	L		RR	Incidents	Remarks
Pressure	Vapour under pressure (in the oil boiler)	Generation of steam that exceeds the design pressure of the boiler	C. Loss of containment	1. Manual valve closed at boiler water outlet	Generation of steam that increases the pressure inside the oil boiler above its design pressure leading to the burst of the vessel and possible fatal impact on personnel	Normal operation	HS3	5	D	SD	CRITICAL	Value position is reported in the Scada system for the operator to move to the boiler room and change the position of the value	Administrative	Prev	1	5	E	SE	HIGH								
											#N/A	Low water level sensor [illegible] which stops the boiler upon reaching a water level of [illegible] No credit given to this control as it is on the same controller as the pressure control which has been given credit for risk reduction	Engineering	Prev	1				#N/A								
											#N/A	Pressure safety valve set at 7.5 barg to open	Engineering	Prev	100				#N/A								
											#N/A	High pressure sensor [illegible] which stops the boiler in case pressure reaches 7.5 bars	Engineering	Prev	10				#N/A								

Figure 2 Example of Rio Tinto HAZID review template

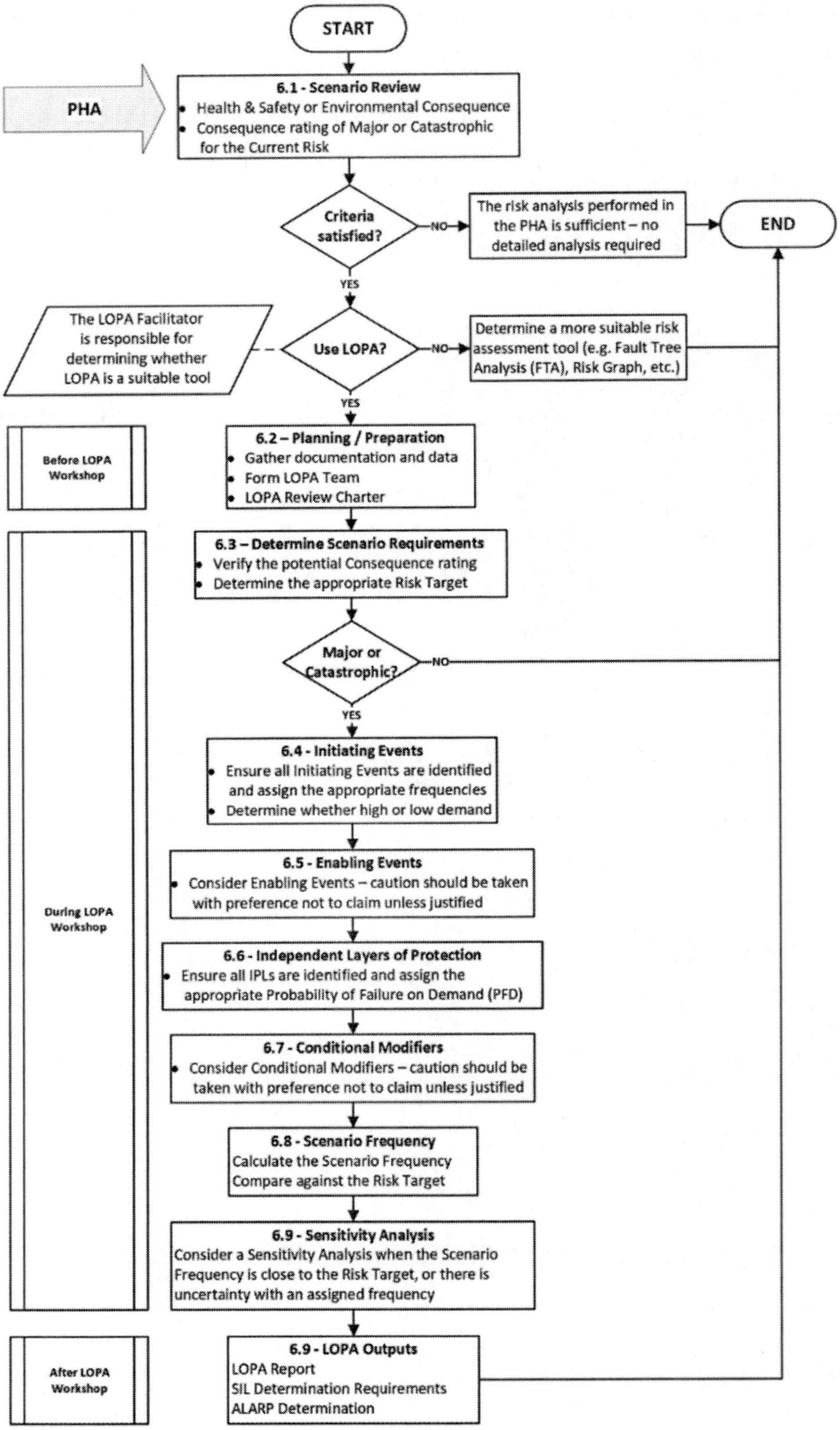

Figure 3 Rio Tinto LOPA Process Overview

If this approach is taken, the following process steps should be followed:

Table 1. Rio Tinto Integrated LOPA Process

Section	Process Step Required	Comments
6.1 – Scenario Review	No	No need to review scenarios as the HAZID/HAZOP Team will trigger the additional LOPA requirements as soon as a LOPA scenario is identified.
6.2 - Planning / Preparation	No	The preparation activities will have already been completed for the HAZID/HAZOP, however, the charter scope must cover the LOPA, any additional documentation and data and LOPA Team requirements.
6.3 - Determine Scenario Requirements	Partially	The Consequence should have been confirmed through the HAZID/HAZOP, however, the appropriate Risk Target will need to be determined.
6.4 - Initiating Event	Partially	The Initiating Events and associated likelihood categories will have been identified from the company's risk matrix through the HAZID/HAZOP; however, the operating mode will need to be determined and numerical values added for the initiating event frequencies.
6.5 - Enabling Conditions	Yes	Enabling Conditions will not have been considered during the HAZID/HAZOP.
6.6 - Independent Protection Layers (IPLs)	No	IPLs and their associated PFDs will have been identified in the HAZID/HAZOP as Controls and RRFs (PFD = 1/RRF).
6.7 - Conditional Modifiers	Yes	Conditional Modifiers will not have been considered during the HAZID/HAZOP.
6.8 - Calculating the Scenario Frequency	Yes	The Scenario Frequency will not have been calculated as a formal numerical value during the HAZID/HAZOP.
6.9 - LOPA Outputs	Partially	An integrated report can be developed to remove the requirement for two separate reports, however, the SIL Determination and ALARP activities shall be completed.

3 Example of LOPA in Mining and Metal Processing Applications

In this section, a hypothetical exercise for a Layer of Protection Analysis is carried out taking into consideration an industry metal processing application. While it does not represent a real or specific situation, the analysis is performed by using the best data available regarding the technology studied.

An exploration of Induction Furnace Process

Induction furnaces melt metals using electrical energy. The specific type of furnace referred to in this section is the coreless type of induction furnace. This furnace consists of a refractory crucible encircled by a solenoid coil excited by a single-phase alternating current supply. The fluctuating axial magnetic field linking the charge within the crucible causes heating within it. The power induced in the charge depends on the physical properties of the material, the magnitude and frequency of the flux linking it, and its geometrical shape. Figure 4 shows a typical schematic of this type of furnace:

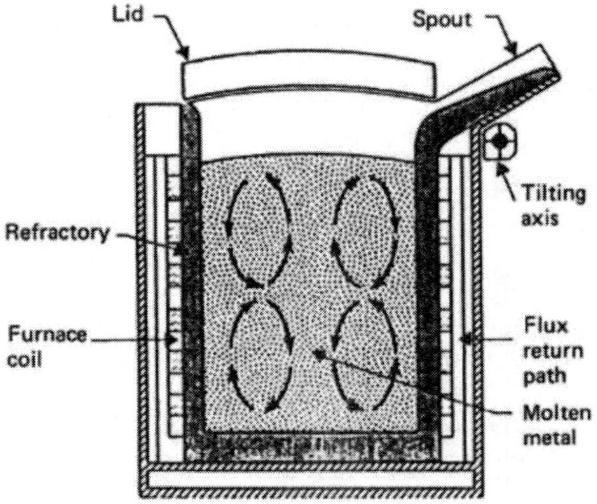

Figure 4. Diagrammatic view of a coreless induction furnace [3]

A layer of refractory material is used to protect the furnace coil from direct contact with the molten metal, which would otherwise cause it to melt. Additionally, the coil requires constant cooling, achieved by circulating water through it and sending it through a heat exchanger.

3.1.1. Process Safety Scenarios to Consider

Several scenarios are considered regarding the operation of induction furnaces. Handling molten metals at extremely high temperatures (above 1000 degrees Celsius) poses a significant risk, leading to severe burns on workers' skin and, in extreme cases, potential fatalities. Workers typically wear protective suits to mitigate such risks in the event of a molten metal containment breach.

Furthermore, the most critical process safety scenario potentially present is associated with the possible interaction between hot molten metal and water, which could result in rapid water phase transition, causing molten metal splashes and/or overpressure waves depending on the circumstances.

3.1.2 How could the molten metal get in contact with the water?

There are generally two ways in which molten metal could encounter water. The less hazardous (though still a concern) scenario involves the introduction of water through the metal parts intended to be melted. Occasionally, moisture might be present in these parts, and when subjected to temperatures exceeding 1000 degrees (much higher than the water's evaporation temperature at ambient pressure), the water rapidly expands. However, as the water is not completely enclosed, the most likely consequence in this case would be a splash of hot molten metal that could reach several meters and could lead to injuries.

A significantly more hazardous scenario may occur when the integrity of the refractory protecting the coils is compromised, allowing molten metal to penetrate the coils. Consequently, the coils may melt, and the water used for cooling inside the coils comes into contact with the molten metal. Refer to Figure 5 for better understanding.

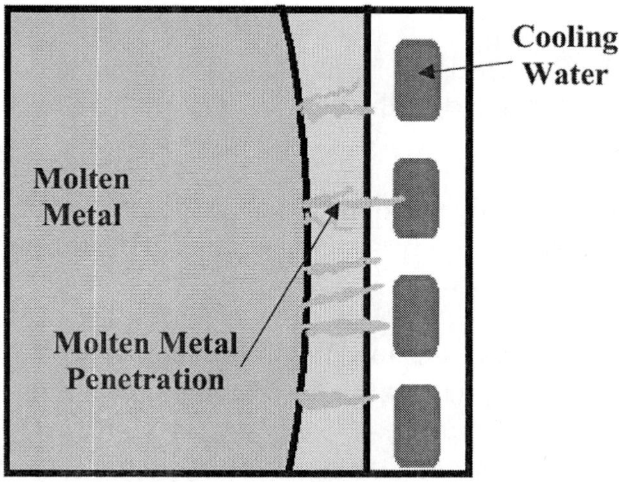

Figure 5. Representation of molten metal penetration through the refractory

In contrast to the previous case where the water was not enclosed and splashing of hot metal was anticipated, here, the water rapidly flashing into vapor becomes trapped inside the coils. This scenario has previously occurred in the industry and has the potential to cause multiple fatalities due to the overpressure generated, along with the release of solid projectiles mixed with hot molten metal. Additionally, the damage to building structures could result in their collapse onto workers. This specific scenario has been considered for the present analysis.

3.1.3 Initiator events

Multiple initiating events are identified that could result in the contact between water and metal. Initially, there is the possibility that the refractory might sustain damage during manufacturing. In such cases, the integrity of the refractory could be compromised, allowing penetration.

Additionally, there is a potential risk of losing cooling water due to various reasons, such as pump failures, human errors resulting from closing water lines, or blockages. This loss could lead to insulation damage of the furnace coils, causing water leaks and subsequent contact between water and hot metal.

The focus of this section of the paper will be on the potential scenario of overheating of the furnace. In this context, the electrical current, if left uninterrupted, has the potential to heat the metal beyond the maximum temperature rating of the refractory, resulting in its degradation and eventual penetration by hot metal.

The initiating events leading to furnace overheating can be primarily associated with human errors in adhering to the furnace operating procedures. The operation of an induction furnace involves procedures encompassing material charging and specific intervals, along with adjustments to the supplied electrical power.

For instance, missing the procedural step to turn off the power after a melting cycle could cause the furnace to overheat, leading to the scenario of concern. Similarly, an error in adding an insufficient amount of charge in proportion to the supplied electrical power could also cause overheating.

Specific scenarios such as "bridging," where charges are not added properly, causing a molten pool with a metal cap forming over the furnace, may result in overheating beneath the cap, leading to refractory erosion.

In all these cases, as previously mentioned, the initiating event stems from human error. However, when attempting to apply standard (e.g. [2][4]) human error probabilities to this application difficulties arise due to:

- The high complexity of the operating procedure.
- A multitude of potential human error combinations that could lead to the concerning scenario.
- The possibility of rectifying a mistake by the same operator or adjacent operators before any protection layers are activated. It has been recorded in the industry that after an initial human error leading to overheating, further consequences were avoided by the same operator or adjacent personnel before any other layer of protection was activated. The continued nature of manual operation of induction furnaces means that operators would be observing the furnace regularly, and since the furnaces are open topped, an experienced operator may easily detect the overtemperature.

Due to these complexities, determining a human error probability value for this application remains challenging following industry standard approaches for human error. A conservative approach can be adopted during the analysis by considering the number of mistakes made in a year that resulted in furnace overheating, disregarding the subsequent correction by the same or other operators.

Regardless of the used frequency of occurrence of the initiator event, some approaches could be applied that may be beneficial such as:

- Enhancing the human interface machine for operating the induction furnace (e.g., incorporating warnings, permissive and action buttons to reduce the probability of human error).
- Reviewing operational procedures and assessing the inclusion of step sign-offs where a human error may lead to severe consequences.
- Undertaking a human factor analysis of induction furnace operational procedures.

3.1.4 Protection Layers.

Through the course of the analysis, various protection layers are visualized based on the technology available. The following sections will describe these layers, accompanied by an examination of the reasons why they could or could not be considered valid protection layers in LOPA. Additionally, recommendations for evaluating potential improvements, if applicable, will be provided.

3.1.4.1 Discrete Temperature monitoring

This proposed protection layer involves the operator detecting high temperatures before molten metal penetrates the refractory and taking action to cut power and empty the furnace, thereby preventing an explosion.

Theoretically, where a human error led to furnace overheating, the same operator or another operator could take steps to recover from the error and prevent an explosion, partly through temperature monitoring. This could give an indication that the barrier can be valid and provide a risk reduction factor.

However, upon basic analysis, it becomes apparent that this layer is not independent of the initiator event, as the same operator who hypothetically committed an error to provoke the overheating is involved. Additionally, even if the procedure required a second operator, questions would arise regarding common cause failures and the true independence of another operator in the same team. No risk reduction credit can be attributed to this layer.

An improvement aiming to enhance this layer's independence could be by incorporating non-contact temperature measurement with an automatic function to prevent the scenario. This would involve not only detection but also a full action independent of the operator to prevent the explosion. The suggested change would include power shutdown, emptying of the molten metal inside the furnace, and triggering an emergency alarm for an evacuation procedure.

3.1.4.2 Energy control system.

Another potential protection layer involves controlling the energy supplied to the furnace, specifically in the form of electrical power. This layer operates based on the weight of the charge intended for melting, incorporating calculations performed by the control system to limit the maximum amount of energy applied during a specified duration of the melting cycle. The underlying concept of this system is to prevent the application of excessive energy, utilizing an automatic system instead of relying on human action (e.g., in case the operator forgets to shut down the furnace), thereby reducing dependence on human intervention and preventing overheating.

When considering this barrier, it would seem like it would offer some level of protection to be accounted for in a LOPA. However, upon closer examination in alignment with LOPA rules, the barrier is deemed non-independent. The primary issue lies in the fact that energy control relies on the weight measurement and input of values by the operator through the machine interface. As discussed previously, since the operator is part of the initiator event, they would not be considered an independent protection layer.

Furthermore, the action of weighing the charge and inputting energy values for a melt cycle is inherent to the operating procedure of the furnace. The failure to perform this action could be viewed as one of the multiple potential human errors during procedure execution, constituting a distinct initiator event rather than an independent protection layer in a LOPA scenario.

A recommendation can be proposed to explore the possibility of implementing an automatic energy control system to prevent the overheating scenario, independent of human intervention.

3.1.4.3 Ground and Leak Detection System.

The ground and leak detection system is another potential layer of protection to be analyzed. This system comprises an electrical circuit connected to the ground with wires spread throughout the refractory material. In the event of molten metal penetration through the refractory, it would be detected upon contact with this system, triggering an automatic shutdown of the furnace power.

The condition of this system would be consistently verified by testing ground conductivity at specified frequencies which can be set by operational needs as frequent as once before every melting cycle.

Although this barrier appears promising in meeting the criteria for LOPA risk reduction and could be considered a safety function, concerns arise about its effectiveness. The final action of the function to avoid the scenario involves an automatic shutdown of the furnace. While shutting down the power stops energy injection into the charge, questions are raised such as: Is it sufficient to prevent an explosion? Without emptying the furnace material, wouldn't the already overheated molten metal continue to penetrate and reach the coils where water is circulating? It's noteworthy that the act of emptying the furnace is not part of the automatic function and would have to be performed by an operator. Considering the emptying of the furnace would reintroduce the issue of the independence of the same operator who may have triggered the scenario with a human error.

After careful consideration, the answer to this question seems to hinge on the original cause of the penetration. For instance, if the penetration resulted from a factory defect in the refractory, such as a small crack, the shutdown of power might be considered effective in stopping the penetration. However, this is purely a qualitative reflection, and no testing or simulation is currently available to confirm it. Conversely, if the penetration was due to a general loss of refractory integrity after overheating surpassed its temperature rating, simply shutting down the power would unlikely prevent an explosion, necessitating an immediate emptying of the furnace to stop the scenario of concern. The protection layer would not be effective in this case and no risk reduction credit could be attributed to it.

Nevertheless, to explore the possibility of accounting for risk reduction for this layer, one suggestion can be made. What if a second operator is included in the operating procedure who would be responsible for emptying the furnace, thereby eliminating dependence on the first operator? Further difficulties are encountered with this idea, including potential common failures within operators of the same team and the challenging detail of the time available to perform the action. It is also important to note that the period between detecting the metal that had penetrated the refractory and the occurrence of the final hazardous event (metal encountering water, producing an explosion) is not clearly defined and can't be easily estimated with reasonable confidence. To even consider a human action for this barrier with an independent operator, this would have to be addressed first. The details of the physics associated with this phenomenon still needs to be better understood.

Finally, a recommendation can be made to evaluate the automation of emptying the furnace contents, coupled with a general evacuation alarm.

3.1.5 Conditional Modifiers and/or Enabling conditions.

Once the considerations about the potential layers of protection and initiator events are concluded, reflection can be done to verify whether if any conditional modifiers and/or enabling conditioners can be present in the scenario that could reduce the frequency of the event.

It is worth mentioning that the operation of an induction furnace would require constant attention and proximity by the operator. Some conditions related to the operation could only be visually verified, and temperature measurements are often manually taken with a probe at specified intervals that can be as frequent as several measures per hour.

Other applications of induction furnaces within the industry are also known where robotics is employed to reduce the operator's exposure, allowing them to be in a safe place while operating the furnace (such as a reinforced control room). However, considering the configuration of the system in question and the heavy reliance on human intervention to troubleshoot problems within our scenario, reducing the risk through probabilities of personnel presence and injury could not be considered.

As for potential enabling conditions, the mode of operation of the furnace could come to mind. For example, there are several modes of operations where the furnace would be either melting, on hold or off. If the furnace is off or on hold mode, the scenario would not be possible. However, these modes of operation are strongly dependent on the operating procedure and the operators themselves would be the ones performing these changes of mode of operation, potentially several times per day. In fact, the frequency of occurrence of the initiator event should already account for this factor and an enabling condition would not be advisable to be considered in this case.

In the end, no conditional modifiers or enabling conditions can be identified.

4 Looking forward

In the mining and metals industry, despite advancements in technology and automation, human and organizational factors play a pivotal role in supporting critical controls within the process safety management system. A significant portion of human errors can be attributed to the industry's significant reliance on human intervention, derived from the need for skilled human operators to execute specialized tasks that are challenging to fully automate.

Given the impact of human factors in all aspects of the industry, Rio Tinto is exploring the implementation of Safety Critical Task Analysis (SCTA), envisioned as a tool to enhance the understanding and management of human errors and their impact on the effectiveness of critical controls. This approach is especially pertinent in situations where LOPA may not be applicable, as highlighted in the previous sections of this paper. By adopting SCTA for critical controls involving human intervention, Rio Tinto aims to achieve a more comprehensive and effective management of process safety risks associated with human factors, aligned with requirements of major hazards legislation like UK Health and Safety Executive Control of Major Accident Hazards (COMAH). Figure 3 presents the HSE COMAH human factors roadmap and the applicability of SCTA to meet its requirements.

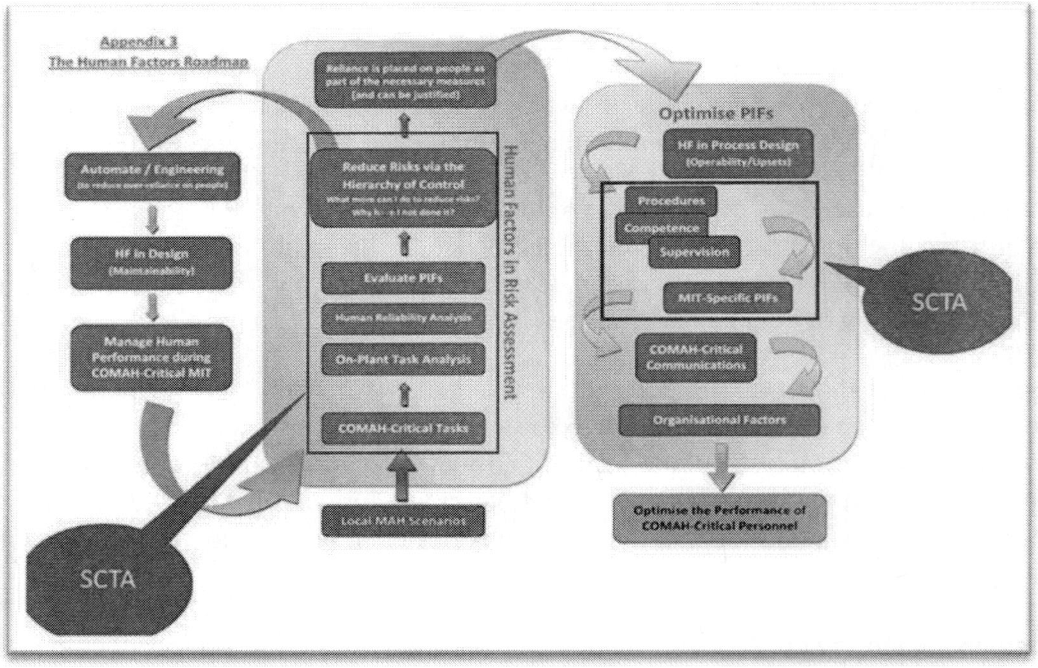

Figure 6. UK HSE COMAH Human Factors Roadmap

Additionally, where LOPA is still an applicable tool, what may be missing in the mining and minerals industry are specific databases on probability of failure on demand of typical layers of protection, and more appropriate guidance on complex initiating events.

Working together with other companies in the industry, through industry groups such as CCPS, would allow a plan to develop relevant databases and guidance, such that there could be mining and metals specific guidance for LOPA in the future.

5 Conclusion

The following conclusions can be mentioned regarding the use of LOPA in mining and minerals melting applications:

- The heavy dependence on human intervention poses a challenge in estimating the risk solely with the LOPA technique. It is widely acknowledged in the industry that LOPA might not be the most recommended technique in situations where many barriers rely on human action. However, for the case studied in the current paper, the identification of these details regarding human dependency became apparent during the analysis. For instance, initially, there was a belief that the leak detection system would solely be an engineering layer of protection, which is not a reality.
- On the application analyzed, several improvements can be proposed to reduce human dependency, including the suggestion of an automatic emptying of the furnace material. Feasibility would need to be explored for the implementation of this improvement measure, defining the proper integrity level for this safety function.
- A comprehensive understanding of the physics of the system is confirmed to be a key factor in providing confidence in the independence and effectiveness of a layer of protection. In comparison to oil, gas, and petrochemical applications of LOPA, where the time periods between initial problem detection and the occurrence of the final event can often be estimated with simple calculations involving process variables or even with specialized software if required, the induction furnace case involves complex physics in the penetration of molten metal into refractory, presenting a limitation for the analysis.
- Use of Safety Critical Task Analysis (SCTA) is envisioned to be a valuable tool for cases where no further improvements can be applied to reduce human dependency despite the best efforts available.
- The advantage of using LOPA is reflected in the in-depth study of the process which leads to the consideration of potential improvements that, if implemented, might reduce the system's human dependency, and effectively lower the risk. Additionally, the authors are left with the impression that the discussions that would be held with the operational team could offer a different perspective on the robustness of their system and all the potential ways it can fail or be compromised, thereby increasing the level of risk awareness of the potential users of this equipment. The authors also gain valuable insights into the limitations and potential of using the LOPA technique in this type of application.

6 Acronyms used.

ALARP – As Low As Reasonably Practicable
CCPS – Centre for Chemical Process Safety (of AIChE)
COMAH – Control of Major Accident Hazards
HAZID – Hazard Identification - a hazard identification tool
HAZOP – Hazard and Operability Analysis – a hazard identification tool

LOPA – Layer of Protection Analysis
PFD – Probability of Failure on Demand
PHA – Process Hazard Analysis
PSM – Process Safety Management
RRF – Risk Reduction Factor
SCTA – Safety Critical Task Analysis
UK HSE – United Kingdom Health and Safety Executive

7 References.

Standards:
[1] International Society of Automation (ISA). Management of Alarm Systems for the Process Industries (ISA 18.2). 2016

[2] Engineering Equipment and Materials Users' Association (EEMUA). Alarm Systems: A Guide to Design, Management, and Procurement (EEMUA-191). 2013

Books:
[3] Laughton MA, Warne DF. *Electrical Engineer's Reference Book.* Elsevier Science; 2002.

[4] Center for Chemical Process Safety. *Guidelines for initiating events and independent protection layers in Layer of Protection Analysis*, New York, 2015

Links:

COMAH Appendix 3: The Human Factors Roadmap http://www.hse.gov.uk/comah/guidance/hf-delivery-guide.pdf

When to Switch from LOPA to Focused QRA

Edward M. Marszal
Kenexis
3366 Riverside Drive, Columbus, OH 43221
edward.marszal@kenexis.com

Keywords: LOPA, QRA, Fault Tree Analysis

Abstract

Layer of Protection Analysis (LOPA) provides an intermediate level risk analysis that is more thorough than PHA, but not as time and cost intensive as a full quantitative risk analysis (QRA). In some cases, LOPA does not provide reasonable results and additional analysis using more quantitative means is desired. LOPA is subject to simplifications that do not allow enough nuance to obtain an accurate estimate of risk. The tools of QRA, such as fault tree analysis, event tree analysis, and dispersion/explosion modeling, can be applied to supplement or replace the original LOPA so that more reasonable recommendations can be implemented.

This paper presents a listing of situations where LOPA should be supplemented with Focused QRA. The discussion begins with guidelines for determining when a LOPA provides results that are "unreasonable". For each situation, the rationale for why LOPA is leading to an inaccurate risk assessment is explained along with presentation of the Focused QRA technique that should be used to address the shortcoming. Each situation will be described using process application examples.

1 Introduction

Process Hazards Analysis (PHA) has evolved since its inception about 50 years ago from a single simple qualitative discussion by a team to a collection of studies with varying amounts of rigor and different objectives and focus. Qualitative PHA in the form of Hazards and Operability Studies (HAZOP) were previously the comprehensive approach employed for conducting risk analysis but are now just the starting points for a string of other activities. The incorporation of Layer of Protection Analysis (LOPA) for critical scenarios – chosen due to their high consequence, risk, or the need for quantitative performance targets – has become the industry standard. This shift occurred in the early to mid 1990's when organizations release standards, like ANSI/ISA 84.01-1996 – Application of Safety Instrumented Systems for the Process

Industries [1] and contemporaneously released guidance documents and books [2] to help practitioners to implement these novel techniques. But even prior to LOPA, PHA studies were supplemented with additional tools such as checklists to address concerns like human factors, plant layout, inherent safety (and more), and the results from HAZOPs were being cascaded into other analyses and reports such as hazard registers and bow tie diagrams.

At this point, LOPA is extremely common with many organizations defining formal criteria that trigger the implementation of LOPA studies. This has led to the unintended effect of scenarios being required to be analyzed in LOPA that LOPA is not suitable to analyze. While LOPA is more rigorous and quantitative (considered by some to be at least semi-quantitative), it is still underpinned by assumptions and simplifications that allow it to be more approachable and efficient than quantitative risk analysis (QRA). For some scenarios, though, the assumptions are violated, or the simplifications are not appropriate, and the LOPA analysis yields results that are inappropriate. Usually, the LOPA results by most organizations will result in a conservative answer, overstating the level of risk and the amount of risk reduction required. Nevertheless, in these situations it is prudent to not necessarily replace but augment the LOPA analysis with quantitative risk analysis techniques in a limited way to overcome the issues that prevent a traditional LOPA from yielding acceptable results. The author's organization refers to this process as Focused QRA, and it includes using the following techniques to supplement the traditional data points that are used in LOPA calculations.

- Dispersion modeling
- Occupancy analysis
- Probability of ignition analysis
- Fire and explosion modeling
- Human reliability analysis
- Fault tree analysis
- Event tree analysis
- Quantitative bow-tie analysis

While these techniques are well documented, further deliberation on the appropriate application of these additional methods necessitates additional discussion. In general, Focused QRA should be applied when an assumption inherent to LOPA or its application generates a calculation parameter that does not reflect reality. These LOPA parameters can include any attribute of a LOPA calculation, including initiating event frequencies, protection layer failure probabilities, conditional modifier probabilities, and enabling event probabilities. Also, violations of simplifying assumptions such as independence between protection layers, can be addressed by more sophisticated techniques. The following list attempts to provide a generalized set of situations where Focused QRA should be considered to replace or supplement LOPA.

- Scenarios resulting in release of toxics/flammables away from potential receptors or at low rates (Dispersion model and occupancy analysis)
- Initiating event is loss of containment (Reliability Centered Maintenance)
- Very complex human interaction with the process (Human reliability analysis and fault tree analysis)
- Many initiating events all with different safeguards and degrees of effectiveness (Fault Tree Analysis)
- Pre-emptive strikes (Fault Tree Analysis)
- Initiating event is a complex sequence of events and conditions
- Shared components between protection layers or initiating events and protection layers
- All or primary safeguards do not prevent loss of containment – mitigative safeguards (Quantitative bowtie analysis or event tree analysis)

This list may not be comprehensive and with time more items will probably be added. Continued improvement of this list should be a goal of industry. The list above provides the starting point for discussion. In the subsequent sections these situations will be described in more detail, and examples are provided.

2 Release of Toxics/Flammables Away from Receptors

While LOPA provides a more rigorous treatment of likelihood of an incident, in general, it does not go into any more detail in treating consequence than the precursor PHA study. Usually, the PHA consequence description and consequence category are carried into the LOPA study without any modification. The determination of consequence category is typically performed by a team, based solely on their experienced judgment, by qualitatively picking a category from a table. These categories often reflect order of magnitude bands of consequence with qualitative descriptions like "potential multiple fatalities" or "likely hospitalization injury". In some cases, organizations supplement the team's judgment with additional tools that will correlate the quantitative parameters to suggested consequence categories. Examples of these tools include degree of overpressure tables that correlate the ration of scenario pressure to maximum allowable working pressure to expected consequence, and release quantity tables which guide users to estimate released quantities and then assigns suggested consequence categories based on the amount of release materials.

Degree of Overpressure Table

Degree of Overpressure	Description of Consequence	Consequence Category
< 115% of MAWP	No consequence	No consequence
115% to 150% MAWP	Flange and instrument connection leaks	Serious injury
> 150% MAWP	Vessel Rupture	Single fatality

Degree of Overpressure Table

Quantity Released	Consequence Category
0 - 10 kg	Minor injury
10 - 100 kg	Serious injury
100 - 1,000 kg	Single fatality
1,000 - 10,000 kg	Multiple fatalities

Examples Only – Do Not Use

Figure 1. Consequence Estimation Tools

Tools like the released quantity estimation tools make assumptions about the extents of dispersion and the occupancy of the area where the release occurs that are generally conservative and intended to overestimate the risk. In some cases, the overestimation of the tools or the team can be excessive to the degree that recommendation for additional risk reduction measures is inappropriate. Team judgement and estimation tools tend to overestimate risk for the following types of scenarios:

- The amount of material released is small
- The release occurs at low pressures, with low momentum
- Toxic materials in the release are at low concentrations
- The receptors – humans for toxins and sources of ignition for flammables – are a significant distance from the release point

When these conditions are present in a scenario it may be prudent to supplement the LOPA analysis with dispersion modeling to determine the true extents of the release in relation to the location of receptors or ignition sources.

Consider a scenario where a toxic gas, carbon monoxide (CO), might be released from the suction of a blower. In this scenario, the fluidized catalytic cracking unit of an oil refinery was equipped with a CO boiler that oxidized the carbon monoxide resulting from incomplete combustion in the catalyst regenerator where coke is burned off catalyst before returning the catalyst to the reactor. The regenerator is equipped with a bypass system, but if the valves that allow bypass fail closed, the flow through the CO boiler will be blocked. The expected response for the blower to stop, and the air flow path from the blower to the boiler will be closed off by shutting a valve. If this valve fails to close, then the CO laden gas in the boiler will reverse flow into the air blower and discharge out the blower' suction.

A conservative PHA team could easily categorize the consequence as a multiple fatality event. After all, a toxic gas was being released to the atmosphere, potentially in very large quantities. A simple quantity of release tool would affirm the decision as a large quantity of toxic material could be released in this scenario. In this example, though, the tool and the judgment of the team might be excessively pessimistic and could result in untenable safety instrumented system (SIS) safety integrity level (SIL) requirements for the isolation of the blower.

More detailed analysis revealed that the probable loss of life for this scenario was actually quite low, and this was confirmed by dispersion modeling and occupancy analysis. Figure 2 shows a side view of the equipment in the scenario, expected occupancy, and the dispersion modeling results.

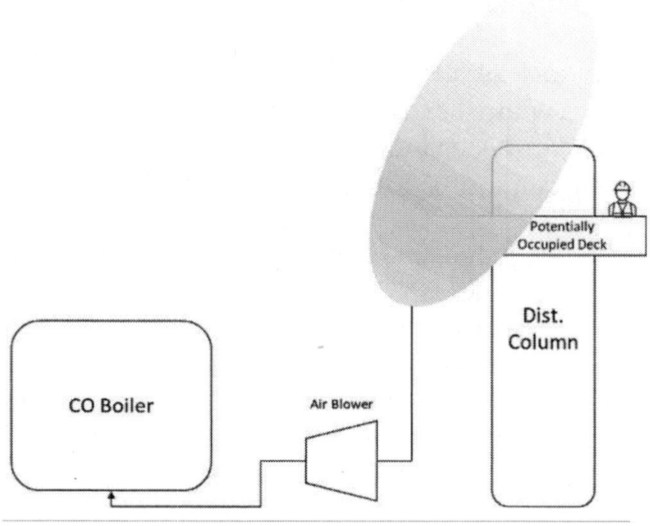

Figure 2. Carbon Monoxide Release Scenario

In this case, the analysis team performed a dispersion model for the reverse flow from the intake of the air blower. The modeling yielded that the gas cloud that was capable of causing harm was actually not very large and would drift upwards and disperse without ever coming to ground level where the bulk of receptors are located. This result occurred for a couple of reasons. First off, the air intake is significantly elevated for this specific reason. Furthermore, the material is released at an elevated temperature, coming directly from the catalyst regenerator. Hot gases released from a high elevation dispersed up and away from the release source, posing little or no risk to personnel located on the ground.

One unexpected result of this analysis was the determination that the gas cloud would, if the weather conditions were right, impinge on a deck of a distillation column that might be occupied. The LOPA analysis that was performed allowed for a conditional modifier of probability of occupancy to be considered. In this case, the probability of occupancy was quite low and the team elected to perform a formal occupancy analysis to justify the figure that was used. An occupancy analysis lists out all of the reasons that a given area would be occupied, estimates that duration and number of personnel for those activities, and then summarizes the results. In this case, the deck would only be occupied by operations inspection of equipment on the deck, engineering inspection of equipment on the deck, or maintenance of equipment on the deck. *Figure 3* shows the results of the occupancy analysis for this scenario.

Task	Duration (hours)	Frequency (per year)	Occupancy Probability
Operator Inspection	1 hour	12	0.00137
Engineering Inspection	1 hour	12	0.00137
Maintenance	16 hours	4	0.00731
Total			0.01

Figure 3. Quantitative Occupancy Analysis

Ultimately, for this scenario, instead of the LOPA analysis using a consequence category that represented multiple fatalities with 100% probability of occupancy, a consequence category of single fatality was chosen and an occupancy conditional modifier of 1% was employed, resulting in much more reasonable and achievable SIS design at the CO Boiler.

While in many cases, qualitative judgment of team and estimation tools overestimates consequence, there are situations where the consequences can also be underpredicted. Consider a scenario where a toxin is released through a vent in an atmospheric storage tank, in this case hydrogen sulfide from sour water in an oil refinery is sent into a "slop tank" that is open to atmosphere. The team judgment did not consider this release to be a significant concern because the slop tank is in a largely unoccupied area of the facility, and the extent of the release are not expected to be very large. The team selected a low consequence during PHA. When the scenario was reviewed during LOPA, the team noted that the release is indeed occurring in largely unoccupied area at the very edge of the plant, but the tank is actually so close to the fence line that if the weather conditions were unfavorable the release could go offsite. Even more unfortunate, the location where the gas cloud would travel is a playground that is occupied by small children on a daily basis. The situation was quickly escalated to management.

3 Initiating Event is Loss of Containment

A requirement to perform LOPA when a HAZOP scenario results in a consequence that is high can often result in scenarios that are inappropriate to be addressed by LOPA. Specifically, scenarios where the initiating event, or cause in the HAZOP, is actually the loss of containment event. The most common of these causes would be a tube leak or a piping leak. If the loss of containment event is the initiating event, the LOPA analysis will not likely yield any benefit because very few safeguards will be effective in reducing risk, because the hazard has already been released. In the situation where the cause is the loss of containment, the only reasonable response is to reduce the frequency of the loss of containment instead of trying to add safeguards that will attempt to respond to that loss of containment. The LOPA process is not suited to developing recommendations that will reduce the frequency of loss of containment. Instead, techniques for recommending changes to material of construction and increasing the type and frequency of inspection of the piping/tubing would be more appropriate, such as the ones described in American Petroleum Institute (API) Recommended Practice (RP) 580 – Risk Based Inspection [3] and 581 – Risk Based Inspection Methodology [4].

Consider where a HAZOP considers a scenario where a leak can cause a significant toxic consequence. An example from the oil refining industry might be where the acid gas generated by an amine recovery unit is cooled in an air cooler prior to being sent to a sulfur recover unit. It is not uncommon for a HAZOP scenario to be generated where the cause is listed as "tube leak" and the consequence is extreme (due to the high concentration of hydrogen sulfide in acid gas and its toxicity) – potentially multiple fatalities. As a result of the high consequence, organizational risk analysis rules might move this scenario into LOPA for additional consideration. During LOPA, the consequence category would generally be carried over. The initiating event would be carried over as "tube leak", and then its frequency would be quantified based on corporate organizational guidance that itself would be based on industry guidance documents like the CCPS book on initiating events and independent protection layers [5]. A commonly used number in LOPA for this "initiating event" could be 1/100 chance per year. The analysis would then consider what other independent protection layers are available and how effective they are in terms of probability of failure on demand. For this scenario, there are no other independent protection layers. After all, the initiating event is the loss of containment, no other actions are available to prevent loss of containment because it has already occurred. One could then consider the implementation of consequence mitigation measures such as gas detection, but as this paper will demonstrate in a subsequent section, consequence mitigation is quite difficult to assess and rarely achieves even one order of magnitude of risk reduction. In this case, which is documented in Figure 4, tolerable risk is not only not achieved, there are no reasonable options that can be incorporated into traditional LOPA that will allow tolerable risk to be achieved. Ultimately, LOPA fixed the frequency of the initiating event, and LOPA protection layers are not effective enough to reduce the risk to a tolerable level. The scenario is stuck with a three order of magnitude gap between estimated and tolerable risk.

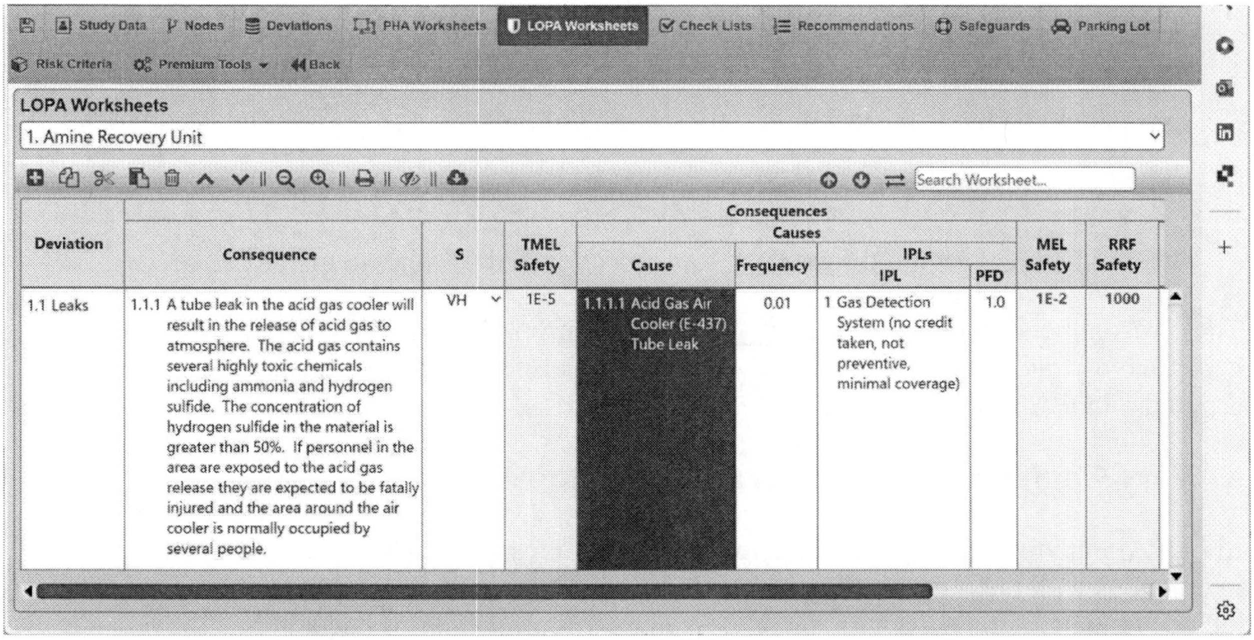

Figure 4. LOPA of Acid Gas Tube Leak

There are two directions that this study can be by implemented more advanced techniques. This first one has already been discussed. This scenario is another case where dispersion modeling could demonstrate the consequence that the team expected is excessively pessimistic. While there is a release of highly toxic material, the release is occurring in the tube sheet of an air cooler with large fans blowing massive quantities of air up into the tube sheet which would be expected to carry any release straight up and allow significant dilution before the release returns to grade level. This scenario would require rather elaborate dispersion modeling using computation fluid dynamics as the driver for the dispersion is the fans, not the ambient weather conditions.

The other quantitative mechanism to improve upon the risk analysis is to better understand the release frequency and the parameters that impact it. In traditional LOPA, the team would simply select the "approved" leak rate from a table in the organizational standard. This figure is certain to be conservative and represent a typical to poor application. As such, the frequency of leaking from unmaintained carbon steel instrument air tubing would be the same as the leak frequency from the exotic metallurgies and frequent extensive testing that is employed for acid gas. This is not realistic or appropriate.

For this scenario, the most appropriate response to intolerable levels of risk is to reduce the frequency at which the leaks occur. A once in 100 years leak for acid gas is not acceptable with no regard of what consequence mitigation is put in place. The best approach is to use more detailed methods to recommend design parameters that will result in leak rates that are more in the range of 1/10,000 to 1/100,000 per year. These techniques and measures are described in API 581 [5] along with methods for quantification. The recommendations from this type of study would include:

- Best option for tubing metallurgy based on service
- Appropriate parameters for design such as wall thickness
- Appropriate methods for techniques for welding/joining
- Inspection methods that should be employed to identify deterioration
- Determination of required frequency for inspection and preventive maintenance

These recommendations are superior to and more appropriate than the recommendations generated by traditional LOPA. These methods will also yield a more appropriate frequency to use in the LOPA and associated HAZOP documentation.

4 Very Complex Human Interaction with Process

In LOPA, initiating events are usually equipment failures, external events, or failed human interactions with the process. Similarly, protection layers include equipment items, but also include human interactions with the process like response to alarms and administrative controls. A typical LOPA scenario will consider a cause, along with the protection layers that protect against that cause. But, in some cases the human interaction with the process can be very complex. Sometimes an initiating event requires multiple initiating events, such as an initial mistake by one person and a failure of the second person to detect that the mistake was made during a checking process. Protective actions performed by human operators might be possible from multiple different people taking actions from different locations using different equipment items. The common LOPA analysis where one operator can respond to one alarm very well might not represent these complex situations. This situation could often benefit from a fault tree analysis that will more elegantly consider the complexity of the situation while allowing for use of more precise and more thoroughly research human error probabilities.

An example of where a traditional LOPA might not provide a sufficient degree of detail is a tank filling operation where material could be simultaneously withdrawn from the tank that is being filled during production operations. In this type of process, filling might involve receiving a shipment by ship transfer of a known volume. An engineer or operator might be required to calculate the volume available in the tank and compare that amount against the volume of the shipment, considering the rate at which material is being removed from the tank, in order to confirm that sufficient tank space is available. Erroneous calculation might result in overfill if the tankage is not capable of receiving the entire shipment. Usually, this type of calculation will be checked by another person, but these checks can often fail. In fact, failure to detect a mathematical error done by another person is quite high.

This leads to the topic of what data is used and its quality. Traditional LOPA will use about three different figures for human failure. The first two are failures in acts of commission, where an action was performed, but improperly. One of these, often 1 in 1,000 per opportunity, is where the person performing the action is trained and following a procedure, and another of 1 in 100 per opportunity where the task is performed on an ad hoc basis without formal training or a procedure. The third number commonly used for human failure in LOPA is a 1 in 10 chance of failure to properly respond to alarms (given a set of criteria where it would be reasonable to consider alarm response would be effective). This last figure is used when a human is a protection layers whereas the others are used when the human is an initiating event. These three figures are representing specific taxonomies of behavior and may be conservative in one situation, while optimistic in another. Ultimately, human failure probability varies by several orders of magnitude depending on the task being performed. Failure to detect a math error in the work of a colleague may be higher than 50%, but the probability that a human will not stop their automobile at a red light is significantly less than one in 1,000,000. When the situations that a LOPA is analyzing are unique, novel, or complex, it is prudent to do more research to determine if the standard three figures are appropriate, or whether other figures should be used. Resources like Human Reliability Analysis from Gertman and Blackman [6] can provide much more detailed discussion of failure probability with data for thousands of different scenarios.

Human response to initiating events can also be complex. In the scenario under consideration, the vessel that is being filled contains three different level measurements – a float and tape continuous measurement, an open path radar level measurement in a stilling well, and a tuning fork type high level alarm if the tank were reaching completely full. Furthermore, the alarms from these three level measurements are annunciated in three separate control rooms – once at the process unit that is tanking material from the tank, second at the tank farm control room, and yet a third time at the marine wharf control room where the transfer is occurring. The multitude of different measurements that are available, the different annunciation equipment available, and the different human operators that can respond are complex and interrelated.

The scenario under discussion would benefit from a fault tree analysis to completely document all of the interrelated conditions and situations. The fault tree can then calculate the expected frequency of the overfill more accurately than a traditional LOPA, especially if more specific an appropriate human error probabilities are researched and employed. An example (simplified) fault tree for this scenario is shown in Figure 5.

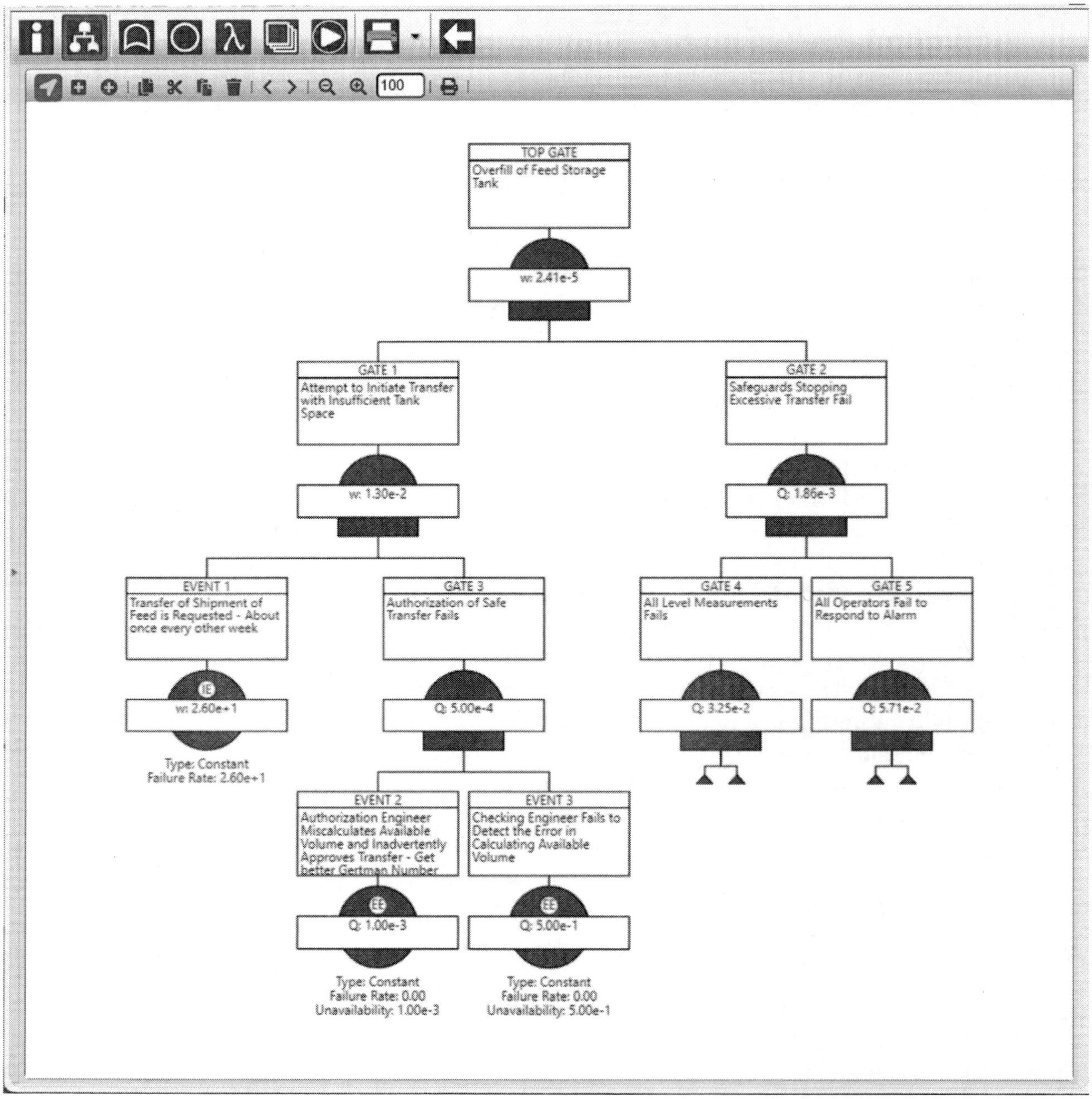

Figure 5. Example Fault Tree Analysis of Tank Overfill Scenario (Simplified)

In the previous paragraphs, a lot of information was presented regarding the departure from traditional LOPA represented in this fault tree analysis. These departures can be summarized as follows:

- Different human error probabilities that are a departure from the "standard three" were researched and implemented.
- The frequency trigger of this scenario is not a traditional "initiating event" but a regular request to transfer product into a feed tank. The actual initiating event is a compound event that includes the request to transfer combined with the probability that the operations team does not determine that it is not safe to do so.
- The IPL of "operator intervention based on alarm" was broken into two pieces to separate the annunciation equipment from the human response
- The human response includes complex redundancy of three operators
- The alarm annunciation equipment incudes complex redundancy of three pieces of equipment

5 Many Initiating Events with Different Safeguards and Effectiveness

Generally speaking, more complex scenarios require more detailed analysis. LOPA is intended to address relatively simply scenarios with simplified techniques based on conservative assumptions. As the complexity of a scenario increases, assumptions inherent to LOPA begin to get violated in increasing relevant ways and the conservatism inherent to LOPA begins to dominate the results to an unacceptable degree. More complex scenarios benefit from movement from LOPA to fault tree analysis where some of the conservatism can be removed by breaking equipment groups, like control loops, into individual components for more accurate analysis. This same technique will also help to overcome the non-conservative error of ignoring that equipment items are shared between multiple protection layers.

Marszal [7] describes the analysis of a hydrocracker runaway reaction using fault tree analysis as a supplement and expansion of LOPA. A hydrocracker can suffer a thermal runaway reaction for a very large variety of causes, and each cause has diverse safeguards that are available. Furthermore, multiple different safety functions that share components to some degree and also may not be 100% effective in returning the process to a safe state. Figure 6 presents a typical hydrocracker flowsheet.

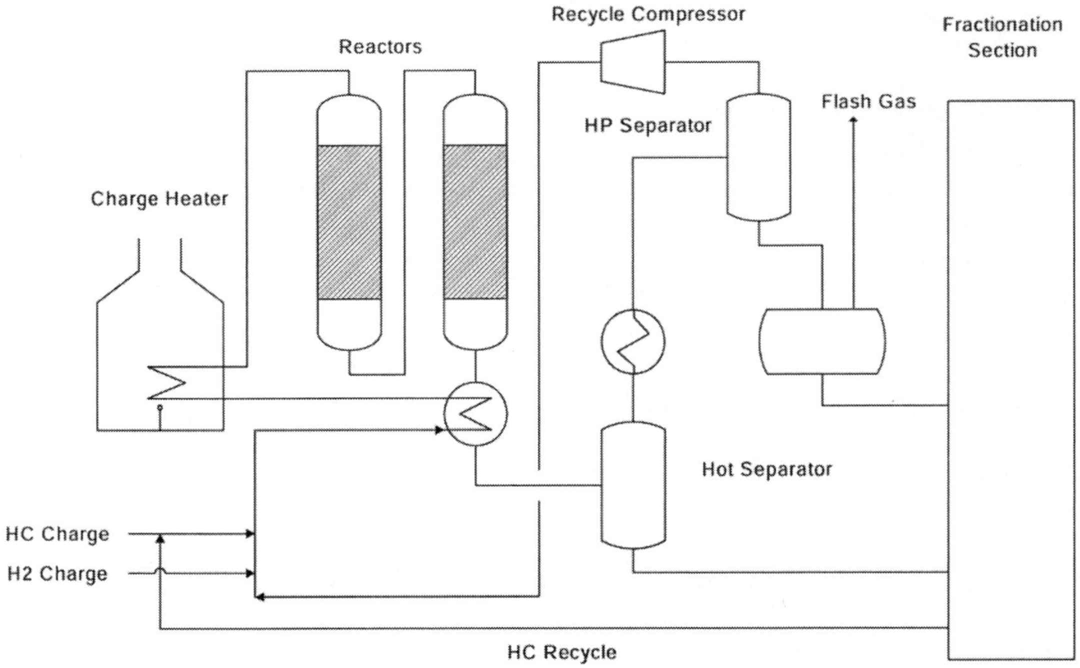

Figure 6. Typical Hydrocracker Flow Sheet

Initial attempts to assess the amount of risk reduction that needs to be provided by the safety instrumented function that detects a high temperature and performs the safety action of depressurizing the reactor system were quite complex. The potential causes of thermal runaway include:

- Recycle compressor failure
- Reactor internals failure leading to preferential flow channeling
- Quench failure
- Plugging and channeling due to coking and contamination
- Improper catalyst loading resulting in channeling
- Bed temperature measurement failure
- Failure of recycle gas flow controller
- Change in feed flow rate and/or hydrogen-to-hydrocarbon ratio
- Failure of fired heater outlet temperature control

For this collection of initiating events, there is a collection of safeguards. Some of the safeguards are applicable to all of the initiating events, others are not. These safeguards are as follows:

- High temperature alarms (various) with operator response – response ranges from adjusting quenches to manual depressurizing activation
- Various SIF that cause low-rate depressurizing, including activation of a compressor or heater shutdown
- High temperature low rate depressurizing activated by the BPCS
- High temperature high rate depressurizing activated by the SIS

Analysis of this scenario and the appropriate assignment of performance targets to the multitude on interdependent protection functions requires departure from traditional LOPA, in this case use of a fault tree analysis. Fault tree analysis is required for several reasons:

- Many initiating events needs to be considered and their risk summed in order to ensure tolerable risk is achieved by the design of the safeguards. Assessing multiple causes simultaneously is possible in traditional LOPA, but can be confusing to achieve and visualize
- The analysis of this single scenario is required to assign performance targets to multiple different safeguards, including SIS, BPCS, and alarms
- The safeguards may share common equipment. For instance, both a BPCS function and an SIS function can activate both the low rate depressurizing valve and the high rate depressurizing valve
- In some cases, a safeguard can be successfully activated, but still not result in a safe state. This concept is discussed in the ISA 84.00.07 [8] technical report as "mitigation effectiveness". LOPA assumes that if a safeguard is successfully activated, then a safe state is achieved, which is not always true.

While some focused QRA techniques supplement a LOPA and provide a more precise parameter, such as a consequence category or safeguard probability of failure on demand. In this case, the LOPA study is entirely replaced with the fault tree analysis which is used not only to confirm that tolerable risk is achieved, but also to confirm that performance targets of all safeguards are both set and achieved.

6 Preemptive Strikes

The discussion of pre-emptive strikes is very similar to the discussion in the previous section from the substance of how the analysis needs to be performed and why fault tree analysis needs to replace the LOPA entirely in these cases. This section simply expands on the definition and purpose of a preemptive strike. A preemptive strike is a type of safeguard that activates in order to prevent an initiating event from occurring. The function by itself does not prevent an incident from occurring directly. Instead, it prevents the situation that will lead to the incident from occurring. The complexity in their analysis lies in the fact that two different safety functions are effectively being used, but the safety instrumented functions share equipment so their analysis cannot be separated as though they were simply two different IPLs. Also, the scenario is complex, making a more detailed analysis prudent.

Consider a batch reaction system that performs a highly exothermic batch polymerization reaction. In this scenario, the reactor is charged with one reactant, and another reactant is slowly metered into the reactor under a control scheme where the flow is controlled with a high temperature override on the flow rate set point. If the second reactant is allowed to accumulate because it is fed into the reactor too quickly, an excessive amount of chemical potential energy will exist in the reactor, which could result in a runaway reaction that generate high pressure, and ultimately will cause a massive rupture of the reactor vessel.

One of the safeguards employed in the reaction procedure is to minimize the amount of first reactant that is charged into the reactor prior to starting the reaction. To accomplish this task, the safety instrumented system includes functionality that uses a weight measurement as a reaction permissive. If the weight of reactant in the initial charging step is too high, the safety instrumented system will stop the batch and prevent the other reactant from being fed into the reactor. When analyzing the scenario, though, one must assume that this first preemptive strike interlock fails and that the second reactant is added into the batch and a runaway reaction begins. If the runaway reaction begins, the high-pressure and/or high-temperature measurements in the reactor can detect the runaway and bring the process to a safe state by adding a killing agent to the reactor.

While some might consider LOPA sufficient in this scenario by treating the two safety instrumented functions as completely independent protection layers, they are not. At a minimum, they employ the same logic solver and a fault tree would be required to elegantly handle the common equipment. Furthermore, in situations similar to this there will probably be more different initiators to the runaway reaction, each having different but equipment inter-related SIF, making the scenario complex and similar to the one that we discussed in the previous section for the hydrocracker runaway reaction.

7 Initiating Event is a Complex Sequence of Events and Conditions

In the discussion of complex human interaction earlier in this paper, we have already described the concept where an initiating event is not a simple failure of a piece of equipment or simply human error to correctly perform a task in a written procedure on which they have been trained. This concept warrants further discussion and another example as there are numerous instances in the process industries where initiating events are not straightforward and require additional analysis outside the boundaries of traditional LOPA.

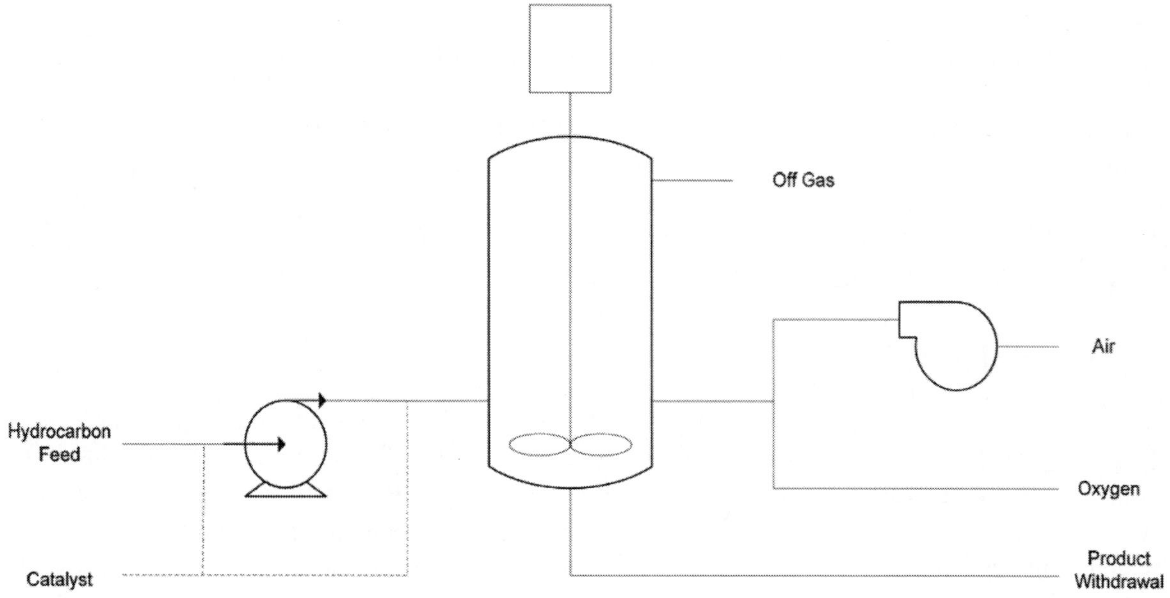

Figure 7. Typical Oxidation Reactor Flow Sheet

Consider a batch oxidation reaction discussed by Marszal [9] that is shown in Figure 7. In this type of reaction, a hydrocarbon feed stock is charged into the reactor – along with other components like liquid catalysts – and then heated to the reaction temperature. At that point, the air blower is started and air inlet valves are opened, sparing air into the fluid which results in the oxygen in the air reacting with the liquid feedstock. The reactor is equipped with an interlock system that will detect that the differential pressure between the air inlet and the reactor is low – indicating that there is not driving force to push the air into the reactor – which causes the valve feed valve to close. In more detail, a double-block-and-pressure arrangement activates where the two valves close and then nitrogen is inserted into the cavity between the two valves providing a positive seal. This type of double block and seal arrangement is employed in situations where leakage through the valve is unacceptable and dangerous even in very small amounts.

Assessment of this scenario using a traditional LOPA will yield very interesting results. Specifically, that the hazard that is detected by the SIF actually has no consequence. When the driving force to push air into the reactor is lost there is a potential for the liquid contents of the reactor to backflow out of the reactor into the air inlet piping. When asked about the consequence of the backflow reactor materials into the piping the team stated that the material would simply pool in the reactor inlet and that the reactor material would not damage the inlet piping either from a perspective of pressure, temperature, or chemical composition.

This situation required further analysis and a process expert provided additional input. While a leak of reaction material into the inlet piping will not create an immediate consequence, it is critical that if a leak does occur, that the inlet piping be disassembled, de-inventoried, and cleaned prior to restarting the process. The oxidation process is highly exothermic, and as long as it is occurring in the reactor vessel, the design of the reactor and its cooling system is sufficient to maintain safe operation. If, though, the reaction were allowed to occur in the inlet piping the reaction would generate temperatures greatly in excess of maximum allowable working temperature of the piping, resulting in explosion and fire.

In this case, the scenario that results in the loss of containment incident is much more complex than simply backflow of reactor contents into the air feed line. The complexity of modeling that a backflow of material had occurred in the past and a reactor restart was going to be attempted to be performed with the operations team failing to disassemble and clean the air inlet system. This complex interaction is better modeled with fault tree analysis due to its complexity.

8 Shared Components

The concept of shared components has already been identified as a portion of a few of the preceding sections, but in this section it will be discussed in isolation using an example that highlights its importance due to every portion of the potential LOPA scenario was interrelated with the same equipment items. For LOPA to be effective the scenario under analysis must have full physical and functional independence not only between the protection layers themselves, but also between the protection layers and the initiating events. In a lot of cases, some degree of overlap between safeguards exists but is ignored because it is minimal. For instance, an alarm that is used as part of an operator intervention IPL might be activated through the same BPCS logic processors whose failure could be the root cause of the failure of the control loop that generated the alarm condition in the first place. When the amount of overlap is small, it can often be demonstrated that the overlap can be ignored in some circumstances [10], but in some cases it is not appropriate to do so.

An example of where components can be shared between every portion of a LOPA scenario, including the initiating event, alarm with operating intervention IPL, and safety instrumented system IPL is provided by Marszal [11]. Figure 8 presents a process segment where a butane storage tank receives shipments of butane via pipeline. In this process, when a shipment of butane is requested to be received, an operator will go to the butane storage sphere and check the level of the sphere at the local gauges. This is done by looking at both of the diverse measurement types presented by LT-101 and LT-102. If there is sufficient available space in the sphere the loading process begins. If the available volume of the tank is exceeded, either because the amount of available space was misread by the operator or because the totalizing controller fails, two protection actions can be taken. First, a high level alarm can be manually responded to be closing valves or manually stopping the transfer system, and second a safety instrumented system can automatically close the inlet valve. The complicating factors is that the initial measurements, the alarm, and the SIF all rely on the same redundant set of measurement devices (i.e., LT-101 and LT-192).

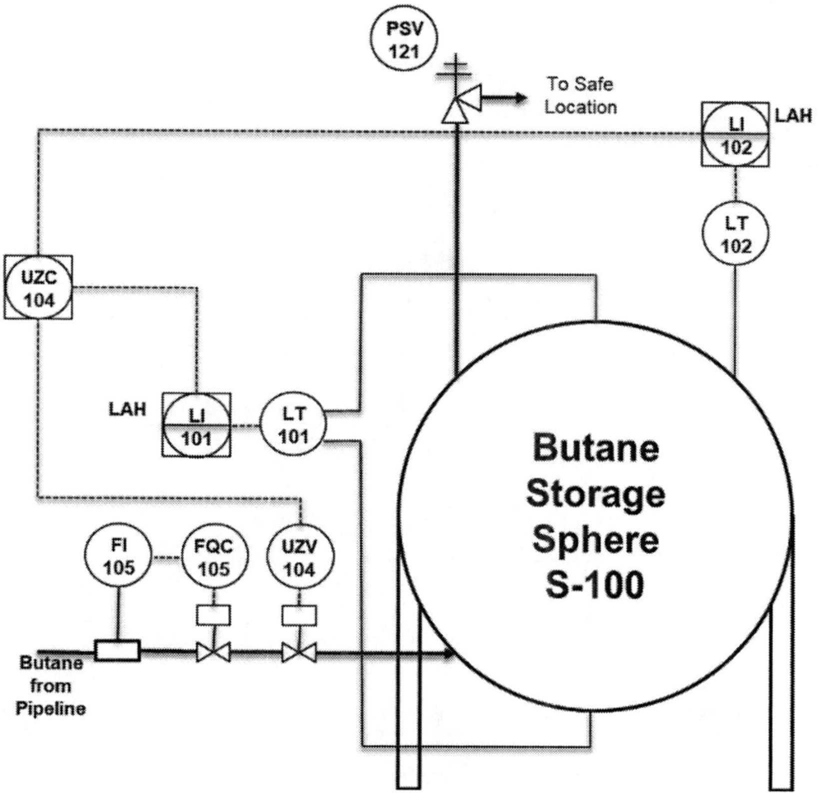

Figure 8. Butane Storage Sphere Flow Sheet

The degree of common equipment in a LOPA scenario cannot be ignored and would typically result in recommendations for use of separate and independent devices for each component of the LOPA scenario. But, if the scenario is analyzed using fault tree analysis instead, the as-designed use of the two transmitters could be justified.

Fault tree analysis can elegantly handle common equipment in multiple independent protection layers because the shared components can be modeled separately and, even though the shared components show up in multiple locations in the fault tree, the minimal cut set calculation methods correctly determine the impact of this sharing on the overall frequency of the loss of containment scenario. This process of addressing the share components of multiple IPL separately is shown in a fault tree in Figure 9.

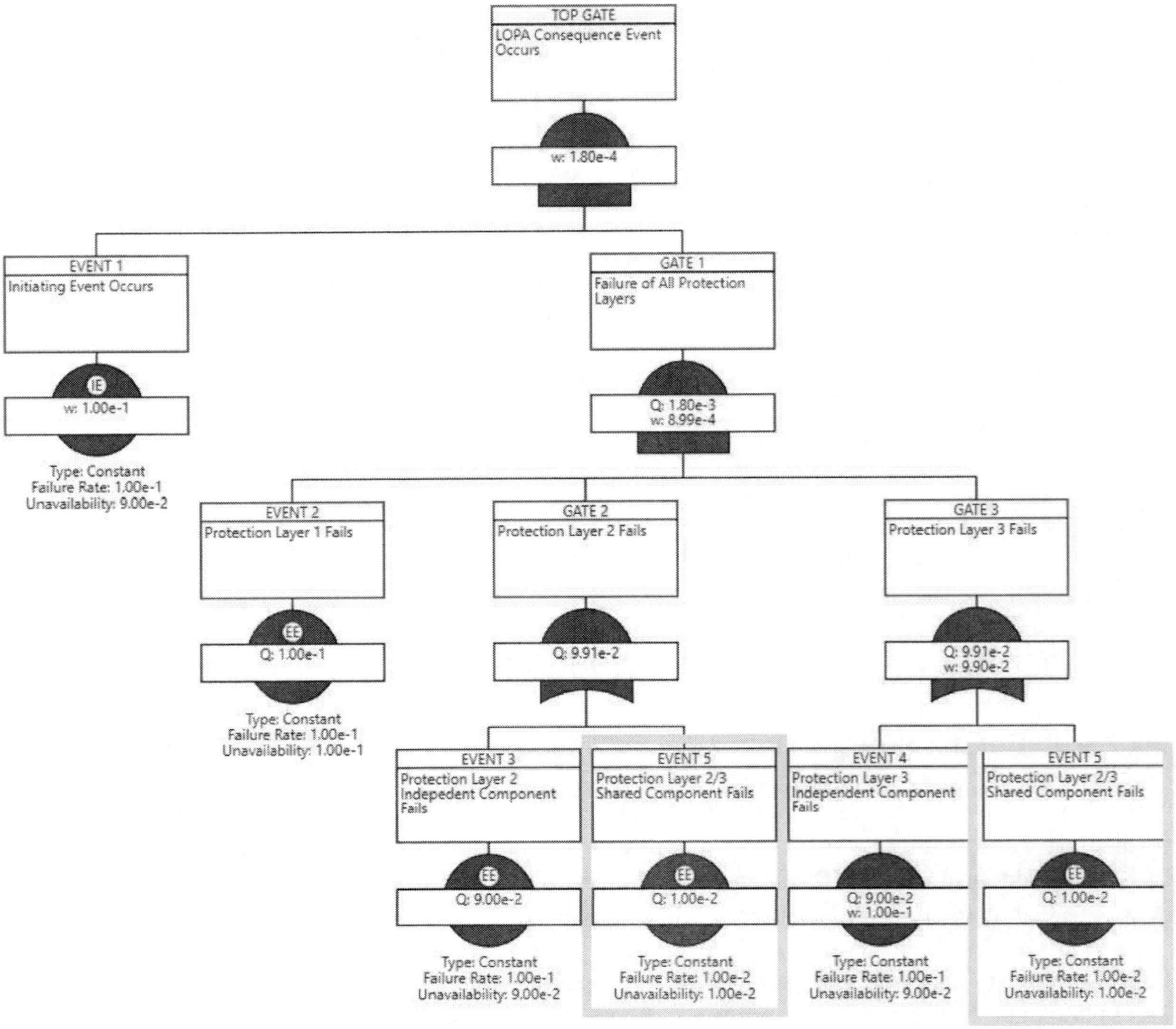

Figure 9. Fault Tree with Shared Components Among Protection Layers

9 Mitigative Safeguards

One of the key assumptions in LOPA is that if an IPL successfully activates, the consequence will not occur. This type of IPL is called "preventive" because it prevents that loss of containment event and thus the consequence. There is an entire category of safeguards that do not prevent the loss of containment from occurring in the first place. Instead, these safeguards detect that the loss of containment has occurred and then take action to reduce the magnitude of the consequence. It is difficult to assess the effectiveness of mitigative safeguards with LOPA, because LOPA analysis is contingent upon a single consequence which may or may not occur depending on whether or not the safeguards are effective. For a mitigative safeguard there is a consequence whether or not the safeguard is effective. If the mitigative safeguard is effective the consequence is smaller. If the mitigative safeguard is not effective the consequence is larger. The analysis is further complicated because the probability that the safeguard is effective is also part of the analysis. Ultimately, when a mitigative safeguard is part of a scenario two consequences need to be considered – the mitigated risk scenario where the mitigation safeguard is effective, the consequence is smaller and the residual risk scenario where the consequence is not mitigated but will occur at a lower frequency. The risk of both outcomes is required to be demonstrated as tolerable. Marszal [12] presents a detailed discussion of these concepts along with a series of examples including the use of a fire detection and suppression system to prevent a boiling liquid expanding vapor explosion (BLEVE) that is the result of a leak in an LPG storage bullet.

While the analysis described above can be simulated by running two separate LOPA scenarios and linking them together, there are more sophisticated tools that will allow both consequences to be assessed in the same scenario. Specifically, event tree analysis and quantitative bowtie analysis were specifically developed to analyze multiple different consequences that can occur as the result of a single loss of containment event. These tools consider multiple different outcomes that can result from a single loss of containment based on activation of mitigative safeguards and other post release conditions such as probability of ignition. *Figure 10* presents a quantitative bowtie assessment of a flammable liquid tank overfill scenario where the mitigative safeguard of tank containment dike is used.

The risk analysis of the initiating events, each of which have two preventive protection layers (i.e., high level alarm and high level SIF). Preventive safeguards are shown on the left of a bowtie diagram as they prevent the loss of containment. It is customary to show the loss of containment event as the bowtie knot. On the right-hand side of the bowtie, you will see two mitigative barriers:

- Containment dike
- Probability of ignition

Since there are two mitigative barriers, four outcomes are possible, as follows:

- Dike is successful, ignition occurs
- Dike is successful, ignition does not occur
- Dike fails, ignition occurs
- Dike fails, ignition does not occur

All four of these scenarios will have different frequencies and different consequence categories. Each outcome should be assigned a consequence category that will define its tolerable risk in the same way as would be done for a LOPA scenario. Similarly, the frequency of each of the outcome is calculated using the same math as LOPA, which is to multiply initiating event frequencies by the probabilities that will result in that outcome occurring. Then, tolerability of all of the scenarios is determined by comparing the estimated frequency calculated for that consequence against the tolerable frequency for that consequence. In the example, all four of the potential consequences were determined to be tolerable.

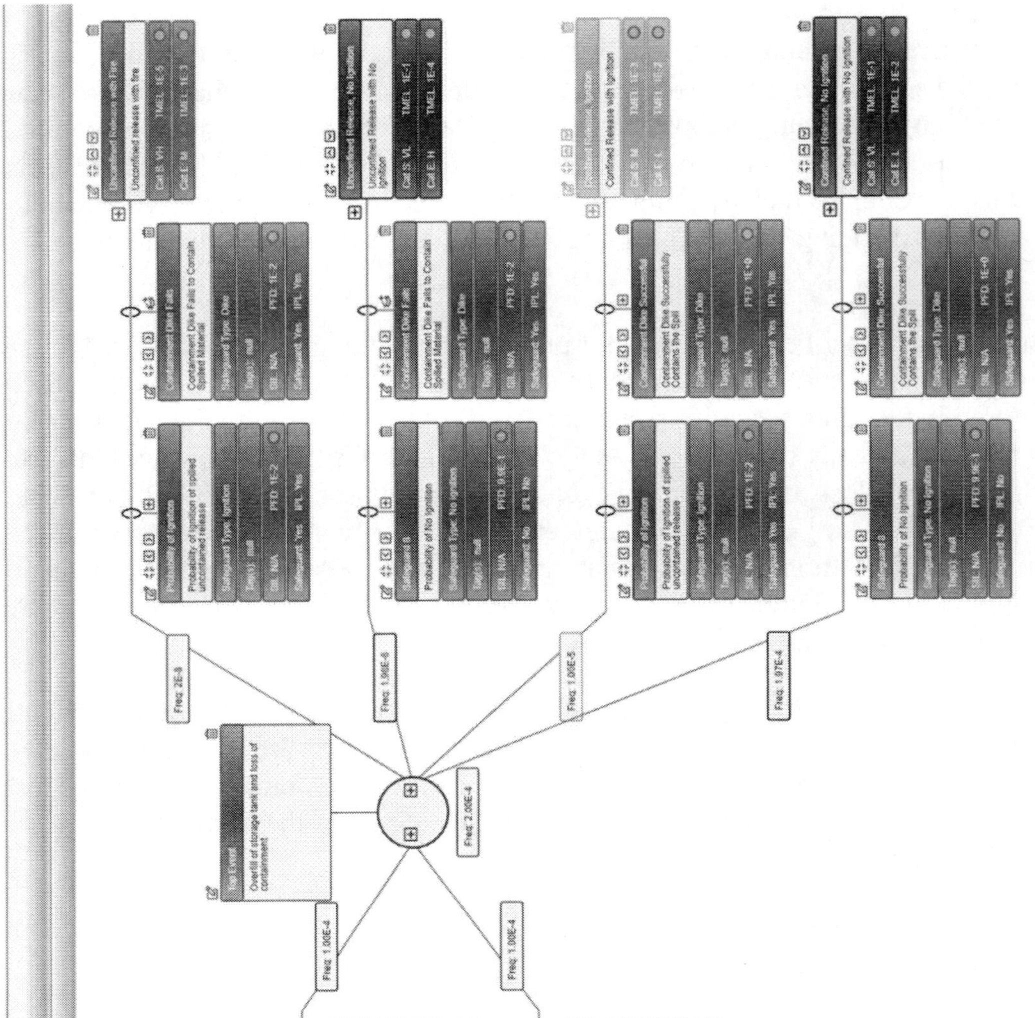

Figure 10. Quantitative Bowtie Diagram of Tank Overfill Scenario

10 Conclusions

LOPA is a powerful tool used to address important scenarios, but in many cases the results of a LOPA analysis are either unsatisfactory or inappropriate. This paper discussed a number of situations where LOPA should be supplemented or replaced with other techniques because the assumptions and simplifications inherent to LOPA are either not true or results in such a degree of excessive conservatism that the results are not usable. Organizations that perform LOPA analysis should be vigilant to detect when these situations are occurring and take additional actions to supplement or replace LOPA as required.

11 References

[1] International Society for Automation, ANSI/ISA S84.01 – 1996 Application of Safety Instrumented Systems for the Process Industries, International Society for Automation, Raliegh, NC, 1997.

[2] Marszal and Scharpf, Safety Integrity Level Selection – Systematic Methods Including Layer of Protection Analysis, International Society for Automation, Raliegh, NC, 2002.

[3] America Petroleum Institute, API RP 580 – Risk-Based Inspection, American Petroleum Institute, New York, NY, 2016.

[4] America Petroleum Institute, API RP 581 – Risk-Based Inspection Methodology, American Petroleum Institute, New York, NY, 2016.

[5] Center for Chemical Process Safety, Guidelines for Initiating Events and Independent Protection Layers in Layer of Protection Analysis, American Institute of Chemical Engineers, New York, NY, 2015.

[6] Gertman and Blackman, Human Reliability and Safety Analysis Data Handbook – 3rd Edition, Wiley-Interscience, New York, NY, 1993.

[7] Marszal and Mitchell, Hydrocracker Safeguarding with SIS – White Paper, Kenexis Consulting Corporation, https://www.kenexis.com/wp-content/uploads/2014/05/HydrocrackerSafeguarding.pdf, accessed 18 Feb 2024.

[8] International Society for Automation, ISA TR 84.00.07 – 2018 Guidance on the Evaluation of Fire, Combustible Gas, and Toxic Gas System Effectiveness, International Society for Automation, Raliegh, NC, 2018.

[9] Marszal and Mitchell, Oxidation Reaction Safeguarding with SIS – White Paper, Kenexis Consulting Corporation, https://www.kenexis.com/wp-content/uploads/2014/05/OxidationReactionSafeguardingWithSIS.pdf, accessed 18 Feb 2024.

[10] Marszal, Protection Layer Commonality in LOPA, 2003 National Conference Proceedings, International Society for Automation, Raliegh, NC, 2003.

[11] Marszal and Applegate, Use Fault Tree Analysis When LOPA Fails, Chemical Engineering Progress Magazine, Volume 6 Issue 2, American Institute of Chemical Engineers, New York, NY, February 2024.

[12] Marszal, Expanding from LOPA to Quantitative Bowtie Analysis, Process Safety Progress, American Institute of Chemical Engineers, New York, NY, 11 October 2022.

Adapting PHA Facilitation Techniques in Response to a Shortage of Skilled Facilitators

Mr. John T. Perez, P.E.
Cognascents Consulting Group, Inc.
14701 St. Marys Lane, Suite 300
Houston, TX 77079
john.perez@cognascents.com

Mr. Jeffrey Miller, P.E.
Cognascents Consulting Group, Inc.

Keywords: American Institute of Chemical Engineers (AIChE), Center for Chemical Process Safety (CCPS), Hazard and Operability Study (HAZOP), Layer of Protection Analysis (LOPA), Occupational Safety and Health Administration (OSHA), Process Hazard Analysis (PHA), Process Safety Management (PSM).

Abstract

Process Hazard Analysis (PHA) is a cornerstone of risk management in the upstream, midstream, downstream, and chemical industries. Traditionally, PHAs are facilitated by highly skilled and experienced individuals whose expertise ensures that potential risks are adequately identified, assessed, and mitigated. However, the industry currently faces a significant shortage of these adept facilitators, prompting a re-evaluation of standard PHA facilitation techniques.

This paper advances the case for modifying traditional PHA facilitation techniques to address the shortfall in expert facilitators, ensuring continued efficacy in risk management practices.

More specifically, this paper will explore the following:

1. <u>Recruiting the next generation of PHA facilitators</u>: Academia and industry are simply not generating adequate talent pools to meet the PHA facilitation service demand. Recruiting techniques are often singularly focused on skillset; however, more effective and creative techniques are available to bolster the thinning ranks of PHA facilitators.
2. <u>Leveraging technology</u>: Given the scarcity of seasoned facilitators, there is a pressing need to capitalize on technology. Advanced artificial intelligence (AI) and machine learning tools can assist in preliminary hazard identification, providing a solid base upon which less experienced facilitators can build.
3. <u>Training and development acceleration</u>: Investing in rapid yet comprehensive training programs can hasten the development of competent facilitators. Modular courses, on-the-

job training, and mentorship programs can all be integrated to bolster the skill sets of budding facilitators.

4. Standardization and templates: To assist less experienced facilitators, the development of standardized procedures and templates can provide a consistent foundation. While customization will still be required based on unique company procedures and protocols, these tools can streamline the facilitation process.
5. Feedback and continuous improvement: Given the shift in facilitation techniques, it is imperative to establish a strong feedback mechanism. Regular reviews and post-PHA evaluations can identify areas of improvement, ensuring that the modified techniques continue to be refined over time. These reviews can incorporate communication packages during management review meetings to afford PHA recommendation review as well as PHA quality assurance.

The shortage of skilled PHA facilitators presents a unique challenge to the industry, but it also offers an opportunity to innovate and adapt. By integrating technology, accelerating training, standardizing processes, and emphasizing feedback, it is possible to maintain rigorous PHA standards even in the face of facilitator scarcity.

The target audience for this paper is anyone whose responsibilities include: (1) leading within an organization required to comply with OSHA 1910.119, (2) establishing effective PHA guidance documents, and (3) participating on PHA teams to identify, assess, and manage process safety hazards.

1 Introduction

Process Hazard Analysis (PHA) is a cornerstone of process safety management (PSM) and risk assessment in the upstream, midstream, downstream, and chemical industries. Traditionally, PHAs are facilitated by highly skilled and experienced individuals whose expertise ensures that potential risks are adequately identified, assessed, and mitigated. However, the industry currently faces a significant shortage of these adept facilitators, advancing the case for modifying the traditional approaches to (1) identifying and recruiting PHA facilitators, (2) conducting PHAs, and (3) communicating PHAs outcomes to all impacted personnel. This paper explores the following in more detail:

1. Recruiting the next generation of PHA facilitators;
2. Leveraging technology;
3. Training and development acceleration;
4. Standardization and templates; and
5. Feedback and continuous improvement.

Most importantly, this paper is a call to action for impacted industries to evaluate their role in ensuring an adequate supply of PHA facilitation talent to meet the ever-present and growing demand for risk assessment expertise.

2 Recruiting and Retaining the Next Generation of PHA Facilitators

The growing demand for PHA facilitation services has underscored a critical issue within academia and industry: the insufficient generation of talent pools to meet this demand. Traditional recruiting methods emphasize technical skillset, neglecting the broader set of qualities that make an effective PHA facilitator. This narrow focus has led to a shortage of qualified individuals capable of effectively navigating the complexities of PHA sessions. As a consequence, industries reliant on PHA facilitation services find themselves grappling with inadequate resources to address essential process safety and risk assessment demands.

Addressing the scarcity of PHA facilitators requires a shift towards more comprehensive and creative recruiting techniques. While technical expertise remains crucial, it is equally essential to recognize the importance of people skills, critical thinking abilities, and adaptability in this role. Incorporating psychometric assessments, situational judgment tests, and scenario-based interviews can help identify candidates with the necessary blend of competencies to excel as PHA facilitators. Moreover, partnerships between academia and industry can play a pivotal role in nurturing talent by designing specialized training programs tailored to the unique demands of PHA facilitation.

1. Collaborative Internship Programs: Companies with PSM-covered facilities should establish collaborative internship programs with academia, specifically targeting students studying relevant fields such as chemical engineering, industrial safety, or risk management. These programs should provide students with hands-on experience in PHA facilitation under the guidance of experienced professionals, thereby fostering the development of future talent pools for the industry.
2. Specialized Training Courses: In collaboration with PSM-impacted industries, academia should develop specialized training courses focused on PHA facilitation techniques and methodologies. These courses should be offered by universities and should cover both technical skills and soft skills essential for effective facilitation. Industries and academia should promote PHA facilitation as a coveted technical career path.
3. Diversified Recruitment Strategies: Employers of PHA facilitators should expand recruitment strategies beyond traditional channels to attract candidates with diverse backgrounds and experiences. While chemical engineers serve the PHA facilitator role well, there simply are not enough of them to meet the demand. Employers should utilize social media platforms, professional networking events, and industry conferences to reach potential candidates who may not be actively seeking employment but possess the requisite skills and interest in PHA facilitation. While technical skill is desired, an effective and successful PHA facilitator has to possess a versatile arsenal of soft skills and leadership acumen – which do not always reside in an engineer.

To bridge the gap between supply and demand in the realm of PHA facilitation, a multifaceted approach is imperative. Beyond refining recruiting strategies, efforts should be directed towards raising awareness about the significance of PHA processes and fostering a culture of continuous learning and development within both academia and industry. By embracing innovative recruitment methodologies and investing in the professional growth of aspiring facilitators, academia and industry can collaboratively address the pressing need for a robust talent pipeline capable of meeting the escalating demand for PHA facilitation services.

3 Leveraging Technology to Safeguard Experience Vulnerabilities

The scarcity of seasoned Process Hazard Analysis (PHA) facilitators presents a significant challenge in industries reliant on robust risk assessment processes. In response to this challenge, there is an increasing recognition of the potential of advanced artificial intelligence (AI) and machine learning tools to complement human expertise in PHA facilitation. These technologies offer the capability to streamline and enhance preliminary hazard identification processes, providing a valuable foundation for less experienced facilitators. By leveraging AI algorithms trained on vast datasets of historical incidents and industry-specific risk profiles, organizations can efficiently identify potential hazards and prioritize areas for deeper analysis, thereby optimizing the utilization of human resources and accelerating the overall PHA process.

One of the primary advantages of integrating AI and machine learning into PHA facilitation lies in their ability to augment human judgment and decision-making. By analyzing complex data sets and identifying patterns that may elude human observers, AI systems can help uncover latent hazards and emerging risks that might otherwise go unnoticed. This proactive approach not only enhances the comprehensiveness and accuracy of hazard identification but also enables facilitators to focus their expertise on interpreting findings, facilitating discussions, and formulating effective risk mitigation strategies. Additionally, AI-powered tools can support knowledge transfer and skill development among less experienced facilitators by providing real-time guidance, recommendations, and feedback based on best practices and industry standards.

1. Implementation of AI-based Hazard Identification Tools: Companies performing or sponsoring PHAs should invest in and deploy advanced AI and machine learning tools specifically designed for hazard identification in process safety management. These tools can analyze vast amounts of data from historical incidents, near-misses, and industry databases to identify potential hazards and assess their likelihood and severity, providing valuable insights to both seasoned and less experienced PHA facilitators. And remember, AI gets smarter with each query. The more queries an experienced PHA facilitator "feeds" into the AI interface, the more opportunities for less experienced PHA facilitators to learn – no matter where they sit in the world.
2. Development of Virtual Reality (VR) Simulations: VR simulations can replicate realistic process environments and scenarios for training purposes. Less experienced facilitators can immerse themselves in simulated PHA sessions, practicing hazard identification and risk assessment in a virtual setting under the guidance of AI-driven feedback mechanisms. This interactive approach allows for hands-on learning and skill development in a safe and controlled environment.

However, it is essential to recognize that the integration of AI and machine learning in PHA facilitation is not without its challenges and limitations. Concerns related to data privacy, algorithmic bias, and the interpretability of AI-generated insights must be carefully addressed to ensure the reliability and trustworthiness of automated hazard identification systems. Moreover, while AI can enhance efficiency and effectiveness, it cannot replace the critical thinking, domain expertise, and people skills that experienced facilitators bring to the PHA process. Therefore, successful implementation requires a balanced approach that leverages technology as a complementary tool to augment human capabilities rather than as a substitute for human judgment and expertise.

4 Accelerating PHA Training and Development

The urgent demand for skilled Process Hazard Analysis (PHA) facilitators necessitates innovative approaches to training and development. By investing in rapid yet comprehensive training programs, organizations can accelerate the cultivation of competent facilitators equipped to navigate the complexities of risk assessment processes effectively.

1. Development of Modular Training Modules: Organizations looking to accelerate the development of PHA facilitators should create modular training modules that cover distinct aspects of Process Hazard Analysis (PHA) facilitation, such as hazard identification techniques, risk assessment methodologies, and regulatory compliance. These courses should break down the fundamentals of PHA facilitation into manageable units. These courses can cover additional topics such as communication skills, cognitive biases, leadership styles, and conflict management. These modules can be completed independently or sequentially, allowing budding facilitators to customize their learning paths based on their existing knowledge and skill levels.
2. Establishment of Mentorship Programs and On-the-Job Training Assignments: Pair budding facilitators with experienced PHA facilitators who can serve as mentors and provide opportunities for on-the-job training during live sessions. Mentors can offer personalized feedback, share best practices, and provide insights into navigating challenges commonly encountered in PHA facilitation. The mentor should personally observe the budding facilitator and provide real-time feedback. This hands-on experience allows less experienced facilitators to observe facilitation techniques in action, participate in discussions, and gain practical insights into managing PHA meetings effectively. Operating companies subject to PSM should consider a co-op arrangement with consulting companies to afford training and development of PHA facilitation talent.
3. Simulation Exercises and Case Studies: Incorporate simulation exercises and case studies into training programs to simulate real-world PHA scenarios. These exercises allow budding facilitators to practice applying PHA methodologies, identifying hazards, and facilitating discussions in a risk-free environment. Feedback from trainers and peers can help them refine their skills and approaches.

The integration of modular courses, on-the-job training, and mentorship programs offers a comprehensive approach to accelerating the development of competent PHA facilitators. By combining structured learning with practical experience and mentorship support, organizations can cultivate a pipeline of talent capable of effectively facilitating PHA sessions and driving continuous improvement in risk management practices. Moreover, investing in the professional growth and development of aspiring facilitators not only addresses the immediate need for skilled personnel but also fosters a culture of learning and innovation that strengthens the organization's long-term resilience and competitiveness in a dynamic business environment.

5 Streamlining PHA Facilitation with Standardization and Templates

To assist less experienced facilitators, the development of standardized procedures and templates can provide a consistent foundation. While customization will still be required based on unique company procedures and protocols, these tools can streamline the facilitation process.

The implementation of standardized procedures and templates is paramount in supporting less experienced PHA facilitators, offering them a structured framework to navigate the complexities of risk assessment. These standardized tools serve as foundational guides, ensuring that essential steps in the PHA process are consistently followed. By adhering to established procedures outlined in templates, facilitators can mitigate the risk of overlooking critical aspects of hazard identification and risk analysis. While customization remains necessary to accommodate the specific nuances of individual company procedures and protocols, standardized templates provide a baseline structure that streamlines the facilitation process and promotes efficiency.

Moreover, standardized procedures and templates foster coherence and alignment across PHA activities within an organization. By establishing a common language and methodology, these tools facilitate communication and collaboration among team members involved in the risk assessment process. Consistency in documentation and approach enhances transparency and clarity, enabling stakeholders to better understand the rationale behind risk assessment decisions. Furthermore, standardized templates serve as valuable training resources for novice facilitators, offering them a tangible reference point to familiarize themselves with established practices and protocols.

1. Creation of Standardized PHA Templates: Companies should develop standardized templates for conducting diverse types of PHA methodologies, such as Hazard and Operability (HAZOP) studies, What-If analyses, and Layer of Protection Analyses (LOPAs). These templates should outline the structure, methodology, and documentation requirements for each type of PHA, providing a consistent framework for facilitators to follow. The templates could be based on common units, scenarios, equipment, or chemistry.
2. Training on Template Utilization: Persons, especially PHA facilitators, who are required to participate in PHAs should receive training on how to effectively utilize the standardized procedures and templates during a PHA. Training should cover topics such as template navigation, data input guidelines, and best practices for documentation. Hands-on exercises and case studies can help familiarize less experienced facilitators with the use of templates in real-world scenarios.
3. Regular Reviews and Updates of Templates: Organizations who own the final PHA deliverable, should establish a process for regular reviews and updates of the standardized procedures and templates to ensure their continued relevance and effectiveness. This process should ensure to solicit feedback from facilitators, stakeholders, and subject matter experts to identify areas for improvement or refinement. Additionally, the periodic review should incorporate updates based on emerging industry trends, lessons learned from past PHA sessions, and changes in regulatory requirements.

Beyond aiding individual facilitators, the development of standardized procedures and templates contributes to organizational resilience and scalability. By codifying best practices and lessons learned into template formats, organizations can institutionalize knowledge and expertise, making it accessible to future facilitators and teams. This institutional knowledge not only promotes consistency in PHA execution but also facilitates continuous improvement by enabling organizations to identify areas for optimization and refinement. Additionally, standardized templates streamline the onboarding process for new team members, reducing the learning curve associated with PHA facilitation and ensuring continuity in risk management practices across different projects and departments.

6 More Effective Sharing of PHA Content and Outcomes with Communication Packages

Incorporating communication packages into management review meetings provides a structured platform for comprehensive PHA recommendation review and quality assurance. By integrating PHA feedback sessions into existing management review processes, organizations can streamline the evaluation of PHA outcomes and ensure alignment with strategic objectives and regulatory requirements. Furthermore, these communication packages enable stakeholders to track the implementation status of PHA recommendations, monitor progress towards risk mitigation goals, and identify any emerging issues or trends that may require attention. Effective communication and collaboration among key stakeholders facilitate informed decision-making and drive continuous improvement in risk management practices.

1. Structured Feedback Forms and Post-PHA Evaluation Meetings: Organizations who employ PHA facilitators should develop structured feedback forms to be distributed to PHA facilitators, participants, and stakeholders after each PHA session. These forms should include specific questions about the effectiveness of the facilitation techniques used, the clarity of communication, and suggestions for improvement. Before dismissing the PHA team after the last session, the team should conduct a post-PHA evaluation meeting to review the execution and outcomes of the PHA sessions. These meetings provide an opportunity to discuss what went well, areas for improvement, and lessons learned.
2. Integration of Communication Packages: To assure more effective sharing of PHA content and outcomes, the PHA process should incorporate communication packages into management review meetings. These packages can include summaries of PHA findings, action item lists, and status updates on mitigation measures. They should also ensure focus on scenarios requiring gap closure with abbreviated, yet detailed, depiction of the scenario from initiating event through consequence and on to proposed closure. More specifically, communication packages can be used to expose unidentified risks by way of plotting high severity scenarios across Process Flow Diagrams or Piping and Instrumentation Drawings (P&IDs). Communication packages help organizations assess and communicate PHA outcomes at all levels of the organization for effective risk management.
3. Establishment of Continuous Learning Forums: Companies where PHA facilitators reside should create forums or discussion groups where PHA facilitators can share their experiences, challenges, and best practices with their peers. These forums

provide a platform for ongoing learning and knowledge exchange, fostering a culture of continuous improvement in PHA facilitation techniques. MS Teams or other similar environments have the functionality to create shared spaces and forums for dedicated content and conversations.

Moreover, a strong feedback mechanism enhances transparency and trust in the PHA process, promoting stakeholder engagement and buy-in. By soliciting feedback from participants and management, organizations demonstrate a commitment to accountability and responsiveness in addressing safety concerns and mitigating risks. Additionally, post-PHA evaluations provide an opportunity to recognize and celebrate successes, reinforcing positive behaviors and motivating continued excellence in risk management efforts. The establishment of a robust feedback mechanism not only enhances the effectiveness of PHA facilitation techniques but also fosters a culture of continuous improvement and innovation within organizations.

7 In Summary

In conclusion, the shortage of skilled PHA facilitators poses a significant challenge to risk management practices in various industries. However, this challenge also presents an opportunity for innovation and adaptation. By re-evaluating traditional PHA facilitation techniques and embracing new approaches, such as recruiting strategies focused on talent development, leveraging technology, accelerating training programs, standardizing processes, and emphasizing continuous feedback, organizations can effectively address the shortfall of expert facilitators. These modifications not only ensure the continued efficacy of PHA in identifying, assessing, and mitigating potential risks but also contribute to the ongoing improvement of risk management practices. Through proactive measures and collaborative efforts between academia, industry, and technology providers, the industry can navigate the shortage of skilled PHA facilitators while upholding rigorous safety standards and enhancing overall risk management effectiveness.

8 References

[1] Venkatasubramanian, Venkat. "The Promise of Artificial Intelligence in Chemical Engineering: Is It Here, Finally?" AIChE Journal. February 2019, Volume 65, Number 2. https://doi.org/10.1002/aic.16489.

[2] Wolbers, Joshua. "An Overview of PHA Templates and Checklists Developed for a Refining System and Lessons Learned." *Process Safety Progress*. 2023. https://doi.org/10.1002/prs.12467.

OPTIMIZING QUALITATIVE HAZARD EVALUATIONS FOR MAXIMIZED BRAINSTORMING

(OR "HOW TO COMPLETE A PHA/HAZOP MEETINGS IN ONE-THIRD THE TIME CURRENTLY REQUIRED WHILE FINDING MORE SCENARIOS")

Authors

William Bridges – President
Process Improvement Institute, Inc. (PII)
e-mail: wbridges@piii.com

Matías A. Massello – Process Safety Engineer
Process Improvement Institute, Inc. (PII)
e-mail: mmassello@piii.com

Abstract

This paper discusses the lessons gleaned from more than 8000 qualitative hazard evaluations completed by the authors and compatriots over the past 34 years. The learnings were from project risk reviews, management of change (MOC) risk reviews, full unit process hazard analyses (PHAs) redos, and PHA revalidations. (All of these evaluations fall within the qualitative category of CCPS's "Hazard Identifications and Risk Assessments [HIRAs]," required in the Risk-Based Process Safety [RBPS] guide [1].) The experience covers all chemical and related industry. The paper shares secrets that will speed up your hazard evaluations while not sacrificing thoroughness. Issues covered include: Should you project your notes during the meetings? Should you use dedicated software? Should you have a dedicated scribe? Should you define the methods and make sections/nodes ahead of the meetings? What methods should you choose? What documentation rules speed up the meeting? What facilitation rules speed up the meetings without crimping brainstorming? What mistakes kill brainstorming and also slow the meetings? Data based on thousands of PHAs is presented, along with a condensed set of optimization rules.

Introduction

PHA optimization is executing the PHA analysis with practices that are thorough and efficient. One key to optimization is optimizing PHA Team Leadership – the PHA leader's ability through training and practice to run efficient meetings while driving the team to make sound engineering- and operationally-based risk decisions. In addition, developing, implementing, and maintaining a PHA management practice with detailed procedures and rules for conducting PHAs is necessary to ensure consistent and efficient implementation. However, PHA optimization is also very dependent on the strength of other process safety management (PSM) practices. The full implementation of these strong PSM elements, over time, will lead to better data needed to conduct better PHAs.

This paper presents PHA team leadership techniques and rules, discusses content of PHA management practice, policy, and procedures, and explains the relationship of PHAs with some other PSM elements where if those elements are weak, can impact the quality of the PHA and increase the PHA meeting or documentation time.

This paper is a significant update of the paper presented, with the same title, in the 5th Global Congress on Process Safety in 2009. [2]

Source of the Optimized Rules

The rules presented in this paper were developed through our efforts in conducting thousands of PHAs, each with slight variations (planned and unplanned experiments). The rules were established with the goal of minimizing the effort it takes to complete a thorough PHA.

The data set for these rules include:

- 34-years of PHAs
- More than 8000 PHAs performed under contract for more than 300 clients
- PHAs were performed by more than 70 leaders and 40 scribes
- Teams were all composed of excellent, required members and often some optional team members
- Leader and scribe used various commercial software packages

Optimized Rules for PHAs

The rules presented in this paper will be new to most of the readers. Some of the rules will be quite surprising (even controversial) and, so at first, may not be readily accepted. Since these rules have been difficult to develop and, until now, have been closely held by a few very experienced PHA leaders and instructors, we recommend that readers give these rules a try.

We also understand that some readers will ask: "Why optimize PHAs? Aren't more meetings and more documentation better for safety?" The answer is no, as will be explained in detail later. In short, inefficient PHAs can burnout teams (lower the brainstorming and causing accident scenarios to be missed). Also, if a PHA takes 3-times longer to complete than it should, then company leaders will be reluctant to invest more in aspects of PHAs where they are currently weak.

Major gap related to PHA's scope

Many companies do **not** perform a thorough analysis of the risk for startup, shutdown, and on-line maintenance modes of operation; the reason normally given is that the analysis of these modes of operation takes "too long." Yet, actually the PHA of the normal mode is taking too long and so the company has no time left for the analysis of procedures for startup and shutdown modes of operation. If these PHAs for the normal mode of operation are optimized, the organization will have time for thoroughly analyzing the non-routine modes (typically discontinuous modes) of operation and the organization will still have a net savings overall! This point is critical since 80% of catastrophic accidents occur during non-routine modes of operation.

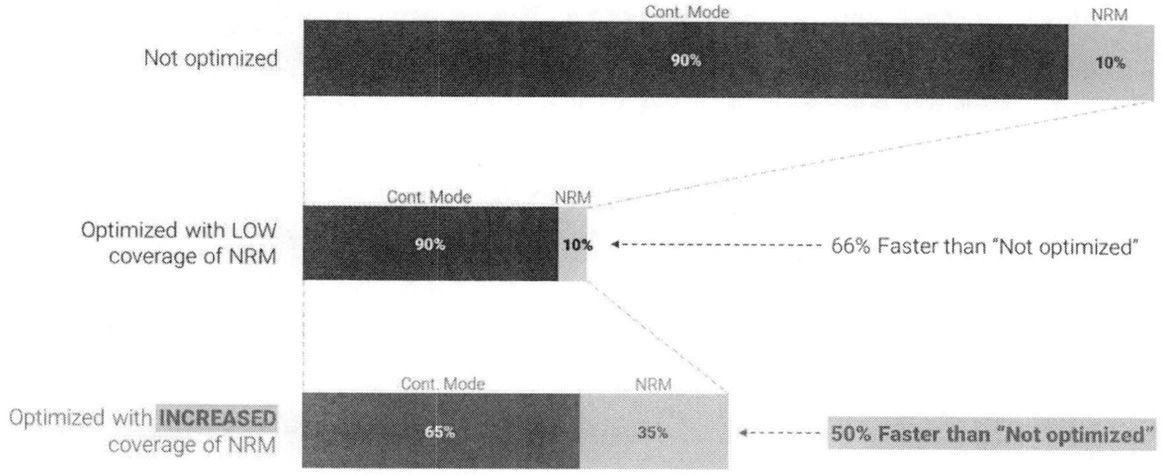

Figure 1. Meeting time (Continuous process)

General rules

This section list best practices for executing efficient PHA meetings.

The rules shown in **Table 1** are considered necessary for any team to be (1) compliant with government and industry standards or (2) are generally understood universally already, and therefore are considered a minimum requirement or common practice and so are NOT considered Optimization rules (but they are rules nonetheless).

There are opportunities for improvement and optimization in each of the the three phases of a PHA (Preparation, meeting and documentation):

- **Meeting Preparation** – Pre-populating the tables with the appropriate deviations will help the leader to quickly transition from deviation-to-deviation within a node, and also transition between analysis nodes or sections.
- **Leader Facilitation** – The PHA leader's ability to run efficient meetings requires training and practice. The Leader must be trained in best practices (not merely what guidewords to ask) and then must be coached through many sessions to practice what he or she has learned. The leader must have at least 10-years of hands-on experience in process operations and plant engineering. Other factors include the elimination of projecting meeting notes for the team to view, which tends to divert the team from brainstorming and instead focuses them on evaluating what the scribe is documenting. As a result, the leader tends to wait for the scribe to "catch up" (since the team members are reading what is projectetd) before leading the team into the next deviation analysis. Additionally, one important rule is for the leader to drive the team to make sound risk decisions as soon as possible in the discussion.
- **Scribing** – Having a separate person to scribe in the meeting (for analysis longer than 4-hours of meeting time) can help tremendously and easily pays for itself. While the scribe is completing the summary of the team's discussion, the leader can move on to the next topic of discussion.
- **Documentation** – Though not as important as the team meetings, documentation of the meeting results is nonetheless critical for the meeting results to be useful to others. There are many styles of documentation, but some styles are more efficient than others. **Table 1** and **Table 2** show the styles that have proven best.

Table 2 lists and describes the best practices for the three phases of a PHA.

Table 1. Regulatory required or universally understood practices (not optimization rules)

Best Practice Rule	**Typical of Current PHAs**	**More details on Best Practice** (i.e., how to optimize for efficiency and thoroughness)
Consolidate Studies – *From Risk Based Process Safety, CCPS 2007, page 230*	Sometimes large plants are broken into smaller units and the PHAs are performed separately for each unit; this reduces efficiency. .	Consider analyzing an entire unit instead of analyzing processes within a unit under separate PHAs. Efficiencies will be gained because the process safety information should overlap, the team members will be almost identical, except maybe for operators. Also, this allows more rapid linking of one discussion to other discussions in other nodes, thereby reducing repetition. Be mindful though that too many consecutive meeting days may require team member substitutions. Following this rule can take one-half of the time of separate, small PHAs that are difficult to link.
Standardize Checklists – *From Risk Based Process Safety, CCPS, page 230*	*Since the mid-1990s, many companies already follow this rule. Therefore, we have not included it in our estimate of potential savings.*	Saves time to have these tools available instead of requiring each leader to develop their own. This is especially true for checklists, such as Human Factors Issues and Facility Siting Issues; these two checklists have been shared across the industry since the mid-1990s. For best efficiency, use these checklists only at the end of a meeting to ensure you have not missed any issues during the pure brainstorming portion of the PHA; using them too early will overwork the checklist issues.

Best Practice Rule	Typical of Current PHAs	**More details on Best Practice** (i.e., how to optimize for efficiency and thoroughness)
Develop and use only a trained and experienced PHA Leader	A majority of team leaders do not know the rules below and therefore do not get the most from their meetings. *We have not included this factor in our estimate of potential savings; instead, we have assumed that each leader will have completed a minimum of 5 PHAs of one or more weeks of meetings, while. Applying the rules after reaching this threshold will give the gains listed.*	A new leader (those who have performed 0-2 PHAs on their own) will take 3-times as much time to cover the same issues and will not document the issues as well. Once a leader has performed 5+ PHAs (or one or more weeks of meetings for each PHA) using the rules below, they will be completing PHAs in one-third the time of other experienced leaders and one-third of a new leader, following the same rules. This gain is likely only possible is the new leader is "coached" during the 5 PHAs by someone who is already expert at using the rules listed in this paper. **See "Elaboration on PHA Team Leader Qualifications" later.**
Make sure you have ALL required team members – such as operations specialist from the area and the process engineer	Most teams are structured correctly, but this has not always been the case in the past. *Since this rule is mandatory for all PHA standards, we have not listed this as an optimization rule.*	The correct team members are critical (paramount) to a thorough meeting, and therefore thorough PHA. *Since this rule is mandatory for all PHA standards, we have not listed this as an optimization rule.* One rule we use in our meetings is: If the senior operator or process engineer from the unit leaves the team meeting (such as to respond to some urgent plant or personal need), the meeting is halted.

Best Practice Rule	Typical of Current PHAs	**More details on Best Practice** (i.e., how to optimize for efficiency and thoroughness)
Follow rules for meeting logistics	Most (but not all) teams tend to follow these rules already, so *we have not included this factor in our estimate of potential savings*	PHA teams have long been told (CCPS 1992, 2008) to meet no longer than 4-5 hrs per day and to take one week off before the next meeting (if possible). Leaders have also been taught to take frequent breaks (every 60-80 minutes) to allow the team members to get out of tunnel vision and refresh their minds and to allow the scribe to ask the leader about one or two confusing issues. It is also best to plan the breaks with hot and cold drinks and with fresh fruit (not pastries) and vegetables, since this will keep team energy higher.

Table 2. General PHA optimization rules

Best Practice Rule	Typical of Current PHAs	**More details on Best Practice** (i.e., how to optimize for efficiency and thoroughness)
Leader and scribe should develop nodes ahead of the team meeting	Many (perhaps more than half) develop the nodes during the team meetings.	A leader/scribe can develop the total nodes before the start of the first meeting. This provides all of the blank spaces for each possible issue, so the scribe can quickly document issues that come up ahead of time. This rule is mandatory if linking is used, as described later. **This rule can save 10% of the meeting time and saves a little after meeting time, as well.**
Make sure to follow node/section definitions that match LEADER or PHA Pro software (see discussion of software later)	Some teams use other software that requires manually choosing each deviation to use for each section/node. This is insanely wasteful.	**Pre-populating nodes with a standard set of deviations or what-if/checklist issues can save 10% of total project time and reduces meeting preparation time by 70%.** This type of optimization is what software should do for us.
Do not project/display (onto a screen) the analysis tables during PHA Team meetings	Some organizations, perhaps more than half, project/display all meeting notes "live" during the meetings.	Only project selected notes at the start of a new meeting-day to allow clarification of a confusing issue. Projecting P&IDs or procedures is fine. **Not projecting the scribe's live notes can save +30% in meeting time while increasing brainstorming.** The brainstorming increases because, if the notes are projected, the students are switching to the editor mode of their brain (while watching notes being typed). There are downsides with not projecting notes, but there are effective ways to overcome these downsides. **See "Elaboration on NOT Projecting Analysis Notes 'Live' during Team Meetings" later.**

Best Practice Rule	Typical of Current PHAs	More details on Best Practice (i.e., how to optimize for efficiency and thoroughness)
Leader should press for decisions as soon as the scenario is understood	Many teams overwork each moderate and major issue.	Pressing for an early judgment of risk can eliminate redundant discussion and can also help to eliminate unnecessary recommendations if the first decision pressed for is: "Is the risk tolerable?" **This can save 10-20% of the meeting time and greatly reduces burn-out, since most of team's energy goes into discussion of scenarios.**
Clarify which safeguards are candidate independent protection layers (IPLs)	Safeguards are a mix of IPLs, non-IPLs, and supportive management system layers.	Safeguards should be noted as either meeting IPL definition or not. This is a major advantage derived from introduction of layer of protection analysis (LOPA [3], [4]) in the late 1990s. **This does not save time, but does not take much if any extra time either.** You can view this as applying the "best" rules from use of risk matrices, which we recommend not using in PHA meetings (see later note).

Best Practice Rule	Typical of Current PHAs	More details on Best Practice (i.e., how to optimize for efficiency and thoroughness)
Perform a PHA of startup, shutdown, emergency shutdown, and certain online maintenance modes of operation	Not performed in nearly all refineries and in most petrochemical plants; the PHAs in these industries tend to focus on continuous mode of operation, even though 80% of the accidents in these industries occur in non-continuous modes of operation.	PHA of all modes of operation, with What-if and 2 Guideword (and rarely 7 Guideword HAZOP) being the methods of choice for non-continuous modes of operation. See section 9.1 of AICHE/CCPS, *Guidelines for Hazard Evaluation procedures, 3rd Edition, 2008* [5] for coverage of this requirement. **This does not "save time" but it does reduce the number accidents by ensuring the process is properly safeguarded for accidents that can occur during non-routine modes of operation.** This also improves compliance with PSM regulations and standards. **This will cost 50% more time than merely doing a hazard review of normal (e.g., continuous) mode of operation.** So, rather than being linked to "saving time," the optimization aspect here is to increase the allocation of time to these modes of operation and therefore find many more high-risk scenarios for the same or lower investment in PHA time (if the other rules are also followed). **So, the benefit of such scope change is a 500% improvement in risk identification and risk reduction.** Choose either 2 Guideword or What-If analysis methods (again, see Section 9.1 of CCPS 2008 for guidance). When documenting, list the data (causes, consequences, etc.) "by exception only" and show in the "deviation during startup," etc., if for a non-continuous mode of a continuous process. Or in a What-if style table of major procedure sub-sections (such as "preparation phase,") if for a batch process (a process that is normally step-by-step).
Use a dedicated Scribe, for meetings longer than 4-total hours	Most meetings do not have a scribe; the reason given is "cannot afford this luxury". However, the savings ratio in staff-hours for the whole team is about 2:1	Use a well-trained scribe to take the documentation load off of the team. **This rule can save 30-50% of meeting time and increases brainstorming** (because the team is not daydreaming as they wait for the Leader to complete the notes). Many junior and senior engineers make excellent scribes with about 1-week of coaching during actual PHAs; attending PHA leadership training can also improve scribe skills. Non-technical secretaries/clerks have been tried as scribes, usually with poor results. [6]

Best Practice Rule	Typical of Current PHAs	More details on Best Practice (i.e., how to optimize for efficiency and thoroughness)
Use efficient Software if you do more than 2-weeks of PHA per year. (Note: Software does not increase the quality of the PHA team meetings.) This is particularly true if you mostly use HAZOP method (versus What-If or FMEA methods)	Many organizations use PHA Works (Primatech), PHA Pro (Dyadem), or HAZOP Manager (Lihou); some organizations do not use PHA software (they instead use MS Excel applications or MS Word).	All of the software vendors are competitors of PII, and so we have no vested interest in any of the existing software packages; however, PII has many years experience with each of the listed commercial products. We recommend using LEADER (by ABS Consulting) if you do a lot of HAZOP of continuous flow processes. **This decision can save 25% in meeting time and can save 80% in preparation time.** Part of the savings is due to the predefined set of deviations for each node/section type (though PHA Pro 7 and above has the same capability of pre-defining deviations easily). Another part of the savings is due to the ability to link from one consequence to the cause of another deviation (even to a deviation in another section/node); LEADER is the only software with this option. Still another part of the savings is being able to decide to "turn off" cause-by-cause mode of documentation when the scenarios are simpler; again, LEADER is the only software with this option. If, however, you perform less than 2-weeks of PHAs per year, then likely MS Word or Excel are your best choices, since they are free and simple and since the benefits above will not be as apparent.
Do not use a risk matrix in the PHA meetings	Use of semi-quantitative risk matrices or even LOPA within PHA meetings.	Only use Risk Matrix within LOPA, and only use LOPA after the PHA meetings for issues that are very confusing to the PHA team (and possibly for multiple-fatality-potential events). **Eliminating forced use of risk matrices (and LOPA) from PHA meetings increases brainstorming and therefore increases the number of scenarios found by 10-20%, and saves 25% in meeting time.** **NOTE:** See the discussion later in this paper if your company forces the use of LOPA or a similar risk matrix tool for each scenario

Best Practice Rule	Typical of Current PHAs	More details on Best Practice (i.e., how to optimize for efficiency and thoroughness)
Use What-if (brainstorming with no guidewords) wherever you believe the risk and complexity is moderate, low, or well understood.	Many (most) organizations require use of HAZOP method in all cases (though a notable few require use of What-If in all cases).	What-If typically only takes 30-40% of the time (saves 60-70% of the time required) that HAZOP of the same system would take. Many see What-If as being too informal or less thorough that HAZOP, but experience of thousands of PHAs has taught us that What-if can be used for selected systems in nearly all PHAs, and in some cases (such as oil terminals and many utility systems), What-If is far superior. There are many PHA leaders who can lead a more thorough PHA with What-If than with HAZOP. **We tend to average a blend of 30% What-if in our PHAs, for a 10-20% savings in total meeting and documentation time.**
Use linking (especially for HAZOP of continuous modes of operation)	Less that 10% of the leaders/scribe use Linking. They instead either use repetition of causes, consequences, and safeguards; or they use "refer to" as a general statement and then attempt to list all safeguards for the entire scenario in one deviation.	Use LINKING to save time and increase speed (both). Use links from a consequence of one deviation to the cause of another deviation, to indicate the scenario path (if the ultimate consequence is of interest). The reader can follow the links forward or backward to find the related causes and safeguards elsewhere in the scenario path. This saves time by eliminating repetition of safeguards. We have also found that, typically, the quality of the final report is greatly improved, since it more accurately reflects how the scenario builds from one deviation to the next. Using linking allows using other clarity rules, such as "only show a safeguard in the deviation where it belongs." **Linking can save 20-30% of meeting time and after-meeting documentation time, and also increases thoroughness, if you doing a HAZOP-based analysis of continuous operation mode. If you are doing mostly What-If, then there is no savings.** Linking is illustrated in **Figure 2.**

Best Practice Rule	Typical of Current PHAs	More details on Best Practice (i.e., how to optimize for efficiency and thoroughness)
Use mostly deviation-by-deviation documentation instead of cause-by-cause documentation	More than one-half of the leaders (including all of the leaders who use PHA Pro, Lihou, and PHA Works) use only cause-by-cause documentation. Note that these three software packages only allow Cause-by-Cause style of documentation; though a power user of PHA Pro can change this at the start of a new PHA	**Cause-by-cause documentation takes 20-30% longer than deviation-by-deviation documentation, and, in most instances, provides little advantage, especially for HAZOP of continuous mode. If you are doing mostly What-If, then there is no savings since cause-by-cause is usually best.** We have found it is best to use a mix; using cause-by-cause documentation style only when absolutely necessary, such as in complex reactor nodes/sections or when LOPA (scoring) of many sceanrios is required by the company policy. Linking (described earlier) can be used in either documentation style. For more description of these styles of documentation, See section 5 of AICHE/CCPS, *Guidelines for Hazard Evaluation procedures, 3rd Edition, 2008;* or contact the authors of the paper. (As mentioned earlier, LEADER is the only software that automates Linking of one deviation to another deviation.)
Follow additional rules as outlined in **Table 3**	Most leaders/teams only follow a fraction of the optimization rules listed in **Table 3**.	**Following the additional rules (or clarification of earlier rules) in Table 3 will reduce meeting and documentation time by about 30%.**

Item Number	Deviation	Causes	Consequences	Safeguards	Actions
2.0 Line – Compressed Air System (dwg: 1)					
2.1	High flow		No consequences of interest		
2.2	Low/no flow	Air compressor failing to start or transferring off	Low pressure in the chlorine railcar (Item 1.6)		41
2.7	High pressure	Operator setting the pressure control valve incorrectly Pressure control valve failing to close or transferring open during unloading operations	High pressure in the chlorine railcar (Item 1.5)	Local pressure indication at the pressure control valve	04

Item Number	Deviation	Causes	Consequences	Safeguards	Actions
1.0 Vessel – Chlorine Railcar (dwg: 1)					
1.1	High level	Railcar overfilled by supplier Reverse flow in the chlorine unloading line soon after a new railcar is connected to the unloading system (Item 3.3) Misdirected flow to the idle railcar from the chlorine unloading line (if the idle railcar is already full) (Item 3.4)	High pressure as a result of liquid thermal expansion (Item 1.5)	Good supplier loading practices	01
1.2	Low level		No consequences of interest		
1.3	High temperature	External fire High ambient temperature	High pressure (Item 1.5)	Concrete crossties of rail spur Dike preventing any combustibles spilled nearby from reaching the unloading rack areas	02
1.4	Low temperature	Flashing of chlorine while using the eductor in the neutralization system to clear the chlorine from the line	Potential brittle fracture of piping, resulting in a loss of containment (Item 1.9)		03
1.5	High pressure	High level with subsequent liquid thermal expansion (Item 1.1) High temperature (Item 1.3) Violent reaction caused by a high concentration of organic contaminants (Item 1.7) High pressure in the compressed air system (Item 2.7) High pressure in the chlorine vaporizer (Item 4.5)	Potential loss of containment (Item 1.9)		04

Item Number	Deviation	Causes	Consequences	Safeguards	Actions
1.0 Vessel – Chlorine Railcar					
1.8	High	Backflow from the neutralization	Loss of containment resulting from	Closed dome on railcar	07
1.9	Loss of containment	Corrosion/erosion External impact Flame impingement Gasket, packing, or seal failure Improper maintenance Material defect Operator failing to close or inadvertently opening a valve to the atmosphere (e.g., a valve at a hose connection) Railcar inadvertently derailed Valve leaking to the atmosphere Low temperature (Item 1.4) High pressure (Item 1.5) Acid corrosion caused by a high concentration of water (Item 1.8)	Large release of chlorine to the atmosphere	Chlorine repair kit Limited vehicular access to area Maintenance/operator response as required, including isolation if needed Operator periodically monitoring the railcar valves while unloading Personal protective equipment in the area Plugs installed in all chlorine valves to the atmosphere when the valves are not in use Relief valve on each railcar for mitigating releases caused by overpressure Supplier maintenance of railcars Video monitoring of the unloading area	10 11 12 13 40

Figure 2. Example: Documentation using linking

Elaboration on PHA Team Leaders Qualifications

A PHA team must be led by a qualified leader. As mentioned, for comparison of gains, we have already assumed that each leader has already completed 5 PHAs of 1 or more meeting-weeks each. There is nothing magic about this number, but after completing a few PHAs, the leader is capable of using ALL of the rules defined herein. The leader may (should) have already learned these rules on the way to completing 5 PHAs. If the leader is NOT qualified, then it does not really matter what rules they follow; the PHA will be poor quality. The rules in this paper only apply to good PHA leaders. What does this really mean? Well, we know a good/bad leader and scribe when we see them, but to put some parameters around the answer, here are minimum criteria we use for qualifying PHA leaders:

Starting Suggested Qualifications:

- 10-years of hands-on experience in chemical or hazardous operations, with special emphasis on hands-on field or plant experience. This is critical to the leader being able to quickly understand the process under review and to be able to quickly understand the aspects of the accident scenario under review. Too many PHA leaders are "academic only" (i.e., their experience is engineering from an office or consultancy). Though we have seen some academic folks become effective leaders, its generally the folks with operations backgrounds that help a team find more of the subtle accidents scenarios during startup, shutdown, and abnormal operations.
- 10-years plus of technical background. This can include obtaining an engineering degree or equivalent experience learned in the field. Some of the best leaders we have observed have been non-degreed staff. With that said, engineers have a natural tendency for learning to be good scribes and leaders (as long as they also have some education in the school of hard-knocks: in the field or operating units).
- Quick learner – since the leader should be independent of the unit/process under review, they must have enough knowledge and experience to be able to quickly learn what the team's concern is for a specific What-If or Deviation and to help facilitate the team through the scenario development and judgment of risk.
- Good writer – the team will need a comprehensive, yet concise/clear record of the analysis, along with easy to understand recommendations. Much of this load can be pushed to a good scribe, but for hazard review of small changes (where the risk review lasts a few minutes or a few hours), a scribe may not be warranted. It is best if the leader serves as a scribe during their training and coaching to become a Leader.
- Inherent facilitator – We look for many traits, which include personable, reasonable, the ability to summarize issues quickly, good at conflict resolution; a good sense of humor is also important.

Once someone meets the Starting Qualifications to become a Certified PHA Leader, most organizations require the candidate to complete the following steps:

1. Attend a 4-5 day course that is approved by the company. There is a lot of bad training out there and many courses are not taught by expert PHA Leaders. The trainer (external or internal) should be carefully selected and approved by the organization. In some cases, the company will only approve

one or two individuals to be their trainers (rather than approving an entire consultancy or internal group). At the end of this first step, there is normally an exam of 30 minutes and, in some cases, a judgment by the instructors on the "potential" for leading a PHA in the near future. The course certificates will therefore say either that you completed the training and passed the exam, or that you are a Candidate PHA Leader. In our experience that only qualifies you for the next step (you are not a leader, yet).

There are exceptions: In one series of multiple PHA Leadership Courses for a client, we trained about 70 individuals, most of whom met the starting qualifications. At the end of the training sessions, each had passed the exam. Of the 70, we (the instructor and the client representatives who helped in the training) concurred that one student was ready to lead or scribe at the end of the 5-day session. This is a rare exception for a rare individual (we have found only 2 in 5000 graduates who could lead effectively after ONLY a 5-day class; even considering the 5-day class was 70% workshop). As a point of contrast, we believed that 30 of the individuals would never make good leaders, but might be good scribes and we believed some, with a little more coaching (1- or 2-weeks during actual PHAs), would make excellent leaders. Likewise, we believed that some would be good leaders eventually, but may never make good scribes. The training was excellent and the structure of the course was right (the client had tried many trainers and courses over a 10-year period before exclusively selecting PII trainers). But, the numbers above are fairly typical results after an excellent PHA Leadership course of 5-days duration.

2. The Next step at the Best organizations is for the "best graduates" to be tutored for their first few PHAs/HAZOPs; this tutoring can be by an already qualified, expert Leader internal to the company or by an already qualified Leader from outside the company. The tutoring can take 1 or 2 weeks or it can take many weeks, depending primarily on the selection of the new leader candidate and how quickly he/she can learn all of the subtle lessons of how to lead and how to scribe. (By the way, it is more difficult to produce a fast and thorough scribe than it is to produce a good leader-only.)
3. The final step is for the new leader to produce his/her first report and have it critiqued by a PHA expert (internal or external). The certifier should review this final report and, if it is good enough, the leader can be certified as a PHA leader; which means they can be trusted to lead and scribe (or work with a certified scribe) to complete a PHA in good and relatively efficient order.

With this background defined, PII staff has trained about 6000 PHA leaders around the world. Of those graduates, we have certified about 280 leaders. But, there are variations across the industry, and our standards may overlap or be different than our client's standards.

Example: In February 2008, we trained 8 candidate leaders from one chemical complex. They practiced on small risk reviews for 8-months. Then, in November 2008, they began to prepare for 4 large PHAs (two were revalidations, which would require significant redo). In January 2009, we completed step 2 of certifying 4 PHA leaders and 3 scribes during 2-weeks of hands-on coaching during actual PHA sessions of the large plants. The Final step will be completed in the next few months, as reports are submitted for critique, revisions, and final issuance. In the meantime, the chemical company considers these individuals certified already because they have met the companies minimum requirement of completing training, passing the exam at the end, and being tutored by a qualified leader for a minimum of 1-week following training. Once the reports are approved, the 4 Leaders and 3 Scribes will also receive a certificate from PII. This approach has been followed at many companies in the past 18+ years.

Please refer to the paper "Building Competencies in Internal PHA-HAZOP Leaders" (2018, PII, presented at GCPS) for details on best practice for developing competent PHA leaders and scribes. [7]

Elaboration on NOT Projecting Analysis Notes "Live" during Team Meetings

Since many readers may be used to projecting meeting notes Live during PHA team meetings, we thought it would be beneficial to compare the pros and cons. (Note that CCPS HEP guidelines of 2008 [5] listed the pros of projecting but did not adequately cover the cons.)

Benefits of projecting notes live during a meeting:

- The team discussions can more easily be kept on track, since the team members can see exactly what checklist item, scenario, etc., is being discussed.
- All team members can see exactly what is being documented, so what is captured by the software can be truly considered as the consensus of the team.
- Documentation errors can be caught by team members and immediately corrected.

Drawbacks of projecting notes live during a meeting:

- Reduces brainstorming by drawing attention away from the discussion of accident scenarios and focusing attention of the text being typed (this causes more scenarios to be missed)
- Projecting the notes is boring for the team (the team's energy is reduced and therefore the team is much less productive)
- Overall, when equivalent teams were used, the teams that projected notes found 15% less accident scenarios for only a marginal improvement in documentation of the scenarios they did find.
- In comparing the speed of meetings with and without projection, the meeting with projection takes 30% longer meeting time, even if the drawbacks of projection are known and attempted to be overcome. This costs applies to all team members (including the team leader, scribe, and participants).

Benefits of projecting other information in the meetings:

- Key information such as P&IDs, procedures, the design intent (PSV specification sheet), plot plans, etc. can be displayed to help everyone focus on the same issue.
- The team leader can make a few notes on the drawing or procedure to highlight a key concept of the scenario, with adding the detail necessary for the scribe to add for the sake of the report.
- Previously completed work (such as during a revalidation, or during the 3rd or 4th phase PHA of a new capital project) can be displayed if it has a bearing on the current discussion or if revalidating results.

Conclusion of leading and documenting meetings with and without live projection of meeting notes:

- The reduction in brainstorming is significant is the team notes are project live during the meetings. All teams miss 10-20% of the accident scenarios (and miss 5-10% of critical, large impact scenarios) that another team will catch, but a team that projects the meeting notes live tends to miss as much as 15% more accident scenarios because they do so much less brainstorming during the same period of time. The teams that watch projected notes tend to digress to analysis and editor modes quickly, instead of fully brainstorming "what can go wrong?"
- The reduction in meeting speed costs more than just additional time for each team member (which takes time away from other critical tasks related to controlling risk). This extra meeting time also leads to more burnout of the team members, and burnout leads to missed scenarios or incomplete analysis of the scenarios found.
- The benefits of projecting can generally be recovered (while *not* projecting) by good and frequent *verbal* summarization by the PHA team leader and by good communication of the leader with a qualified and dedicated scribe. Also, the leader or scribe can, on exception, project the notes that are confusing to them; this seems to work best at the start of the next session (e.g., the next morning).
- Projection of meeting notes appears to speed up a revalidation of a previous PHA, if not much brainstorming is required, so for revalidations we are generally in favor of projecting notes live during PHA meetings.

Projecting of meeting notes may be necessary if the leader and the scribe are both lacking in experience or are not quick learners, because the team will need to compensate for their inadequacies; however, we have already stated that the efficiency rules shared in this paper are for experienced, "good" leaders (not poor, inexperienced leaders).

Other optimization rules

Table 3 summarizes these additional rules in a "simplified" PHA analysis table format to assist the reader in "seeing" the rule in context. These rules apply to analyzing deviations for a particular process section or node. They include specific documentation rules for certain deviations, as well as more general rules; again all of these rules are targeted as achieving high efficiency with little or no sacrifice of thoroughness. Following these rules will reduce meeting and documentation time by about 30%.

The Notes in the following subsections will help the reader understand the entries in each of the columns of **Table 3**. The Notes are presented in the order in which they should be filled:

1. **Deviation**
2. **Consequences**: Complete first for the deviation. **Go to consequences first and make sure the consequence is of interest to the organization** (not below the scope assigned to or agreed to by the PHA team). If the consequence is too low, then state "No consequence of interest" in the consequence column and do not list any causes or safeguards or recommendations. This rule alone can save 35% of the meeting time.

3. **Causes**: Complete second for the deviation, if there is a consequence of interest
4. **Safeguards**: Complete third after stating the consequence for the deviation and stating the causes
5. **Recommendations**: The last column to complete for a deviation, if additional protection against the scenario is necessary

Deviation column

A – Make sure you document each deviation that has a consequence of interest. For HAZOP, document even deviations that do not have consequences of interest.

B – Discuss Loss of Containment for each node; do this FIRST as stated above

C – Reverse flow is usually a credible scenario, even if there is a check valve in the line. (If there is a check valve, and if it is inspected and tested about every 4-5 years, and if it usually passes these checks, then it can be listed as a safeguard against reverse flow.)

Consequences column

D – Use LINKING to save time and increase thoroughness (both). Link from a consequence of one deviation to the cause of another deviation to indicate the scenario path (if the ultimate consequence is of interest). The reader can follow the links forward or backward to find the related causes and safeguards elsewhere in the scenario path (see also **Figure 2** above).

If high pressure links to loss of containment, and if one or more PSVs are safeguards against loss of containment, then one consequence listed in high pressure must be "PSV opens on demand, releasing _____ to ______".

Causes column

D – Use LINKING to save time and increase thoroughness (both). Link from a consequence of one deviation to the cause of another deviation to indicate the scenario path (if the ultimate consequence is of interest). The reader can follow the links forward or backward to find the related causes and safeguards elsewhere in the scenario path.

E – "Thermal expansion, if liquid blocked in" is shown as a cause in loss of containment (and not shown as a cause anywhere else).

F – Do not show external fire as a cause of high temperature; it is instead shown as a cause of loss of containment (since it is not a "process deviation" and since flame impingement is also a concern).

G – Tube leak/rupture cannot cause misdirected flow, but it can leak (and therefore link) to the node that relates to the "interchanged" stream, such as linking to high concentration of contaminants.

H – High pressure causes high flow (in a line) and not the other way around. So, "pump over-speed," if credible, is a cause of high pressure and then high pressure would be linked to high flow (assuming all of this leads to a consequence of interest).

I - Low pressure causes low flow (in a line) and not the other way around. So, "pump off," is a cause of low pressure and then low pressure would be linked to low/no flow (assuming all of this leads to a consequence of interest).

J - PSV opening is not a cause of misdirected flow.

Safeguards column

K - TRV (thermal relief valve, for thermal expansion) are only shown as safeguards in loss of containment and not in high temperature, since thermal expansion is almost never an issue during normal, process deviations (but instead is usually an issue during shutdown)

L - PSV is not a IPL if it is too small for the scenario listed, but can be listed in the Safeguard with a "non-IPL; not capable" tag.

M - Safeguards are only listed for the specific deviation and for the specific node to which they apply (flow safeguards for a line are only listed in a flow deviation; they are not listed in level deviations or pressure deviations).

N - Do not add the label "IPL" if a safeguardt is associated with a cause or another safeguard (in or linked to the same deviation). In other words, all IPLs must be truly independent of all other safeguards and of the causes of a scenario (we use the same definition as an independent protection layer (IPL) as in the LOPA and IPL guideline books).

> **Example:** If the flow control valve (FCV) failing closed is a cause, then you cannot use the flow indicator or low flow alarm (FAL) as an IPL for the same accident scenario, if all are from the same instrument loop, because they are not independent

O - Do not list any safeguard with the lable of "IPL" that has not been properly tested, maintained, and assured. For example, interlocks must be on a reasonable check/test plan; check valves must be on an inspection/text plan with 5 years of less test interval (in clean service); critical SOP steps must be emphasized in training and routinely practiced in the field.

P - A PSV is not a safeguard against high pressure, but instead is only listed in the safeguard column of loss of containment, and can be labeled "IPL" only if the PSV is sized large enough for the scenarios listed and only if it is inspected/tested according to industry standards.

NOTE: See the textbook on *"Guidelines for Initiating Events and Independent Protection Layers,"* 2015, CCPS for a deeper understanding if IPLs and see the paper by PII on *"Identifying IEs and IPLs During PHAs"* [4], [8].

Recommendations column

Q - Judge risk before moving forward to recommendations. Only list recommendations if, in the judgment of the team and/or by performing a LOPA, the risk is too high. If you ask your team "Do you have any recommendations?" then 9 out of 10times they will give you a recommendation. But, many times these are not necessary since the risk is already tolerable. The team should always first answer the question "Is the risk tolerable or not?" and then, if the risk is too high, discuss recommendations.

R - Make one recommendation for ALL the minor corrections to the SOPs (but only for minor corrections, not for "changes"); have one person on the team (such as an operator) keep track of all minor fixes for a procedurr and reference in the recommendation who has the official copy on minor corrections (who has the procedure that was marked up with minor corrections during the PHA). Then, then PHA team leader can make notes for recommended "changes" to the SOP and forward this recommended markup along with the PHA report to the unit for resolution of the recommendation for procedure changes.

S - Make one recommendation for ALL the fixes/corrections (not the physical changes that may be recommended) to the P&IDs and other system drawings. Then, have one person on the team (such as an engineer) keep track of all fixes (on the drawings and referenced in the recommendations), and have that person maintain the official copy of the marked up drawings from the PHA. (Any necessary changes to the system must have their own recommendation.)

T - Make sure you designate which recommendations are Operability only.

U - Have a formal method for closing OPEN items before meetings (cycle) ends.

Table 3. Additional detailed rules for team facilitation and documentation

Deviations	Causes	Consequences	Safeguards	Recommendations
A – Make sure you document each deviation that has a consequence of interest.		Make sure you ask for consequences "first." If the consequence is not of interest, then state "No consequence of interest" and move to the next deviation or concern.		
B – Discuss Loss of Containment for each node	**D –** High Pressure **E –** Thermal expansion "if liquid is blocked in" **F –** External Fire		**K –** TRV **L –** PSV (do not list as a safeguard if the PSV is sized too small)	**Q –** Judge risk before moving forward to recommendations. **R –** Make one recommendation for all minor procedure changes.
High Flow	**H –** High Pressure			
Low Flow	**I –** Low Pressure	LINKING	**N, O –** IPLs must be independent and proven by testing; safeguards can support IPLs or be candidate IPLs (if gaps are closed) **P –** PSV are not safeguards against high pressure; they are safeguards against Loss of Containment	
C – Reverse Flow			Check Valves	**S –** Make one recommendation for all fixes/corrections to P&IDs.
Misdirected Flow	**G –** ~~Tube Leak/Rupture~~ **J –** ~~PSV Opens~~		**M –** In general, list safeguards for the specific deviations and for the specific node they apply	**T –** Make sure you designate which recommendations are operability.
High Temperature	**F -** ~~External fire~~			
High Pressure	**H –** High Flow **H –** Pump overspeed	**D –** Loss of containment **D –** PSV opens on demand, releasing (material) to (location)	**P –** PSV are not safeguards against high pressure; they are safeguards against Loss of Containment	**U –** Have a formal method for closing Open Items before meetings (cycle) ends
Low Pressure	**I –** ~~Low Flow~~ **I –** Pump off			

Specfic case: Optimization rules when use of Risk Matrix or LOPA is mandatory

As described in **Table 2**, the use of Risk Matrix or LOPA in a PHA is NOT recommended because it impacts negatively on the brainstorming and adds 2 to 10%% additional meeting time. However, there are some companies in which the use of a risk matrix or LOPA is mandatory on all or most of the scenario risk judgments. For those cases, there are some specific rules that can streamline the meeting while still using a Risk Matrix or LOPA:

- Use LOPA rules (for IPLs and their PFDs and Initiating Events and their Initiating Event Frequency (IEF) for the risk ranking of scenarios with highest consequences (Multiple and Single Fatality, and the equivalent in environmental and financial damage). Use qualitative judgement (with the matrix) for lower consequences. **Never use risk matrix scoring without using the rules of LOPA.**
- Repeat IPLs in the deviation where the final consequence (and the risk ranking) is recorded. In **Table 2** it was mentioned that Safeguards / IPLs are only recorded in the applicable deviation; however, if risk ranking / LOPA is performed, it is better to show all the applicable IPLs in the deviation / row with the final consequence (for clarity).
- Applicable Safeguards that do not meet IPL criteria should be clearly labeled as "Non-IPL" along with the explanation why it is not (not independent, not capable or big enough, not tested or documented, etc.)
- Use Deviation-by-Deviation documentation style and use Cause-by-Cause to group "equivalent" Initiating Events. Score (perform LOPA) on the the worst-case cause-consequence pair for a set of related sceanrios in the same deviation.

  Example: A vessel has two major causes for High Pressure leading to Loss of Containment: 1) Gas blowby from the vessel upstream, 2) runaway reaction. There might be several Initiating Events leading to Gas blowby (recorded in Low Level) or Runaway reaction (recorded in High Temperature), but the LOPA scoring will be performed to worst case cause-consequence pair for gas blowby and for the worst case cause-consequence pair for runaway reaction.

PSM implementation and its relation to efficient and thorough PHAs

The PSM elements are very dependent upon one another. These interrelationships determine the management of process safety risks. While some benefits are realized from implementing a single process safety management practice, the implementation of process safety elements is not intended to be ala cart. A missing PSM element or major weaknesses of one PSM element will most certainly weaken the implementation of other PSM elements; the overall impact is a weak PSM program. The relationship between some PSM elements are stronger (or more dependent) than others. Understanding the relationship between the elements is critical for optimization of any PSM program. Missing PSM requirements, for instance, if the MOC system and MI program are not fully designed and implemented, will lead to unnecessary team discussions and a larger number of recommendations generated, which adds to meeting time.

While optimizing PSM will help to optimize PHAs, we realize this is global to all PHA leaders and scribes. There is nothing the Leader can do if the organization has a poorly implemented PSM system, but he or she will feel the effects in both less efficient and less thorough PHAs. So, for completeness we have included a summary (below) of the relationship of PHA to the rest of the organizations PSM (or RBPS) systems.

Identify weak management practices by analyzing types of PHA recommendations

One way to recognize weaknesses in the implementation of PSM elements is to categorize PHA recommendations from previous PHAs. After several cycles of PHAs for a given process, expect the number of recommendations to decrease as the process safety design is tweaked and PSM management practices are strengthened and fully implemented. **Table 4** summarizes some typical categories for PHA recommendations. If there are many across multiple PHAs, then the management practice warrants review.

Table 4. Typical categories for PHA recommendations

Category	Description
Process design	Indicates a possible weakness in engineering or process safety engineering standards
Standard Operating Procedures Changes/Updates	Indicates a possible weakness in the procedure writing process. If there are sizeable areas of weakness, a global recommendation to rewrite the operating procedures may be necessary. Poor procedures can greatly affect the quality of the PHA. If the weakness is widespread, the PHA of the procedures should be halted.
Admin/Policy/ Procedure Verification Items	Every PSM element, including PHAs, requires policies, authority, procedures, check and balances, and verification steps to work effectively. Weakness in these can lead to lower reliability of safeguards and impact the PHA.
Mechanical Integrity	These may simply indicate weaknesses in documentation or communication of these systems. But, perhaps critical tasks are missing or perhaps the frequency of a task is wrong.

Category	Description
Process Safety Information (PSI) Deficiencies/ Availability	These recommendations indicate that PSI is not up-to-date or available for the team to use to evaluate the safeguards and/or define the consequence of interest. Poor PSI can greatly affect the quality of the PHA. If the weakness is widespread, the PHA should be halted.

Process Design

Recommendations for process design changes may increase or decrease in number for a variety of reasons. As operational experience is gained, the reliability of safeguards and "lessons learned" from incident investigations could lead the team to make better risk evaluation decisions. This is related to quality of incident, near miss investigations, and documentation discussed elsewhere. Ultimately, recommendations related to improving designs indicate areas for improvement in engineering and design standards used by the organization; if these are amended to account for the recommendation, then perhaps the deficiency will not show up in the next (similar) design.

Standard Operation Procedures (SOPs) Update/Corrections Recommendations

Many PHA recommendations are the result of using incorrect SOPs. As a result, these inaccuracies slow the hazard evaluation process. As the team is analyzing procedures (all procedures for batch operations and non-routine procedures for continuous processes) the leader guides the team in understanding the consequences of not performing a step or performing the step incorrectly. The team reviews the process safety information, such as operating limits and consequences of deviation, which are stated in the procedure. If the procedures are incomplete, the team must then take the time to discuss the SOP inaccuracy and take the time to recommend, in general, how the SOP should be updated. Simple recommendations like "add a warning..." or "change the operating limit to reflect the current process..." are typical but would be unnecessary if procedures were written using best practices for procedure writing, SOP changes recognized as needing approval through the MOC system, and if the MOC system recognized the need to update SOPs as a result of a process change (if the change impacted the SOP content). One ultimate solution would be developing and implementing a system for writing effective operating procedures. See PII's paper "Best Practices for Writing Operating Procedures and Trouble Shooting Guide" for more details on best practices. [9]

Administration/Policy/Procedure Verification Recommendations

A PHA procedure is necessary to define roles and responsibilities, PHA leader qualifications, data to gather for the preparation of PHAs and for reference during the meetings, team membership requirements, and documentation requirements. Such a procedure should also include best practice rules, such as those stated in this paper.

Mechanical Integrity (MI) Recommendations

Many recommendations are the result of the team not having information on the reliability of engineered safeguards. Since the leader must drive the team to a risk decision for a scenario, the team should know or be able to reference certain information to help in deciding if the safeguards in place are meet the definition on an IPL and if they collectively are adequate to control the risk at an acceptable level. Many recommendations are made to determine the frequency of a safeguard's inspection, testing, and preventive maintenance tasks, or to improve a safeguard by establishing an inspection, test, or preventive maintenance for the safeguard. Recommendations related to MI may simply indicate weaknesses in documentation or communication of these systems. Perhaps, though, critical tasks are missing or the frequency of a task is wrong. Many safeguards are not IPLs because of these deficiencies.

Process Safety Information (PSI) Recommendations/Verifications

Process safety information is necessary to identify valid consequences of interest for a process deviation and to evaluate the risks (the adequacy of the safeguards to prevent causes and mitigate/prevent the consequences). If the information is missing or is poorly documented, or otherwise unavailable for reference by the team, the team will not be able to fully analyze the hazards.

Inefficiencies also arise if there is an incomplete understanding of the scenario because of missing or incomplete PSI. As consequences are discussed, if the team does not fully understand the extent of consequences, they may understate or overstate the consequence or the safeguards needed to control the risk. For example, consider process chemistry. If the team does not fully understand the consequences of adding an incorrect amount of catalyst, then the consequence may not be fully documented (understood); however, there is the possibility of overstating the consequence and discussing a scenario that is not credible or has no consequences of interest. Information on reaction rates, operating limits, process chemistry, chemical properties, chemical volumes, equipment design, and others, assist in determining the consequences. **If information is missing on the safeguards, then the adequacy of the safeguards to prevent the cause or mitigate the consequence may not be properly evaluated**. Often, team members cannot recall the exact alarm set point or interlock trip. This results in an "open" item where the information is located over lunch or at the end of the day. The team must then re-examine the scenario, based on the information now available. If the information cannot be located, then a recommendation must be written to develop the PSI and to reassemble the team to evaluate the scenario using the PSI.

P&IDs – P&IDs are referenced throughout PHA to verify connections, valve locations, etc. As team members realize that the P&IDs are incorrect, a discussion begins, adding to the meeting time. **Ensuring that P&IDs are up to date prior to the PHA meeting helps to minimize meeting time.**

Management of Change

During PHA revalidations, MOC reviews can take considerable time. If MOC documentation is incomplete (key information the revalidation team needs to ascertain that the change process included a health and safety risk review) or is of poor quality, then the PHA team will have to spend more time discussing each process change. Designing the MOC system correctly will facilitate MOC reviews during the revalidations. Thorough documentation of the risk review – what questions are asked and what process is used to determine the level of risk review required for the type of change. Documentation of the risk review including the deviations, consequences, safeguards, and any action items will be necessary for the PHA team to incorporate the process change into the PHA analysis and to allow the PHA tables to be updated to reflect the process change. Also, documentation confirming exactly what was installed and implemented is necessary so any design changes that are safeguards (or IPLs) can be added to the PHA tables.

In the past 15-years, while conducting PSM audits, assessments, and PHA revalidations, the MOCs reviewed at many of the 40+ sites did not include information on the risk review. Some MOC forms had only a checked box indicating that the risk review was required or not required. Some MOC procedures contained a short questionnaire to evaluate the risk category of the change, and depending on the category, determined the type of hazard evaluation method required to further evaluate the risk. In many of these risk categorizations there was no supporting documentation of the risk review. Were these risk reviews completed? Were the risk review teams of the same composition as a PHA team? If not, can these MOCs be reflected in the revalidated PHA without a new risk review with proper team composition? Obviously, the MOCs at the mentioned sites had insufficient risk review documentation and this will greatly reduce the efficiency of the PHA Revalidation system.

Certain risk-based questions must always be asked, so the reviewers can confidently say that the proposed change does not introduce a new risk or increase the current risk. That in itself is a level of risk review, and those questions, and their responses should be documented to help the PHA revalidation team when reviewing the MOC.

Incident Investigations

It is best practice and required by OSHA's PSM standard to evaluate previous incidents related to the process under review. Poor documentation of incident reports can add hours to a PHA meeting, depending on the number of reports under review. Inefficiencies can occur at the meeting preparation phase, as well as during the PHA meeting. Electronic incident report generation and storage can make it easier to retrieve reports, but can also be a timely process depending on how the system is designed and how incidents are categorized. Some companies have a "process safety" category, but typically process safety incidents are categorized under equipment damage, environmental, loss time/injury, operational, or fire. Having to go through each category, in search of process safety incidents, is time consuming. Having a clear definition of what a process safety incident is, and properly categorizing these incidents, will make the reports easier to access and will save preparation time.

In order for the PHA team to analyze incident investigations, the report must clearly describe the incident, the root causes, recommended corrective actions, and the status of the corrective actions. If the incident investigation management practice doesn't define investigative leadership and team requirements or, documentation and implementation requirements, then poor quality investigations can result.

If the root causes are missed or incorrectly identified, then incident investigation corrective actions will not adequately prevent the near miss or incident from recurring. (Some corrective actions are engineering/administrative controls and others are changes in the management systems to directly address the root cause.) Longer discussion times for poorly documented incidents, and most likely the generation of recommendations to address the causes or probable causes, will increase meeting time.

Conclusions related to PSM management systems and their impact on PHAs

When reviewing PSM management practices, think in terms of how the information generated from these management practices is used in PHAs and PHA revalidations, to ensure the policies and procedures are designed to facilitate efficient process hazard analyses.

Summary of savings and other benefits

Following the rules for PHA preparation, leadership, and documentation as described in **Table 1** and **Table 2** can result in PHAs meetings being completed in one-third the time of typical PHAs today. The improvements percentages listed in these tables is not necessarily additive and some rules are dependent upon others. However, we have watched some PHAs, conducted by "experienced" leaders, which took 5-6 weeks of meeting time using the wrong rules, and yet an identical unit with an identically structured team took only 6-days using the optimized rules presented here. When the documentation of the two analyses are compared, the one that took less time, but used the optimized rules, found more accident scenarios and documented them more clearly. We do not expect 80% improvement as typical. However, when all the rules are followed, we have seen dramatic improvements in both meeting speed and number of meaningful accident scenarios discussed.

For many highly experienced leaders, the meeting time may only be cut in half, instead of cut by two-thirds, if these rules are followed. All of the rules were learned through experiments (planned or unplanned), and in all cases we tried to hold the qualifying analysis and clarity of the results to the highest standards. In most cases, 15-20 deviations (or What-if questions) per hour average speed is achievable by the rules listed here for meeting facilitation. If all of the rules are followed, even slightly higher "per deviation" speed is possible. Additional efficiency gains are possible within meetings if the proper mix of hazard evaluation methods are used and the proper changes are made as to choice of software. Finally, the optimal efficiency and thoroughness for preparation, meeting, and documentation phases of a PHA is achievable if all of the rules are followed. **The typical reduction in overall PHA labor cost is two-thirds, when compared to typical, un-optimized PHAs**.

The tables below provide two comparisons of following the optimization rules to not following these rules. Comparison 1 is for the same scope, but following all optimization rules and assuming an average speed for what we have observed for well trained leaders who do NOT follow these rules. Comparison 2 (**Table 6**) is the same as Comparison 1 (**Table 5**), except analysis of operating procedures for uncovering accident scenarios during non-routine modes of operations has been added to Only the Optimized PHA column to provide comparison of un-optimized to fully-optimized PHAs. These tables provide a similar comparison to what was shown in **Figure 1** in page 4.

Both comparison are for a basis of 100 nodes of continuous HAZOP sections (and all related operating procedures for startup and shutdown, for Optimized case only) and consider 4 full-time PHA team members.

Table 5. Comparison 1: LOW coverage of Non-Routine Modes – Potential savings when following optimization rules

		Optimized (Using rules + dedicated Scribe)			Un-Optimized (No rules + No Scribe + No NRM)	
PHA Phase	**Use of optimization rules**	**Leader labor**	**Scribe labor**	**Team members** (4)	**Leader labor**	**Team members** (4)
Preparation	Predefining nodes and section Pre-populating deviations and What-If questions	10 h (5% of total time)	10 h (5% of total time)	-	30 h (~15% of total time)	-
Meeting	Not projecting the meeting notes live Pressing for decisions on risk soonest Use a dedicated scribe Use efficient software (must save time) – must support different documentation style and Linking (if you do HAZOP of continuous modes) Do not use risk matrix in qualitative PHAs Use What-if wherever possible Use linking for continuous-mode HAZOPs Use mostly D-by-D style of documentation Following detailed rules in Table 3 Clarifying which safeguards are candidate IPLs Performing analysis of startup and shutdown modes of operation	60 h (50% of total time)	60 h (50% of total time)	240 h	200 h (~65% of total time)	800 h
Documentation	After-meeting benefits directly relate to meeting optimizing rules, but when using optimization rules, more load is shifted to the after-meeting effort	50 h (45% of total time)	50 h (45% of total time)	-	40 h (~20% of total time)	-
	Total staff hours	440 h			1070 h	
	Accident scenarios found	**500**			**400**	

Table 6. Comparison 2: INCREASED coverage of Non-Routine Modes – Potential savings when following optimization rules

		Optimized (Using rules + Dedicated Scribe + NRM)			Un-Optimized (No rules + No Scribe + No NRM)	
PHA Phase	**Use of optimization rules**	**Leader labor**	**Scribe labor**	**Team members** (4)	**Leader labor**	**Team members** (4)
Preparation	Predefining nodes and section Pre-populating deviations and What-If questions	15 h (5% of total time)	15 h (5% of total time)	-	30 h (~15% of total time)	-
Meeting	Not projecting the meeting notes live Pressing for decisions on risk soonest Use a dedicated scribe Use efficient software (must save time) – must support different documentation style and Linking (if you do HAZOP of continuous modes) Do not use risk matrix in qualitative PHAs Use What-if wherever possible Use linking for continuous-mode HAZOPs Use mostly D-by-D style of documentation Following detailed rules in Table 3 Clarifying which safeguards are candidate IPLs Performing analysis of startup and shutdown modes of operation	90 h (50% of total time)	90 h (50% of total time)	360 h	200 h (~60% of total time)	800 h
Documentation	After-meeting benefits directly relate to meeting optimizing rules, but when using optimization rules, more load is shifted to the after-meeting effort	75 h (45% of total time)	75 h (45% of total time)	-	40 h (~20% of total time)	-
	Total staff hours	720 h			1070 h	
	Accident scenarios found	**600 – 700 (but the gain is risk reduction is 500%)**			**400**	

Some organizations may be reluctant to implement the changes to accommodate use of these rules. We encourage them to think through the potential savings for all team members time. Also, improvement in meeting efficiency improves brainstorming and increases the number of significant accident scenarios found. Finally, the savings obtained can be used to fully analyze hazards during non-routine modes of operation, finding even more scenarios and more importantly finding the scenarios that are likely least safeguarded.

Closing

Although more than a million PHAs have been performed, few organizations perform these in an optimized fashion. The goal of a PHA is to identify potential accident scenarios, describe the scenario fully, and qualitatively judge the risk of the scenario. A goal of any business practice is efficiency. Both goals are readily achieved if rules for optimizing the PHA effort are strictly followed. The rules in this paper are based on thousands of PHAs, with many variations across the set. Therefore, we believe these rules represent at least a first cut at PHA optimization. Following these rules can lower the labor costs significantly for completing PHAs and can, at the same time, improve thoroughness.

Acronyms

CCPS: Center for Chemical Process Safety
CM: Continuous Mode
COI: Consequence Of Interest
CSB: US Chemical Safety Board
EPA: U.S. Environmental Protection Agency
HAZOP: Hazard and Operability study
IE: Initiating Event
IEF: Initiating Event Frequency
IPL: Independent Protection Layer
L+S: PHA Leader and PHA Scribe
LOPA: Layers Of Protection Analysis
NRM: Non-Routine Modes
OSHA: US Occupational Safety and Health Administration
PHA: Process Hazard Analysis
PFD: Probability of Failing on Demand
PSM: Process Safety Management
RBPS: Risk Based Process Safety
SOP: Standard Operating Procedure

References

[1] CCPS/AIChE, Guidelines for Risk Based Process Safety, Wiley, 2007.

[2] W. Bridges and R. Tew, "Optimizing qualitative hazard evaluations for maximized brainstorming," in *AIChE/CCPS 5th Global Congress on Process Safety*, Tampa, FL, 2009.

[3] CCPS/AIChE, Layers of Protection Analysis (LOPA), Wiley, 2001.

[4] CCPS/AIChE, Guidelines for Independent Protection Layers and Initiating Events in Layer, Wiley, 2012.

[5] CCPS/AIChE, Guidelines for hazard evaluation procedures, Wiley, 2008.

[6] M. Massello and W. G. Bridges, "The art of PHA Scribing: The invisible role," in *AIChE/CCPS 14th Global Congress on Process Safety*, Orlando.

[7] W. Bridges, "Building Competency in Internal PHA/HAZOP Leaders - lessons learned in 40 years of doing so," in *AIChE/CCPS 14th Global Congress on Process Safety*, Orlando, FL, 2018.

[8] W. Bridges and A. Dowell, "Identify SIF and Specify Necessary SIL, and other IPLs, as part of PHA/HAZOP - or - Why it is not necessary to "Boldly go beyond HAZOP and LOPA"," in *AIChE/CCPS 12th Global Congress on Process Safety*, Houston, TX, 2016.

[9] W. Bridges and R. Tew, "Best Practices for Writing Operating Procedures and Trouble-Shooting Guide," in *AIChE/CCPS 13th Global Congress on Process Safety*, San Antonio, TX, 2017.

Enhancing PHA Facilitation through the Elicit-Provide-Elicit Method: A Novel Approach Inspired by Motivational Interviewing

Brandon Morris
Nutrien, Inc.
bl.morris@nutrien.com

Keywords: PHA, Hazop, EPE method, motivational interviewing, communication

Abstract

Effective facilitation of PHA studies requires not only technical expertise but also skilled communication; however, conventional PHA facilitation training has largely overlooked the teaching of communication methods. This paper presents a novel approach to bridge this gap, integrating the Elicit-Provide-Elicit (EPE) method from Motivational Interviewing into PHA facilitation.

The EPE method, a central technique in Motivational Interviewing, fosters engagement and collaboration by eliciting an individual's understanding, providing tailored information, and subsequently eliciting their response. Adapting this method to PHA facilitation offers a promising way to enhance team dynamics, critical thinking, and engagement.

The paper outlines the fundamental principles of PHA facilitation, the EPE method, and the noteworthy absence of communication techniques in traditional PHA training. It then provides a detailed methodology for incorporating the EPE technique into PHA facilitation, supported by theoretical frameworks and practical guidelines. By combining motivational psychology with process safety, this paper offers fresh insights into PHA facilitation and contributes to a more participative and effective hazard analysis process. It calls attention to a significant gap in current practice and paves the way for a more holistic approach to risk management.

1 Introduction

Communication skills are often listed as a critical skill for PHA facilitators. Job postings typically have it as a skill requirement as a facilitator. Experts in the field list communication as vital for a PHA [1]. Training available in the field is geared toward technical communication rather than interpersonal communication. Publications and training related to communication skills within the process safety industry focus on safeguarding against team member personality traits [2], conflict management [3], and promote using scare tactics in safety communication [4]. This paper offers a particular motivational interviewing technique, called the elicit-provide-elicit method, as a tool that can be used in PHA facilitation to navigate roadblocks in risk perception and willingness to implement process changes due to risk ranking.

Motivational interviewing is an evidence-based approach used in the medical field to achieve behavioral change with respect to risky behaviors. The communication methods used in motivational interviewing focus on exploring the person's own reasons for their resistance to change. Listening to a person's own perception of risk and resistance to change behavior allows the user to provide personalized direction. Motivational interviewing requires the user to maintain a sense of humility while engaging in the conversation. It is essentially the opposite from providing unsolicited advice and confrontational conversation. Motivational interviewing also requires the user to remain respectful of the person's reasons for resistance to change. The outcome of a motivational interviewing based conversation is a clear path forward built upon a person's intrinsic motivation to change [5]. This paper focuses on the elicit-provide-elicit (EPE) method in motivational interviewing.

2 Motivational Interviewing in the Process Hazard Analysis Setting

Effective communication is a critical skill in PHA facilitation. Facilitators coordinate with diverse, multidisciplinary teams to identify hazards and risks related to a chemical process. Direct communication is used when giving clear messages to apply the set of rules defined by the process safety management program. This language consists of clear, explicit instructions or information to avoid errors in risk analysis. Hazop teams use open ended questions to encourage creative discussion and participation from all team members to uncover hidden hazards. Assertive communication is used when stating facts related to the process and its operation. Non-verbal communication including body language offers insight to the conversations at hand.

There are times at which a team or team member has a substantial disagreement with the risk related to a certain part of the process. Team members may also be reluctant to write recommendations for risk mitigation due to any number of reasons for not wanting to change the operation of the process. There are two assumptions the facilitator can make in these situations. The person exhibiting resistance to change has a knowledge deficit or is competent. Motivational interviewing research in the medical setting shows that outcomes are improved when using the assumption that the patient is competent [6]. A PHA facilitator using motivational interviewing should approach the conversation with the assumption that the other person genuinely cares about process safety and is competent. Approaching a disagreement with the assumption that the person is competent results in a more collaborative conversation. Motivational interviewing requires active listening. It requires the user to remain humble. Facilitators should avoid language that disrupts listening:

- Agreeing/Disagreeing
- Selective attention/Response preparation
- Technical jargon/Complex language
- Absolutes
- Overly complex/Lengthy explanations
- Persuasive explanations

Conversations grounded in empathy and understanding can improve team engagement and result in better recommendations/corrective actions that are more likely to be implemented. By feeling understood, team members might be more willing to actively participate in the PHA sessions. Using motivational interviewing in the PHA setting can drive the team towards collaborative, well-supported PHA outcomes.

The specific language used by a facilitator can either encourage or discourage conversation from the PHA team. A generic example of the impact of subtle differences in language can be seen in the "what else?" and "anything else?" questions. Asking the team "what else?" encourages additional conversation about the topic while asking "anything else?" commands a yes or no response. Motivational interviewing research indicates that certain language use can steer a client either towards or away from decisions to change health behaviors [7]. The following table shows language adapted from the Motivational Interviewing Skill Code version 2.5 with examples provided that relate to PHA facilitation [8]. The language codes are separated into two categories. Facilitator language that is consistent with motivational interviewing, and language that is inconsistent with motivational interviewing.

Code	Description	Example
Motivational interviewing consistent language codes		
Advise – with permission	Facilitator provides advice or expertise after obtaining permission. Permission from the PHA team can take several forms including verbal and non-verbal communication. It could be in the form of the facilitator asking "can I provide input for a potential safeguard?" In the case of non-verbal communication, the PHA team stopping and looking at the facilitator for guidance on the subject would be non-verbal permission.	"Can I suggest something?" "I have an idea for a potential recommendation." "This may or may not work for your site, but I've seen other sites successfully use a high temperature trip on this type of equipment."
Affirm	Facilitator says something positive or complimentary to the PHA team. This can be related to the previous work the team has put in to improve the plant.	"You have been resourceful making this equipment operate as safely as possible with the tools you have been provided." "You have made major improvements with the older style

		boiler system. Have you considered upgrading the BMS?"
Emphasize control	Facilitator acknowledges the team's autonomy and the personal responsibility of the individual groups that the PHA team represents (operations, engineering, electrical & instrumentation).	"It's up to the engineering team on how to design this system." "It's your decision on how to re-write the operating procedure to avoid this hazardous scenario."
Open question	Questions that leave room for a response.	"How do you want to handle this scenario?" "What would you do to change this system?" "What would have to happen to consider making changes?"
Raise concern with permission	Facilitator points out a potential problem with the PHA teams plan or intention with permission.	PHA Team: "What do you think about adding a DCS output limit to ensure the valve doesn't close all the way?" Facilitator: "Well, it concerns me. The DCS output limit doesn't keep it from going closed due to other failure modes."
Simple reflection	Facilitator restates or slightly rephrases something the PHA team said. This conveys understanding but does not change the content of the statement.	PHA Team: "Operations has been talking about wanting a new alarm for this." Facilitator: "They want a new alarm."
Complex Reflection	Facilitator restates or slightly rephrases something the PHA team said while adding content to the statement.	PHA Team: "Operations has been talking about wanting a new alarm for this." Facilitator: "They want a new alarm, but they are comfortable with it as is?"

Motivational interviewing inconsistent language codes

Advise – without permission	Facilitator provides unsolicited advice.	"You should add a high temperature trip for this."
Confront	Facilitator speaks with an implied uneven power relationship along with disagreement or negativity. There is facilitator over-ride of what the PHA team says.	"You knew about the risk and operated the plant anyways?" (criticizing) "Sure you did. Right." (Disbelieving, sarcastic tone) PHA Team: "This equipment is safe; we operate it all the time without it failing." Facilitator: "So you're risking everyone's safety by continuing to operate like that?"
Direct	Facilitator gives a direct order.	"You have to install a SIL rated trip." "You can't operate the pump like that."
Raise concern without permission	Facilitator points out a potential problem with the PHA teams plan or intention with permission.	"That's not going to work." PHA Team: "What do you think about adding a DCS output limit to ensure the valve doesn't close all the way?" Facilitator: "Nope. That's not an IPL."
Warn	Facilitator gives a direct warning or threat.	"You will hurt someone by running this equipment."

Other language codes

Facilitate	Other language to keep the conversation going with no relevant meaning.	"uh-huh" "yeah"
Filler	Other language that does not fit other categories.	"Hello" "Good afternoon"

		"Another day in paradise!"
Giving information	Facilitator makes a statement without advising.	"The daily safety briefing said the plant is down today."
Closed question	Facilitator asks a question that has a simple yes/no answer. This also includes asking an open question but ending it with a closed question.	"Will you be installing this proposed safeguard soon?" "What would have to happen to consider making changes? More funding?"
Support	Facilitator makes a statement that sides with the team.	"You make a good point."
Structure	Facilitator gives information about how the meeting is held or about meeting transitions.	"We will be discussing checklists next." "Last week you brought up a concern about this other system, but you needed to gather more information. Are you ready to discuss now?"

3 Elicit-Provide-Elicit Method in PHA Facilitation

Motivational interviewing contains several core techniques and communication frameworks that each contribute to its overall effectiveness in the clinical setting. The elicit-provide-elicit (EPE) method allows the user to provide tailored information to a person while acknowledging the person's experience and resistance to change. It should be used in a conversation once the facilitator becomes aware they have information or facts that may help continue the conversation. Rather than offer unsolicited advice, this framework assists in the delivery of information. It is a three-part communication framework centered around information sharing while remaining neutral in the conversation. The user elicits the person's existing beliefs, understanding, or viewpoint on the subject. Information is then provided in a way that respects the person's viewpoint and is tailored to their response. It is important that the information is provided in a balanced manner that avoids disruptive language. The last step is to elicit the person's response to the information provided. This method can be used when encountering significant opposition to a risk ranking, if there is resistance to creating recommendations, or if the team is using biased logic.

Elicit – Ask what they know.

"What has your experience been?"

"Can I share industry information?"

Provide – Neutral statement of fact, information sharing.

"Industry standard is…"

"Other sites have had success with…"

Elicit – Get their interpretation.

"What are your thoughts?"

"Considering the information provided, what would be the ideal safeguard implementation?"

Conversations in PHA sessions frequently become complex. Technical jargon is used. Explanations become lengthy and sometimes overly complicated. Debates occur often and are necessary for hashing out accurate risk analysis. Disagreements happen, but communication breaks down when team members stop listening and begin waiting for their turn to speak. This method is intended to halt this breakdown in communication and return to listening.

Poor communication can significantly delay PHA studies. To provide an accurate and timely risk analysis, PHA teams are required to discuss complex scenarios while relying on their experience and expertise. Along with discussing complex technical scenarios, teams must communicate their own perception of risk. Risk perception varies among individuals and is dependent on a sense of vulnerability. Vulnerability and perceived risk are dependent on the combination of several factors, including knowledge, experience, and process safety culture [9]. The EPE method can be used in several scenarios to navigate conversations around risk perception and an individual's or PHA team's resistance to change. The rest of this section consists of example conversations and how the EPE method can be used by a PHA facilitator.

Recent incident bias: Recent incidents are viewed as more likely to occur.

PHA Team: "This pump just failed recently and caused some major headaches. Maintenance is still going through the cause map. The failure frequency is at least once per year."

Facilitator (Elicit): "What do you know so far about the failure? What concerns do you have regarding the cause map process?"

PHA Team: "A trip valve downstream of the pump went closed and caused a leak in the pump casing. Maintenance is busy getting this pump repaired and reinstalled and don't have as much time to investigate the root cause right now."

Facilitator (Provide): "I understand the frustration when getting bogged down with repairs in the plant. In this process, other sites have had success setting a DCS minimum limit on the minimum flow control valve, so it doesn't deadhead when the trip valve closes. What do you think about doing something similar here?"

PHA Team: “That sounds promising. We can explore that.”

Facilitator (Elicit): “With a potential solution in mind for preventing additional failures with this pump, what should the typical failure frequency be for this equipment?”

Anchor bias: The risk matrix severity and/or frequency discussed first can anchor the conversation.

Facilitator (Elicit): “I notice we started out talking about a high severity scenario that may not accurately portray the risk for this scenario. Do you feel like this is indeed the worst credible consequence?”

PHA Team: “Yeah, maybe we went to the extreme with this scenario.”

Facilitator (Provide & Elicit): “I typically see this scenario ranked as a lost time injury. Looking at this scenario with a new perspective, what do you think the worst credible consequence would be?"

PHA Team: “Let’s use the single LTI.”

Survivorship bias: Wrong conclusion drawn from incomplete dataset.

PHA Team: “This pump has been running for the last 20 years and has never given us any problems. Our operators catch any issues early on and we have never had any incidents from this pump.”

Facilitator (Elicit): “It’s great to hear it has been running smoothly, and it’s certainly a testament to the work operations and maintenance has put into this pump over the years. When you say operations has caught issues early on, can you share more about what kinds of incidents they have prevented? Are we covering all relevant information, including near-misses?”

PHA Team: “We always report any incidents, but if an operator prevents an incident, I’m not sure it would be reported.”

Facilitator (Provide): “That’s a common approach, but it might lead us to overlooking lessons learned from the smaller issues or near-misses, which can be warning signs of bigger problems. Statistics show that incorporating data from near-misses into analysis improves risk analysis and prevents major incidents. Would it be helpful to get info on how other sites handle near-miss reporting?”

PHA Team: “Yes, that makes sense. I’d like to see how they report near-misses.”

Facilitator (Elicit): “Based on what we know now, how do you think we can incorporate your current knowledge of the near-misses on this system into our analysis?”

Loudest voice in the room: The most-listened-to voice in the room can rule the conversation.

A PHA team meeting where one member frequently dominates the conversation, potentially overshadowing quitter team members' contributions.

Facilitator (Elicit): "I've noticed we have a lot of valuable expertise around the table. I want to make sure we're capturing all perspectives. Can anyone share how they feel about the discussions so far? Are there thoughts or concerns you haven't had a chance to express yet?"

PHA Team Member (Quieter Voice): "Yeah, it's been hard to find a moment to chime in. I think we have some risks that haven't been fully explored yet."

Facilitator (Provide): "Thank you. It's important we hear from everyone to get a comprehensive view of our risks. To facilitate this, we might adopt a round-robin approach for sharing ideas or use other structured communication techniques. How does the team feel about trying some of these strategies?"

PHA Team (Louder Voice): "I see your point. I didn't realize I was overpowering the discussion. I'm open to trying a new approach if it helps us cover all our bases."

Facilitator (Elicit): "It's great to hear a willingness to try new methods. Given our goal to include all perspectives, what are some ways we can structure our discussions to ensure everyone feels comfortable and empowered to share their insights?"

Unknown information: Unknown information is often viewed as unimportant.

PHA Team: "We use a rental skid to mix chemicals and fill up our product tank. We don't have P&IDs for this equipment. The company we get it from does their own PHA, so all the hazards are already covered. We don't need to cover it in our PHA."

Facilitator (Elicit): "How do you connect to it? What do you know about the risks associated with this equipment?"

PHA Team: "We connect to it with hoses. It is a contained trailer mounted unit with a cover over the piping. Any leaks would be contained by the cover, and it would drain to the containment sump. We don't have any concerns with it."

Facilitator (Provide): "It sounds like you have done a great job mitigating the risks that you are aware of. You set this up over a leak containment sump to mitigate the risk of any leaks that might happen. I'm concerned though about how this equipment is a "black box" of unknown piping connections. It appears that your process chemical and potable water are both connected to this system. I can show you some examples where similar systems have had the process chemical backflow into the potable water system."

PHA Team: "We can check into that. We should get the vendor drawings and confirm the design and their safeguards."

Facilitator (Elicit): "I agree. Would you like to schedule another session to cover this additional scope?"

The impacts of using the EPE method in PHA sessions can be difficult to measure. In the medical field, the impacts of using motivational interviewing and the EPE method can be analyzed through quantitative heath monitoring measurements to evaluate its efficacy. There have been some efforts in the medical field to codify motivational interviewing language used in client interactions. This research shows additional studies are needed to further evaluate the efficacy of motivational interviewing [10]. The efficacy of motivational interviewing in the medical field is dependent on the unhealthy behavior targeted for change along with several other factors. Motivational interviewing is particularly useful in treating substance abuse since it is an economical and effective method [11]. In the PHA setting however, facilitators are not treating any kind of medical issues. The consequences of failures in using motivational interviewing and the EPE method are negligible compared to that of the medical setting. When communication breaks down in the PHA setting, the main consequence is a delay in study due to poor communication. The facilitator might call for a meeting break and later regroup to continue the conversation to finish the analysis. The benefit of using effective communication methods in the PHA setting is time savings.

Motivational interviewing and the EPE method have the benefit of being able to be learned and implemented after a three-day interactive training session [12, 13]. Motivational interviewing is not the only evidence-based treatment available. The part that makes motivational interviewing effective outside the medical field and in the PHA setting is that it focuses on communication structure rather than the information provided [10]. The communication structures within motivational interviewing promote reflective listening, which helps the user develop a deeper understanding of the client's reluctance to change [14]. Reflective listening is taught in other courses including leadership courses, conflict resolution workshops, communication skills seminars, team building retreats, and several other courses. Motivational interviewing and the EPE method, however, provide a structured approach to communication. The structured approach and practical guidelines of both motivational interviewing and the EPE method allow for adaptability across diverse professions. This versatility ensures they can be effectively integrated into various professional contexts.

4 Conclusion

Facilitation of PHA studies can present unique challenges in interpersonal communication. While communication is a critical skill for facilitators, training courses in the field of process safety often lack content on these skills. Poor communication in the PHA setting can lead to delays in sessions and can impact the quality of PHA results. There are several courses that offer training in leadership and conflict resolution, but few have a concise framework like the elicit-provide-elicit method.

This paper presents the EPE method from motivational interviewing as a communication technique to use in PHA sessions to enhance team engagement and improve collaboration. The EPE method allows facilitators to elicit an individual's understanding, provide tailored information, and subsequently elicit their response. It encourages active listening, humility, and empathy in conversations, which can improve team engagement and lead to better PHA outcomes.

The paper provides a methodology for incorporating the EPE method into PHA facilitation. It highlights the impact of language on communication and offers examples on how the EPE method can be used to navigate differences in risk perception and resistance to change.

While the impacts of using the EPE method in PHA sessions can be difficult to measure, it is evident that effective communication can significantly enhance the efficiency and outcomes of PHA studies. The paper acknowledges the need for further research to evaluate the efficacy of motivational interviewing and the EPE method in the PHA setting but emphasizes the adaptability of the communication technique across diverse professions.

Overall, this paper calls attention to the gap in current PHA practice regarding interpersonal communication skills and paves the way for a more holistic and comprehensive approach to risk management. By incorporating the EPE method, PHA facilitators can foster collaboration and engagement, leading to more effective hazard analysis and ultimately improving process safety.

5 References

[1] aeSolutions. 5 Facets of an Efficient Process Hazard Analysis (PHA). https://www.aesolutions.com/post/5-facets-of-an-efficient-process-hazard-analysis-pha . Accessed January 20, 2024.

[2] AIChE Conference Presentation. Blah, Blah, Blah!!! - Effective Individual and Team Communication Lead to Effective PHAs. https://www.aiche.org/conferences/videos/conference-presentations/blah-blah-blah-effective-individual-and-team-communication-lead-effective-phas . Accessed January 21, 2024.

[3] PrimaTech. Leadership Skills for PHA Facilitators - Managing Conflict. https://www.primatech.com/technical/pt-notes/263-leadership-skills-for-pha-facilitators-managing-conflict . Accessed January 20, 2024.

[4] Larkin TJ. Use scare tactics to communicate major incidents. *Chemical Engineering Progress* 2018; **August:** 39-45.

[5] Miller RW, Rollnick S. Motivational interviewing: preparing people to change addictive behavior. New York, NY: The Guilford Press; 1991.

[6] McCain J. To heal the body, get into the patient's head: motivational interviewing: to improve adherence. *Biotechnology Healthcare* 2012; **9(4):** 10-12.

[7] Apodaca TR, Jackson KM, Borsari B, Magill M, Longabaugh R, Mastroleo N, Barnett NP. Which individual therapist behaviors elicit client change talk and sustain talk in motivational interviewing?. *Journal of Substance Abuse Treatment* 2016, **61:** 60-65.

[8] Houck JM, Moyers TB, Miller WR, Glynn LH, Hallgren KA. Motivational interviewing skill code (MISC) version 2.5. 2010. https://casaa.unm.edu/assets/docs/misc25.pdf. Accessed February 21, 2024.

[9] Subbarao NV, Senthilraja M, Sangeetha B. Improving process safety culture through developing and maintaining the sense of vulnerability at all levels in the organization. *Process Safety Progress.* 2024; **43(1):** 80-84.

[10] Hurlocker MC, Moyers TB, Houck J. Can a pure motivational interviewing intervention be manualized and still efficacious? A test of feasibility and initial efficacy. *Psychotherapy* 2021, **58(2):** 196-205.

[11] Bischof G. Bischof A. Rumpf HJ. Motivational interviewing: An evidence-based approach for use in medical practice. *Deutsches Arzteblatt International* 2021, **118(7):** 109-115.

[12] Lundahl B, Moleni T, Burke BL, Butters R, Tollefson D, Butler C, Rollnick S. Motivational interviewing in medical care settings: a systematic review and meta-analysis of randomized controlled trials. *Patient Education and Counseling* 2013, **93:** 157-168.

[13] Motivational Interviewing Network of Trainers. A Brief Introduction to MI: Training Options. https://motivationalinterviewing.org/training-expectations . Accessed 1/20/2024.

[14] Miller WR. The evolution of motivational interviewing. *Behavioural and Cognitive Psychotherapy* 2023, **51:** 616-632.

"Selling" Process Safety Concepts for Non-PSM Applications

Edward Gaither, P.E., MBA, CCPSC
PROtect, LLC
ed.gaither@protect.llc

Keywords: Process Safety Management; Risk-Based Process Safety

Abstract

Many production plant facilities are not governed by OSHA PSM or have areas that lie outside a self-determined PSM boundary. That doesn't mean those processes are inherently safe. It doesn't mean those plants do not have hazards. It quite literally means that those plants (or areas of the plant) merely do not exceed the Threshold Quantity of any Highly Hazardous Chemical. Since the inception of PSM, many facilities have been reluctant to apply PSM elements to processes that lie outside the defined PSM boundary.

The process safety industry rightfully focuses on PSM facilities. This presentation makes the case that non-PSM facilities also deserve attention and offers some personal experience-based propositions to help bring Risk-Based Process Safety (RBPS) principles to non-PSM facilities and areas. The adoption of even just some of the RBPS elements can improve safety, operability and maintainability of the process.

Particular attention will be given to food and pharmaceutical industries. They may have some PSM areas or combustible dust hazards and are likely to benefit from adopting PSM principles facility wide. The existence of U.S. Food & Drug Administration (FDA) food safety and current Good Manufacturing Practices (cGMP) programs at these facilities also put them only a few small steps away from improving process safety by implementing RBPS elements in full or even only in part.

1 Introduction

Thousands of facilities in the United States fall under the scope of U.S. Occupational Safety and Health Administration (OSHA) Process Safety Management (PSM) regulations as defined in the Code of Federal Regulations, 29 CFR 1910.119. Many thousands more than that do not fall under PSM, even for a portion of the facility. Those facilities would benefit from adopting at least some PSM elements.

The two most cited OSHA elements for PSM covered facilities over the past decade are Process Safety Information (PSI) and Mechanical Integrity (MI). If the facilities required to have PSI and MI programs are lacking, how much more room for improvement is there in facilities that do not have such OSHA requirements. Other elements such as Incident Investigation, Hot Work, and Emergency Planning are effectively required by other regulations and many sites have some version of these regardless of their PSM status. Although the importance of all PSM elements is recognized, this paper will focus on the benefits of implementing PSI and MI. Both of these are also very closely linked with Management of Change (MOC), the proper performance of which relies on PSI and MI for inputs while also providing information back to them after the change is implemented.

Asset Integrity and Reliability, the RBPS corollary to MI, is as critical to profitability as it is to process safety. Process Knowledge Management, the RBPS corollary to PSI, supports Hazard Identification and Risk Assessment (HIRA) and Asset Integrity and Reliability, as well as directly enabling efficient operations and project development. Terminology throughout this paper includes both PSM and RBPS language to emphasize the interchangeable and complimentary nature of the two standards.

The content of this paper will reflect the author's experience in food production and pharmaceutical industries. Several other industries would be just as relevant to the discussion. The argument can be made that any industry would benefit from implementing elements of PSM and CCPS' *Risk-Based Process Safety (RBPS)*[1]. Much discussion could also be made regarding other PSM and RBPS elements. Publications like CCPS' *The Business Case for Process Safety*[2] have presented compelling evidence and testimonials for implementing PSM.

Lastly, this paper will discuss tactics for persuading stakeholders to realize the benefits of implementing a PSM or RBPS program. A question repeatedly asked by PSM engineers is: "How can we best influence executives and senior management?" This paper will present some ideas regarding that question.

2 Asset Integrity and Reliability

2.1 Overall Equipment Effectiveness

Overall Equipment Effectiveness (OEE) is the product of equipment availability, performance, and quality. It is a common metric of the Total Productive Maintenance (TPM) system originally developed in the 1960's and widely used in industry.[3] Availability is another term for uptime or production runtime. The performance score represents throughput. Quality reflects the out of spec product scrap or rework rate. For example, a process with equipment that is available for

90% of the targeted uptime, performing at 85% of their throughput capacity, and making 97% good product has an OEE of 74.2% (90% x 85% x 97%).

The effect of equipment availability is the most obvious benefit of maintenance. Downtime is unpredictable and costly. Yet many non-PSM facilities are performing minimal preventative maintenance activities and are in a near-constant state of reactive repair activities. This causes strain on the business.

In the world of consumer-packaged food with low inventories and frequent product delivery to retail outlets, loss of run time quickly leads to inability to fulfill commercial contracts. The retail outlet may find another supplier to provide a product that fills that shelf space. When shelf space is lost to a competitor, it is difficult to reestablish your product.

Equipment failures may not take an asset completely out of production but reduce its production rate instead. Production may be able to proceed but with a lower OEE score due to a performance rate less than the normal throughput target or substandard quality. An MI program should seek to keep each piece of equipment running at its target OEE, not merely keep it from breaking down entirely. Running with one or more unit operations at low performance rates forces operations to compensate in other ways and negatively affect the cost of goods sold.

2.2 Plant Staff Safety

Tank, piping, and pressure vessel failures can result in a sudden loss of containment or release of energy. These acute occurrences have high potential to injure.

Poorly maintained equipment still in operation, but with issues (or deficiencies) perceived to be "minor" by plant staff, can foster a normalization of a deviation from the design basis for safety.

- A packaging machine that requires frequent readjustment can be a hazard. An operator may try to perform a minor correction while the equipment is in motion, possibly resulting in an entanglement injury. A maintenance tech may become tired of going through the lockout tagout process and rely on safeguards like access door interlocks to positively deenergize the equipment while working on it.
- Frequent steam or water hammer can be seen as merely an annoyance by staff who are unaware of the damage being caused internally to the equipment. The eventual sudden failure is an injury hazard.
- Pipe erosion and corrosion are often met with an attitude of "keep an eye on it and we'll fix it when the leak gets too bad". In the author's experience, few non-PSM facilities perform non-destructive testing (NDT) on piping.

2.3 Asset Value

A piece of equipment that has been well maintained will retain value on the secondary market. Resale of equipment can take some of the sting out of the capital expenditure to replace an old unit with an upgraded model. There is an active used equipment market among co-manufacturers and co-packers. Used equipment that has been retired by the original owner (the "buyer" in the co-manufacturing agreement) is often desirable to the co-manufacturer because it is already set up to make the product that they are under contract to produce.

2.4 Public Perception

While not unique to the food and pharmaceutical industries, brand value can be jeopardized by poor quality product hitting the shelves, or worse still, a product recall. Asset Integrity is a primary line of defense against biological, adulteration, and contamination risks in food and drugs. Storage bins and tanks are good initial targets to be managed with a "PSM-level" PSI and MI program. In partial-PSM facilities, raw material storage and receiving operations are frequently outside the PSM envelope.

The integrity of the production building shell is, in effect, a quality control device whose failure can negatively affect a product and public perception. Leaking roofs have caused many food recalls. An infamous food safety incident involving a leaking roof caused substantial brand damage to a popular peanut butter brand. This issue led many food and pharmaceutical producers to ensure that building maintenance received the same level of MI attention as critical process equipment. Public perception has always been critical to brand value and the company's bottom line profit.

2.5 Insurability

Asset Integrity and Reliability are obviously key to property insurance coverage. The potential effect on injury and business interruption insurance is just as important. Claims filed due to equipment breakdowns, vessel failures, etc. inevitably lead to greater insurance premiums. This is often the most apparent return on investment (ROI) argument for a comprehensive MI program. It also tends to be more tangible to executives and senior management unfamiliar with the business risks of process safety incidents.

3 Process Knowledge Management

3.1 Process Knowledge

Call it "design basis" or "process history" or "process knowledge" or "tribal knowledge" or "knowing where the bodies are buried." By any name, the understanding of how and why the process works without safety incidents and mechanical breakdowns is essential to running a facility. Generational shifts in the workforce are emphasizing the importance of knowledge management and knowledge transfer. Frequent staff turnover or mobility within the company also puts pressure on process knowledge management.

Many non-PSM facilities do not have current revisions of even the most fundamental process design documents: design basis, mass & energy balance, and Process Flow Diagram (PFD). A surprising (in the author's opinion) amount do not have current Piping and Instrumentation Diagrams (P&IDs) or facility layout drawings. It is not easy to convince complacent plant staff that these are necessary when they may have been running a "successful" plant without them. Reverse engineering all these seminal process design documents requires substantial effort. The value in reproducing these documents is not always tangible. The perceived low ROI of a document update project is usually the end of the discussion. It is challenging to communicate the "What" and "Why" if leadership does not have an engineering background.

Insufficient process knowledge can manifest itself with dramatic results. Examples from the author's experience include:

- A biological fermentation process had a severe process upset that greatly reduced production for weeks. The company lost a substantial market share of downstream food producer clients. Market loss and resulting future forecasting made the ROI on an ongoing large (40% increase) expansion project undesirable. Attempts to improve project ROI by cutting budget and scope when the project had already begun construction greatly complicated that project.

  The problem was that the microbiological culture from one fermenter supplied the seed for a subsequent fermenter, so an invasive microbiological infection was presumed to be the cause. The plant operated on the mantra of "if you lose the micro culture, you are in big trouble."

  Several process knowledge management gaps came to light while pursuing that line of investigation. The plant had a long-tenured staff in key operations, quality and engineering positions, so they relied on the memory and experience of those individuals. Problems were addressed by those individuals meeting together, then coming to a consensus and correcting this issue. That method had been very effective for many years, so they became complacent about managing historical data. Laboratory results of process sampling were difficult to cross reference to Basic Process Control System (BPCS) trendlines. Laboratory Inventory Management System (LIMS) data was not accessible to production, engineering and maintenance staff. Substantial effort was needed to retroactively cross reference years of LIMS data with other minor fermentation upsets. Although the data needed to investigate the upset did exist, a method to efficiently cross reference LIMS data with BCPS historical data was not in place. The QA/QC team ended their investigation without concluding that the presumed microbiological culture was the definitive cause of the upset.

  The process remained running at a lower production rate throughout the upset. It eventually recovered, but the facility had a recurrence two years later. Even if a structured process knowledge management system were in place, the problem still may have remained a mystery, but a good system would have made the investigation process much less labor intensive.

- A project to replace two legacy bioreactors with larger, more automated bioreactors was complicated by the loss of key personnel. The project suffered major cost overruns, potentially because the approved budget assumed a relatively brief design phase. When key personnel left the company, their knowledge left with them. The new project team was capable of performing the project but required more input from the remaining plant staff. Knowledge that had been only possessed by the staff who left had to be relearned. Plant standard specifications needed to be updated. Process knowledge of clean-in-place (CIP) requirements and of the multiple products intended for the bioreactors was not readily available. Plant management and corporate program management staff began to emphasize better PSI systems going forward.

3.2 Equipment Information

Non-PSM facilities that have made an effort to maintain PSI databases often fall short of what a PSM engineer, project engineer, or Inspection Testing and Preventative Maintenance (ITPM) program requires to perform at a high level. The vast majority have implemented some form of a Computerized Maintenance Management System (CMMS). Implementation is often initiated and performed with asset value accounting as the focus. Data management can also be upset by migration to a new system, again often driven by someone other than maintenance or engineering. This can motivate users to avoid a central database and instead use a spreadsheet-based system that will remain entirely under their control.

Recurring ITPM activities performed by plant staff are typically scheduled and tracked. The various maintenance disciplines (mechanical, electrical, or automation) rarely use a CMMS with equal effectiveness. Predictive maintenance and third-party inspection data is often managed outside of the CMMS for the convenience of the engineering and maintenance staff.

Several types of equipment information are lost in the shuffle of ineffective data management. Examples from the author's experience include:

- Corporate project management culture often emphasizes commissioning and production ramp-up at the end of a project. While this, in itself is not problematic, it results in a lack of resources available during project closeout to update PSI with as-built updates to drawings. This policy has a negative effect on future projects. Projects kick off with a herculean effort to produce as-built P&IDs, layouts, electrical one-lines, and structural drawings in short order. A key driver in food industry projects is speed-to-market for a new consumer product. Schedule concerns often outweigh the budget. Companies that see the value in using slack time between projects to update PSI usually reap the benefit by having good information during pre-project planning and project justification for future projects. Plus, they are better able to accelerate schedules and have better speed-to-market for new products.

- In many non-PSM facilities, historical data of third-party API and ASME compliant inspections can be difficult to find or entirely unavailable. Indeed, those inspections may not have been performed for decades. Facilities have claimed ignorance of the ASME requirement for pressure vessel inspections. In other cases, management simply refuses to fund ITPM activities that are not explicitly required in an OSHA standard or not required by their insurer.

4 The Sales Pitch

The question of how to influence management to develop an appreciation of process safety is as old as process safety engineering itself. Whether the targets of your persuasive efforts are at the upper levels of your company's organization chart or if they are clients you are trying to influence as a third-party consultant, the struggles are the same. Many of these arguments seem maddeningly like common sense conclusions to those of us living in the process safety world. That inclination needs to be suppressed when presenting to executives and senior management. They know that we know what we are talking about. That's why our jobs exist. But their decision-making process has a different set of constraints than an engineer's. If we are giving them a pitch, we need to meet them more than halfway.

4.1 Know Your Audience

4.1.1 Personalities Matter as Much as Positions

Obvious management targets are operations, engineering, safety, and risk management leaders. While that list is likely to be the people that PSM engineers approach, they need not be the only ones. Others have reason to embrace PSM principles, too. Benefits of people focused PSM elements of Employee Involvement, Training, and Incident Investigation, as well as RBPS elements of Process Safety Culture, Stakeholder Outreach, and Conduct of Operations should be recognizable to corporate development or human resources management. When approaching that side of the organization chart, emphasize how PSM holistically influences positive culture, not how important it is to have up-to-date P&IDs. Asset Integrity and Reliability, Compliance with Standards, and Operational Excellence align more with financial executives who concentrate on balance sheet items and financial/commercial risk management implications.

For the obvious targets, an approach should fit their personality. There are engineers who are engineers, and they cannot stop being engineers, even when they are in executive positions. They are influenced by facts and data… probably even more so by raw data that they can dig their fingers into and analyze. Then there are engineers or business managers who prefer to stay out of the weeds and see the bigger picture. They want to see how PSM and RBPS "move the needle." The project performance impacts of MI & PSI discussed above are good to discuss with them.

Maintenance and operations staff tend to have a "doer" personality. Try to find arguments that demonstrate how PSM and RBPS make their job easier to do. Point out how a PSI program enables troubleshooting or how an appropriate ITPM program should let them be more proactive than reactive. Figure out how to make the operating procedure revision process smoother for them.

Of course, in order to speak to their personalities, you are going to need to get to know them as people. My apologies to the introverts.

4.1.2 What's in it for them?

Human nature makes all of us evaluate a proposal by wondering what it means to us as an individual. It could alleviate a pain they have had to tolerate. It could provide them with a new benefit or help them achieve a performance goal. It could not really affect them at all. The first two of those are the levers to pull when validating the value of PSM and RBPS.

The examples given in the sections above demonstrate some of the pain points people may have experienced or continue to experience throughout their careers. This paper focused on MI and PSI not only because they are among the most frequent OSHA citations, but also because they can address many different pain points.

The most tangible benefits involve the move from reactivity to proactivity, as well as a more engaged employee and contractor culture. That may take some foresight to see past the initial pain of change. A corporate culture where engineers or managers frequently rotate through sites or positions can make this tactic more challenging. The long-term benefits of funding a robust PSI program can be a more difficult sell if stakeholders are not likely to be around to benefit from it.

On the other hand, investments in mechanical integrity may pay immediate dividends by improving OEE. An effective argument may be to emphasize that unplanned downtime caused by mechanical failure or unexpected complications during scheduled maintenance is typically more costly than performing ITPM well. Unplanned repairs may require costly expediting of parts, sudden contractor mobilization, or overtime for maintenance crews. The predictable planning afforded by a well-managed ITPM program also provides a solid return on investment.

4.2 Be Patient, Pleasant, and Persistent

A positive response to asking for adoption of PSM or RBPS elements where they aren't explicitly required by law or dictated by an insurance underwriter may take some time. Change is difficult at the individual level. It is exponentially more difficult at the organization level. You may quickly win over an executive champion, but unless that person can unilaterally implement new policies, you will probably have to help them win over their peers.

Maintaining patience can be difficult, especially when you know hazards are present in the workplace that could cause a loss or injury at any moment. You may have to keep motivational notes for yourself commemorating every minor win along the road to the end you envision. Be happy with taking small bites. Selling PSM to non-PSM sites is definitely an elephant you need to eat one bite at a time.

The attitude of a speaker has a profound impact on a listener. Unless there has been an unfortunate incident to increase immediate pain on a company, a pleasant demeanor will be more effective than a frustrated or, worse yet, condescending demeanor.

Use every opportunity you have to jump on a point of emphasis. The breadth of the PSM and RBPS elements provides you with many opportunities to make relevant points regarding whichever elements are most applicable to the situation. Concrete examples of the benefits are most impactful when they can be highlighted in real time.

5 Conclusion

Non-PSM facilities are not non-hazardous facilities. There are many reasons pertaining to hazards and also pertaining to profitability for all facilities to adopt elements of PSM and RBPS. Many industries fall under the requirements of other codes and standards with adjacent purpose and effect to PSM. Steps taken to adopt PSM elements are often not a paradigm shift from a facility's current management programs. Industries like food and pharmaceutical production have several opportunities to improve their operations and reduce their process safety risks.

The implementation of PSM-level Mechanical Integrity and Process Safety Information programs provides several benefits. Overall equipment effectiveness, plant staff safety, asset value, public perception, and insurability are areas that are likely to benefit from a structured MI program. The process knowledge and equipment information database of a robust PSI program also provides substantial benefits.

Persuading senior management to invest in implementing some or all of the PSM or RBPS elements is not an easy endeavor. The sales pitch must contain information that addresses the pain points that PSM can cure and also the tangible benefits gained from implementation. A PSM engineer attempting to influence management must be patient, pleasant, and persistent with stakeholders having a broad range of decision-making constraints.

6 Reference

1. Center for Chemical Process Safety, "Guidelines for Risk Based Process Safety," AIChE, John Wiley & Sons, Inc., Hoboken, NJ (2007).
2. Center for Chemical Process Safety, "The Business Case for Process Safety, 4th Edition," AIChE, 2018
3. Nakajima, S., "TPM Development Program: Implementing Total Productive Maintenance," Productivity Press, New York, NY (1989)

SPRING24
+20TH GCPS
A Joint AIChE and CCPS Meeting

AIChE Spring Meeting and Global Congress on Process Safety

CAPTUER – Using Digital Technology for Effective and Efficient Procedures

Elliot Wolf-Stokes
Global Process Safety Leader
The Chemours Company
elliot.wolf@chemours.com

Rob Fisher
President and Director of Operations
Fisher Improvement Technologies, Inc.
rob.fisher@improvewithfit.com

Prepared for Presentation at
American Institute of Chemical Engineers
2024 Spring Meeting and 20th Global Congress on Process Safety
New Orleans, LA
March 25 – March 28, 2024

CAPTUER – Using Digital Technology for Effective and Efficient Procedures

Introduction

Most organizations believe that their operating procedures are good, they just need people to follow them. There is now enough science (Deming[1], Reason[2], Dekker[3], Conklin[4], Rasmussen[5], et.al.) that proves that simply is not true. Possibly an underlying reason that drives these beliefs are decisions to allocate time toward more perceived "value-added" tasks which may include cost savings, capacity release, new product trials, and process improvements. Process safety professionals often find themselves in this dilemma competing for time among technical and operational staff, thus the value proposition for process safety procedure improvements must be highly compelling and well understood. Historically, procedures have not typically been the most exciting and rewarding projects to work on. The authors have been part of the development team for the ***C***ompletely ***A***utomated ***P***rocedure ***T***ool for ***U***ser ***E***rror ***R***eduction, ***CAPTUER*** which uses latest technology to search for, identify, flag, and recommend changes for many of the error drivers that appear in operating and technical procedures. This electronic tool resolves many of the priority conflicts for procedure updates and results in more effective process safety by automating part of the review process using an interactive search add-in for Microsoft Word™.

Through multiple site-level pilots of the application, the authors have demonstrated a 40-60% time savings required to review the procedures with a significant increase in identifying the procedure error drivers that could result in a process safety incident. By reducing the burden and time commitment, especially for long detailed operating guidance, this add-in enables people to reduce the probability of bad process safety outcomes and to be more effective and efficient in their roles. Another direct benefit of the add-in is that it increases engagement between frontline workers and technical staff when discussing the identified procedure issues which leads to operational improvements and standardization. Instead of procedure reviews being a superficial "check the box" activity, they can be a collaborative exercise that increases process safety, teamwork, and reduces field errors. Procedure improvements are not limited to process safety but also quality, production, effectiveness, efficiency, and cost.

Some Procedure History

Rob Fisher has been involved in human-factored procedure writing and best practice development for over 40 years. Process safety-related procedures have always followed the development technology from typewriters and white-out to desktop computers, to laptops, to tablets. All of these changes improved how we able to see and access procedures, but not how the procedure actually drove the critical work. In other words, we impacted the visibility but not the usability. As computing power increased so did our ability to manage the information in technical procedures.

In the early 1990s, the concept was introduced that guidance in process safety procedures cannot just be technically accurate, it also has to be usable for the person performing the tasks. High-risk industries began creating "procedure writing guides" that helped the developers of written guidance understand the parameters and methods of writing good procedures, and how they were different than good technical writing as taught in universities. In other words, a great engineer may not be great at writing effective procedures. Something had to be done, especially with an increased focus on process safety.

After an incident or near-miss, often procedures were discovered to be technically accurate. Management then posited that "we have good procedures, if we could just get people to follow them, we would be ok." While this made managers feel good about their procedures, and the actions they could take to increase "accountability", it did nothing to discover what the potential drivers were of the incident or near miss related to procedure quality. As incidents were analyzed deeper into understanding human performance and human factors there were systemic error drivers that were embedded in the procedures that made "people do what they did, at the time that they did it, for reasons that made sense to them at the time." [3]

It became increasingly clear that there are five (5) error drivers that appear most often, and with the most impact, on field performance.[6] As procedure and process writer training started to include the definition and understanding of these error drivers, field performance improved. Organizations started to develop writers' guides and procedure quality checklists and grading tools to help identify and rectify the error drivers before they get to the field. Organizations that paid attention to procedure and process error drivers saw improvements to the safety aspects, as well as almost immediate improvements in quality, production, and standardization. Field performance and procedure adherence improved somewhat, but the administrative burden on procedure writers and reviewers was immense.

The Top-5 Error Drivers are:

Field Decisions – Forcing a decision without guidance increases user error rates to as high as 50%.
To Reduce:

- Use conditional terms like 'IF-THEN' or 'WHEN-THEN.'
- Add conditional clarification like (e.g. 'as required to prevent water hammer')
- Provide a decision-making table for more complex decisions.

Vague Terms – Ambiguous phrases or terms with multiple meanings lead to higher error rates from misinterpretation or misunderstanding.
To Reduce:

- Use terms consistently and within the organization's requirements and guidance for developing procedures.
- Clarify the need for the actions or the desired outcome - how are they supposed to be sure if they did the right thing?
- Add clarifying information (e.g. Slowly fill to prevent over pressurization).

Ineffective Verbs – Actions that could mean multiple things or different things to different people increase the probability of misinterpretation or use.
To Reduce:

- Avoid words like 'check', 'ensure', or 'verify' without clarifying the information.
- Use the terms consistently within the writer's guide recommendations and do not interchange terms.
- Avoid words like 'use', 'attempt', 'review', 'evaluate' without clarifying the information.

Permissive Terms – Use of 'shall', 'should', and 'may' type terms drive users to make inappropriate conclusions due to the use of passive voice.
To Reduce:

- Write in active voice and tell the user the action to take (e.g. use action verbs).
- Describe the action instead of the outcome.
- Convert shalls, shoulds, mays, etc. into the actions required to achieve what is being described.

Multiple or Embedded Actions – Steps or actions that are hidden in notes, cautions, warnings, precautions, or limitations. Any time there are 3 or more actions in a step there is an increase in error rates.
To Reduce:

- Move actions into steps and leave notes, cautions, and warnings as "informational" not operational.
- Put the actions that appear in precautions and limitations into the section of the procedure where the actions should occur.
- Limit actions in a step to one when possible, and at the most two (and only if those actions are closely coupled by the time they need to be performed)

For decades, one of the authors dreamed of a way to automate the review for some of the more common traps. This would reduce the burden on writers and improve the capabilities of the reviewers and approvers to identify error potential before it becomes a problem. The ultimate predicted outcome would be better procedures in the field which drive fewer errors and process safety incidents. The problem was that they did not have the computer expertise or financial wherewithal to develop software that would do the work right.

Value Proposition – The Need

One of the biggest untapped resources in many organizations are the frontline workers. Their knowledge and awareness of how work really gets done in the field is essential input to create guidance that is correct and useable. Over time gaps may arise in how work is planned versus how work is performed due to changes in personnel, equipment, procedures, and processes. This is where engagement with frontline personnel is key so that accepted variability (adaptability) and drift in the work is acknowledged in written guidance. Understanding why certain tasks are difficult or how workers need to adapt should be continually evaluated as part of the procedure review process, especially for procedures that impact process safety. The CAPTUER add-in makes that engagement process easier for both the writers and users of the procedures because it reveals tasks within steps that are either mentally difficult or contribute to error-likely situations.

Top 10 Value Propositions to Focus on Procedure Improvements

1. Correct, readable, and human-factored procedures prevent incidents.
2. OSHA PSM regulation 1910.119 requires procedures to be "correct and accurate".
3. To understand how to practically reduce error rates in the field.
4. To improve teamwork and communication.
5. To have accurate and usable process safety procedures necessary for implementing digital strategies.
6. To learn new and better ways of doing work safely and producing safer outcomes.
7. To train new employees on the right way to do tasks.
8. To increase standardization which leads to more consistent outcomes.
9. Current technologies (such as ChatGPT) cannot write procedures for us…accurately!
10. Procedure writers spend up to 40-60% too much time on review for format and too little time on usability.

Industry spent decades trying to ideate a tool that would result in improving written guidance less reliant on the individual writing, reviewing, and approving procedures, and more aligned with current technological capabilities (and able to grow with the technology). Taking a page from our quality friends, we needed to learn to BUILD quality into our procedures, instead of trying to REVIEW it in. If review was going to work, it would have been working for years!

Developing the CAPTUER Tool

This first-generation tool consists of a Microsoft add-in application that identifies several of the Top 5 Error Drivers in procedures and processes and assigns a performance metric to the document. Over a period of two (2) years, most of the possible combinations of words, phrases, traps, and document designs were fed to people who know how to program Word Add-ins. It was tested against procedures known to contain the error drivers. People would perform manual procedure error driver assessments and review the results of running the tool on the same procedure. In each of the test runs, the tool found the error traps faster (it only takes about 15 seconds to run) and more effectively (the people missed many of the error drivers that the tool identified). Subsequently, after providing the tool to multiple clients, there were identical results. The tool saves time and increases the probability of finding problems that could impact process safety.

The original loading process and criteria for an add-in were fairly complex, but the process has been refined and defined so most people who use Word can perform the loading and use tasks. The tool can be used on any document that is in ".docx" format. Unfortunately, to date we have not been able to crack the Apple™ code, so it only works on PCs and laptops.

At this time this tool has a limitation that it cannot find multiple or embedded actions. This must still be done manually; however, procedure writers tell us, and it has been our experience, that these drivers are relatively easy to find and fix when you don't have to look for and identify the fixes for all of the other error drivers.

CAPTUER Details

Once the add-in is loaded as an add-in to Word it appears in the upper ribbon as an option. Opening CAPTUER reveals the opportunity to "Score" the procedure. In other words, the add-in runs and counts the total number of error drivers in the document, identifies them in the document by type, and flags them with the error driver label. This allows the developer or reviewer to identify and fix the error drivers in the procedure quickly and efficiently.

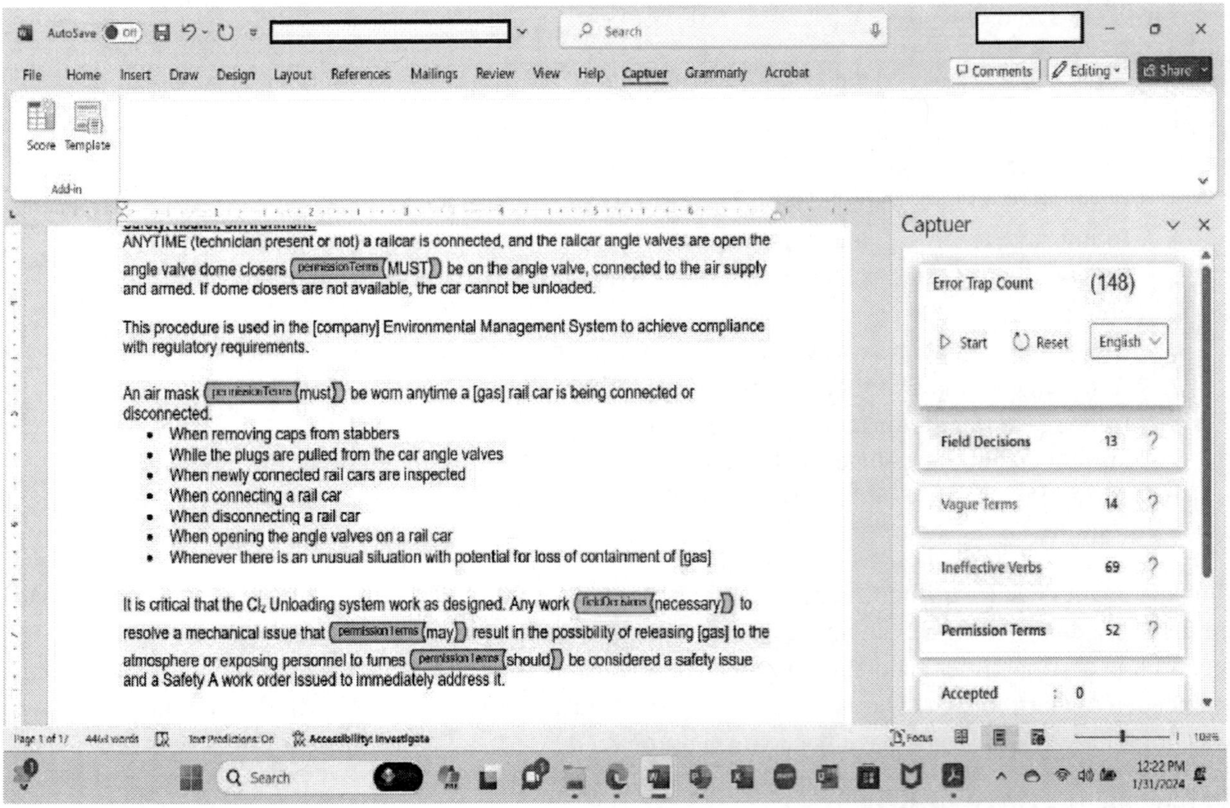

As you can see from the example, this procedure had 148 known error traps or drivers identified. The writer or reviewer can then address each one and determine if the identified and flagged statements either (1) need to be fixed, or (2) are used appropriately and can be disregarded. If it is discovered that one of the identified drivers or trap statements is actually used correctly, the user can remove that phrase from future reviews using the tool. This allows the flexibility to ensure that all are identified, but you can still justify certain uses of terms that would normally be an error driver.

You can use the add-in on any procedure, process, or document in .docx format.

Future generations of the app will utilize natural language processing (NLP), generative AI, and gamification to reward the user for high quality, error-proof procedures. Think of it as your own personal writing assistant which finds errors and offers suggestions to fix them. This add-in tool can be the first step for organizations toward sustainable improvement, usability, and standardization of process safety-related operating and maintenance procedures.

The business case for error-proofing written guidance in procedures goes way beyond process safety…using digital technology and tools are Operational Excellence in nature. Improving the integrity of procedures can save lives, protect the environment, improve Quality, Cost, Reliability, Productivity, and Effectiveness in organizations.

References

1. Deming, W. E., Out of the Crisis, 1982, MIT Press
2. Reason, J., Managing the Risks of Organizational Accidents, 1987, Ashgate
3. Conklin, T., The 5 Principles of Human Performance, 2019, Amazon Press
4. Dekker, S., The Field Guide to Understanding Human Error, 2002, Ashgate
5. Rasmussen, J., & Vincente, K., Coping with human errors through system design, 1988, Riso National Laboratory
6. Fisher R., Understanding Mental Models, 2022, Amazon Press

Improving PHA Outcomes to Better Manage Critical Controls in Mining Industry - from PHA to Verification in the Field

Christophe Catala,
Chief Advisor Process Safety
Rio Tinto
1190 Avenue des Canadiens de Montreal,
Montreal, QC, H3C 1B3, Canada
Christophe.catala@riotinto.com

Laura Anato[1], Luis Carrero[1], Catherine Morar[2]
Principal Advisors Process Safety
Rio Tinto
[1] 1190 Avenue des Canadiens de Montreal, Montreal, QC, H3C 1B3, Canada
[2] 155 Charlotte St, Brisbane City QLD 4000, Australia
Laura.anato@riotinto.com, luisjavier.carrerorivero@riotinto.com,
catherine.morar@riotinto.com

Keywords: PHA, facilitation skills, competency, leadership

Abstract

Process Safety Management in the mining and metals industry is relatively new compared to other high hazards process industries such as oil and gas or chemicals. Practices are less mature and are developing, transferring from other industries, and using International Council on Mining and Metals (ICMM) guidance to manage critical controls.

This paper shares how Rio Tinto, a leading mining and metals company has implemented its Critical Controls Management practice, the challenges faced and recent improvements. The focus is on how the company has improved the quality of its Process Hazard Analyses (PHAs) and critical controls management activities through a series of actions. This covers clarification of the methodology used, development of training packages, revision of the methodology used to identify critical controls, development of a PHA facilitation approval process and a PHA facilitation approval committee, implementation of a training and coaching program for internal candidates and development of assurance activities to monitor effectiveness of the process.

These actions have resulted in very encouraging results in terms of overall quality of the PHAs, effectiveness of the PHA approval process to support the new methodologies introduced and the development of internal facilitation capabilities.

1 Introduction

1.1 What is different about the mining and metals industry when it comes to process safety?

Process Safety Management for the mining and metals industry is relatively new compared to other high hazards process industries such as oil and gas or chemicals. There are no globally recognized major incidents such as Bhopal or Piper Alpha associated with this industry; however, high process safety risks exist and need to be managed, with process safety being the most relevant program to address these.

The mining and metals industry is different in the sense that the process safety risk profile across the sites varies widely (a mine presents typically a low profile versus a smelter) and each site can present a unique technology and type of risk. Also, the industry presents unusual hazards such as molten materials whose production heavily relies on humans to perform critical activities, hence mostly administrative types of controls.

While most of the practices developed for oil and gas and chemicals industries are transferable, there are clear limitations with some practices (e.g., LOPA, lagging performance indicators). The mining and metals industry, while implementing process safety, has not developed customized practices.

1.2 A brief history of Process Safety management within Rio Tinto

Rio Tinto started its Process Safety management journey in the early 2010s with the development, issue, and implementation of our Process Safety Management standard (Reference [3], developed based on international practices issued by CCPS or Energy institute) at 32 of our operating sites across the world. The paper describes some of the challenges identified while implementing the standard since 2015.

1.3 Development of critical controls management process by the ICMM

At the same time as Rio Tinto issued its Process Safety Management standard, the International Council on Mining and Metals (ICMM) issued in 2015 its Health and Safety Critical Control Management Good Practice guide [References 1 & 2]. The intent of this document is to provide the industry with a reference on how to identify and manage critical controls that can either prevent a serious incident occurring in the first place or minimize the consequences if a serious incident were to occur. It provides specific guidance on:

- identifying the critical controls
- assessing their adequacy
- assigning accountability for their implementation
- verifying their effectiveness in practice.

This process is an adaptation to the mining and metals industry of the Safety Critical Elements management process developed in the UK for the oil and gas industry. It is summarized on Figure.1 below.

According to reference [1], a critical control is "*A control that is crucial to preventing the event or mitigating the consequences of the event. The absence or failure of a critical control would significantly increase the risk despite the existence of the other controls. In addition, a control that prevents more than one unwanted event or mitigates more than one consequence is normally classified as critical.*"

And critical control management (CCM) is "*A process of managing the risk of Material Unwanted Events that involves a systematic approach to ensure critical controls are in place and effective*".

Figure 1 Critical Control management process developed by ICMM (extracted from Reference [1]

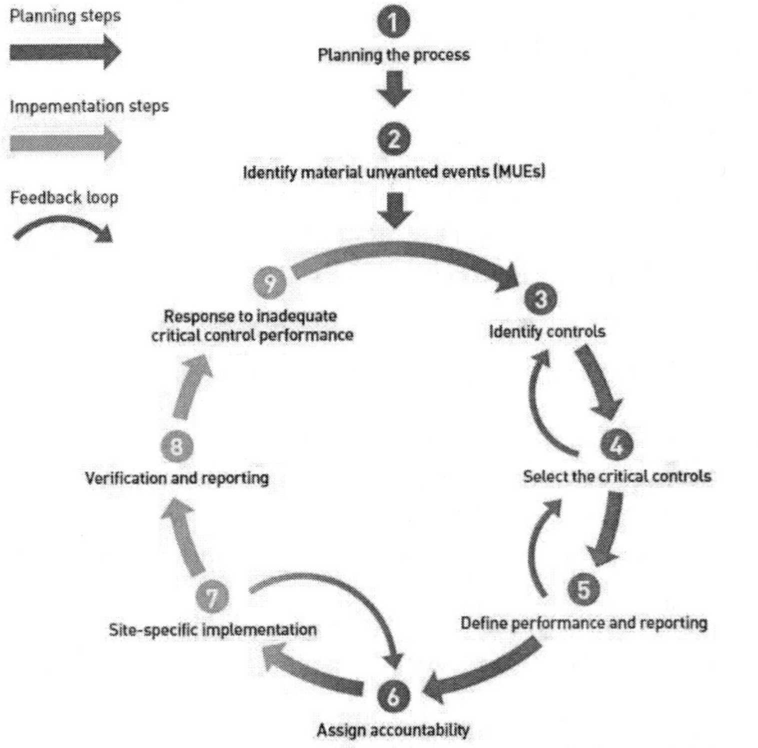

	STEP	TARGET OUTCOME
Planning steps	1	A plan that describes the scope of the project, including what needs to be done, by whom and the timescales.
	2	Identify MUEs that need to be managed.
	3	Identify controls for MUEs, both existing controls and possible new controls. Prepare a bowtie diagram.
	4	Identify the critical controls for the MUE.
	5	Define the critical controls' objectives, performance requirements and how performance is verified in practice.
	6	A list of the owners for each MUE, critical control and verification activity. A verification and reporting plan is required to verify and report on the health of each control.
Implementation	7	Defined MUE verification and reporting plans, and an implementation strategy based on site-specific requirements.
	8	Implement verification activities and report on the process. Define and report on the status of each critical control.
	9	Critical control and MUE owners are aware of critical control performance. If critical controls are underperforming or following an incident, investigate and take action to improve performance or remove critical status from controls.

2 Implementing the critical controls management process

The implementation of the critical controls management process was a key part in the initial deployment of the Process Safety Management standard. The process starts with the identification of what is called a Material Unwanted Event (MUEs) using hazard analysis techniques and then using Bow Tie to identify critical controls.

2.1 Hazard identification and risk assessment methodology used within Rio Tinto

The Rio Tinto Process Safety Management Standard defines the hazards to be included in the management program and an appendix provides threshold quantities that can be used to determine if a given hazard is part of the scope or not. Based on a preliminary hazard identification (HAZID), hazards that would meet our criteria for major hazards are identified followed by hazard analysis to treat risk to ALARP. Then a Bow Tie review is performed to identify risk scenarios and controls put in place to reduce risk.

2.2 Critical controls identification

Once the Bow Tie analysis is completed, critical controls are identified using the requirements from Rio Tinto Process Safety Management Standard (Reference [3]) which are aligned with ICMM guidance (Reference [1] & [2]) and ISO 31000 [6].

The approach is simple and consists of a set of criteria to be met. If:

1) The control by itself stops the threat; or

2) The control applies to more than one threat: or

3) The control is the only control for that threat.

,then, the control is considered critical.

2.3 Critical controls verification in the field

Once critical controls have been identified, the next step is to determine critical controls' objectives, performance requirements and how performance is verified in practice. This is achieved by developing a Critical Activity Data Sheet (CADS) that provides information on the critical control as well as performance requirement and responsibility for verifying performance of the control. An example is provided in Figure.2 below.

Figure 2 Example of Control Activity Data Sheet (CADS)

RioTinto		CADS # 160
PSM CONTROL ACTIVITIES DATA SHEET (CADS)		
PSM Hazard	Combustible gas	**Hazard Owner: Fred**
Critical Control	**Gas detection with interlock**	
Risk Scenario	LOPC combustible gas	
Consequence	Gas explosion	
Critical Control Owner	Maintenance Superintendant - Plant B	**Bob**
Control Objective *(What the control is expected to do?)*	To detect gas release inside building and shut-down the process (source of gas) before it accumulates at the LEL	

N°	**Activities required to achieve the control objective**	**Done by**	**Verification methods and frequency of the activity**	**Verified by**
160.1	**Calibration of gas detectors**			
160.11	- Plant A: calibration of fixed gas detectors as per SAP PM # 552233 (includes detectors #A33, A34, A35, A36)	Electro-Technician	- Confirmation of work executed as per procedure and specs. - Review of calibration report - Confirmation of detectors in working condition 1 month	Mtce Supervisor (Donald)
160.12	- Plant B: calibration of fixed gas detectors as per SAP PM # 448855 (includes detectors #B33, B34, B35)	Electro-Technician	- Confirmation of work executed as per procedure and specs. - Review of calibration report - Confirmation of detectors in working condition 1 month	Mtce Supervisor (Rod)
160.2	**Functional testing of the interlock shutting down the process in the event of gas detection**	Electro-Technician		
160.21	- Plant A: Functional testing of the interlock shutting down the process in the event of gas detection at 100ppm from either detectors #A33, A34, A35. Test as per SAP PM#669988		- Confirm the process shuts down required equipment on a 1000 ppm gas detection. Refer to functional description #A11 for specifications. - Review functional test reports 6 months	Mtce Supervisor (Donald)
160.22	- Plant B: Functional testing of the interlock shutting down the process in the event of gas detection at 100ppm from either detectors #B33, B34, B35. Test as per SAP PM#784512		- Confirm the process shuts down required equipment on a 1000 ppm gas detection. Refer to functional description #B11 for specifications. - Review functional test reports 6 months	Mtce Supervisor (Rod)

Additionnal information :	
Performance Criteria:	
Green	Inspection/tests completed as per schedule. 100% Detectors calibrated within specs and functional. Functional testing successful. Evidence available of the inspection/tests.
Yellow	n/a
Red	Calibration/testing not performed, Or 1 detector or more not calibrated, OR functional testing failed at shutting down required equipment.

Clear responsibilities are assigned for the management of the hazard and verification happens in the field to verify the performance of the control remains as expected.

The performance of each control is then reported using a simple color code (green, yellow, or red). This is then reviewed at various levels within the organization and actions are taken should a critical control not be meeting its performance criteria.

2.4 PHA facilitation

Rio Tinto standard (reference [1]) requires "*Facilitators of process hazard analysis must be deemed independent and competent in the application of the approach and capable of application in the hazard environment under consideration*".

2.5 *Identification of areas of improvement for the initial approach*

During the first years of implementation (2015 to 2019) of this program, focus was on getting quick results and getting the basic standard implemented. The next round of improvement (since 2020) focused more on developing guidance and reaching sustainable practices. The areas of improvement identified are summarized in Table 1 below together with the proposed improvements.

Table 1 Key areas of improvement identified with the initial practices.

Topic	Potential improvement	Solution identified
PHA methodology	• No group guidance and specific requirement for the performance of process hazard reviews • Variability in hazard identification practices across the group • How to decide which hazard analysis technique to be used versus complexity of the process	• Issue and implementation of 2 group procedures to specify requirements for HAZID, HAZOP and LOPA reviews performance. • Development of training packages to clarify roles and responsibilities, expectations and provide templates and examples of what good looks like
Critical controls identification	• Identification process to suit specifics of process safety and aligned with recognized practices[1] (e.g., Independent Protection Layers) • Too many critical controls identified and verification activities overload in the field.	• Preparation of a critical controls identification flowchart and alignment with CCPS guidance on identification of Independent Layers of Protection [4, 5]
PHA facilitation	• Variability in the experience and training of facilitators	• Implementation of a PHA facilitation approval process • Training and coaching program for new candidates • Implementation of a PHA facilitation approval committee • Development of assurance activities, increased support to sites and regular review of PHA reports

Topic	Potential improvement	Solution identified
Verification in the field	• Verification focused mainly on compliance and less on effectiveness	• Definition of group controls library and group controls performance criteria and performance reporting[1] • Review and improvement of front-line verification activities[2]

3 Implementing changes to the critical controls management process

While the overall critical control management process presented in Figure.1 has remained unchanged, the improvements presented in Table.1 have been implemented and are detailed in the following sections.

3.1 Defining PHA methodologies requirements

Two group procedures have been developed and issued to define the specific requirements associated with the performance of Hazard Identification (HAZID), Hazard and Operability reviews (HAZOP) and Layers of Protection Analysis (LOPA).

These have been aligned with best industry practices, especially introducing semi-quantitative techniques for the performance of HAZID and HAZOP. This is a highly rigorous approach to risk assessment that allows to determine directly critical controls and their theoretical effectiveness on reducing risk during the review.

An example of HAZOP template is provided in Figure5 below.

[1] Not detailed in this paper.

Figure 3 Example of HAZOP review with new methodology

Node: Boiler 5 Outlet

Drawings / References: 06072-M-0104, Sheet 2

Design Conditions/Parameters: Outlet Temperature - 350 F
Outlet Pressure - 300 psi
Outlet Piping Size - 10 inch

Deviations	Causes	Consequences	CAT	Inherent Risk			Effective Controls	CAT	RRF	Residual Risk			Recommendations		Incidents	Remarks
				S	L	RR				S	L	RR	Recommendations	RRF		
1. High Pressure	1. Closed or partially closed valves on 10" Boiler 607-BLR-5 outlet	1. Restricted or no flow leading to high pressure in Boiler 607-BLR-5, resulting in overpressure and equipment damage.	PER	4	C	Critical	1. Pressure relief system composed of 2 x PSVs	Preventive	100	4	E	High	1. Implement periodic testing of safety functions identified in the HAZOP to ensure safeties will be available as required and provide functionality status to the operator.	5		
							2. Boiler Trip: i) boiler high pressure; ii) boiler high temperature; iii) boiler outlet low flow	Preventive	5				2. Install a boiler high water pressure alarm to alert operators and provide adequate reaction time for operators to take action to avoid a high high pressure condition and potential rupture at electrical boilers (Boilers 607-BLR-1 and 607-BLR-5).	5		
							3. Operator intervention to regulate boiler outlet flow;	Preventive	10							
2. No / Low Pressure	1. No credible causes identified															

A similar approach has been used with the HAZID methodology; however, the structure of the spreadsheet is aligned with the structure of a Bow Tie (Hazard, top event, threat, consequence, controls). The aim being to optimize efforts with the development of Bow Tie post review.

Figure 4 Example of HAZID review with new methodology

Hazard category	Hazard	Hazardous scenario	Central event	Causes	Consequence	Operational state	CAT	Inherent risk				Controls				Current risk				Recommendations	RRF	Residual risk				Incident	Remarks
											RR	Controls	Type	CAT	RRF				RR						RR		
Pressure	Vapour under pressure (in the oil boiler)	Generation of steam that exceeds the design pressure of the boiler	C. Loss of containment	1. Manual valve closed at boiler water outlet	Generation of steam that increases the pressure inside the oil boiler above its design pressure leading to the burst of the vessel and possible fatal impact on personnel	Normal operation	HBS	5	D	5D	CRITICAL	Valve position is reported in the Scada system for the operator to move to the boiler room and change the position of the valve	Administrative	Prev	1	5	E	5E	HIGH								
											#N/A	Low water level sensor [illegible] which stops the boiler upon reaching a water level of [illegible] No credit given to this control as it is on the same controller as the pressure control which has been given credit for risk reduction	Engineering	Prev	1				#N/A								
											#N/A	Pressure safety valve set at 7.5 barg to open	Engineering	Prev	100				#N/A								
											#N/A	High pressure sensor PI112 (9532) which stops the boiler in case pressure reaches 7.5 bars	Engineering	Prev	10				#N/A								

A specific guidance note has been developed on how to perform the risk assessment with these methodologies and in particular which data to use to determine initiating event / threat frequency and control effectiveness as well as the rules to be followed by the review team during the sessions.

Finally, a LOPA guidance has also been developed to determine the specific requirements to apply the method within Rio Tinto supported by the approval of group acceptance targets. The LOPA would be used where required to complement the HAZID or HAZOP and determine with more accuracy the likelihood of selected events.

3.2 Changing the approach to identify critical controls.

Methodology to identify critical controls (see Figure.7) has been refined and aligned with recognized practices in the process safety area, i.e., the ones defined in CCPS books (Reference [4&5]). This ensures consistency between any HAZID/HAZOP and subsequent LOPA review and allows reference to international practices.

Focus has been put on ensuring that any control claimed as critical meets the criteria for Independence, Functionality and Auditability. Also, considering the reliance on administrative controls within the Mining and Metals industry, specific guidance has been developed on how determine if such a control can be claimed as critical (Reference [5]). This considers the following:

- Presence of a written operating procedure
- Clear means to detect a problem and clear indication even in emergency situations. The operator is fully available and is not flooded with alarms and he/she can clearly distinguish the alarm among others.
- The information displayed to the operator is easy to understand / analyze and decide. There is no need to do calculations or use graphs to respond when quick action is required.
- Operator clearly knows what action needs to be taken according to the situation.
- Operator has been trained on how to perform the task with regular, documented, drills/tests.
- All needed materials or tools for the task are available to the operator including appropriate personal protective equipment (PPE) and the human factors associated with the action such as communication or work environment are properly addressed.
- Operator has sufficient time to observe the condition or alarm, to diagnose problems and analyze what should be done, and to correctly perform the task. It is considered as a minimum that operator must be able to complete the action in half the time required for the consequences associated with the process safety scenario to occur.
- The operator response is tested or audited regularly to ensure the performance of the control is maintained over time.

Figure 5 Revised flowchart to identify critical controls.

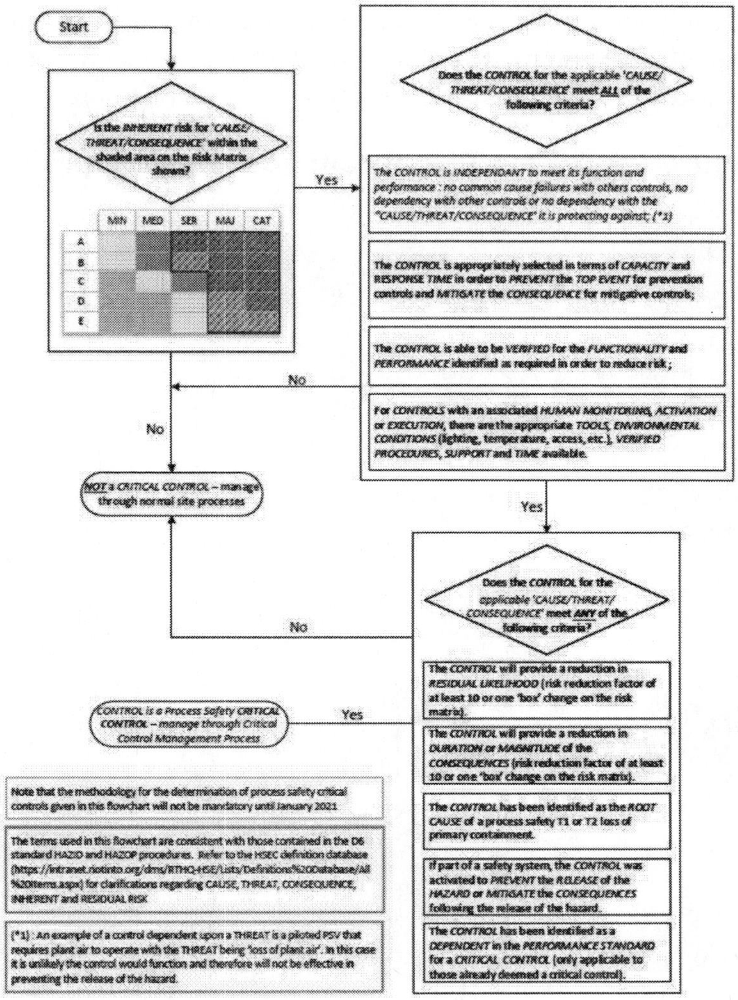

3.3 Improving PHA facilitation leadership and skills

The introduction of new PHA methodologies was an important change for the organization and improving facilitation skills was a critical success factor to reach a consistent and sustainable quality of the PHAs.

We have been working on implementing a robust PHA facilitation approval process that would support the quality of the PHA and the development of facilitation of the skills of our employees. For this we have set up a program that includes:

- A PHA facilitation approval process with
    - a PHA facilitation approval committee
    - a properly maintained and accessible list of approved PHA facilitators
- A skills development program for Rio Tinto internal candidates
- An assurance program to monitor the effectiveness of the program.

3.3.1 Setting up a PHA facilitation approval process

With the new PHA methodology being implemented we wanted to set up a clear process for internal and external candidates to be approved, using transparent assessment criteria, and recognized external assessors. The process is described on Figure.8 below.

Figure 6 Rio Tinto PHA facilitation approval process

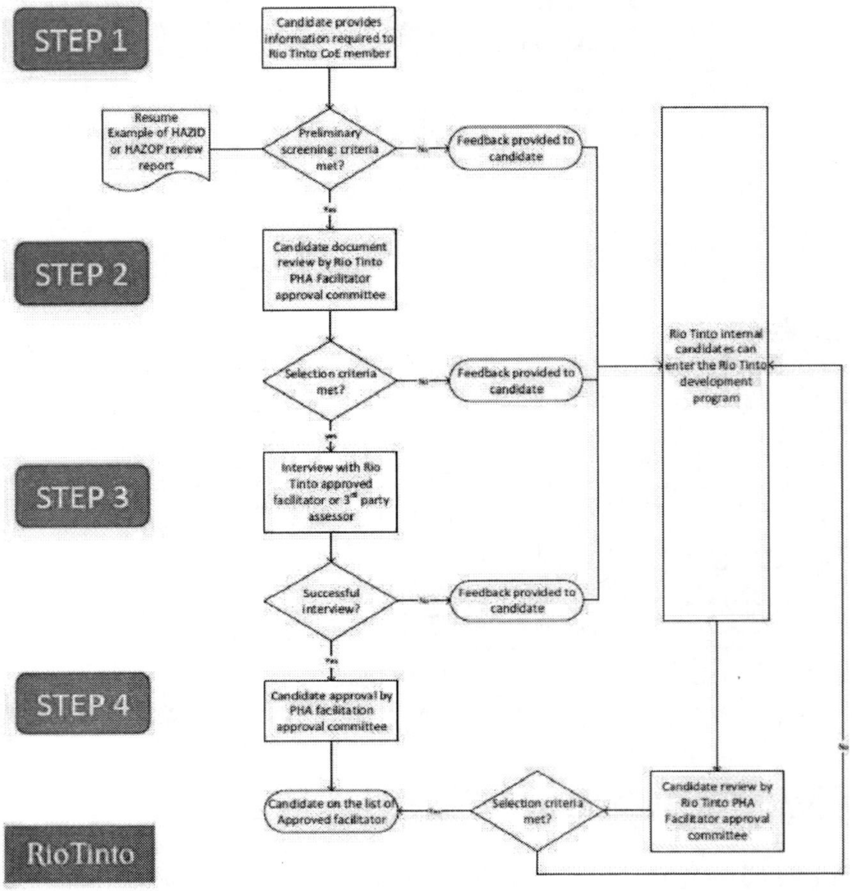

To initiate the process, a candidate must provide minimum information to a member of the Center of Excellence (CoE), and must satisfy basic requirements (minimum years of experience, evidence of formal training in risk assessment techniques, evidence of PHA reviews facilitated, example of report of PHA review facilitated). At any stage, internal candidates that do not meet the minimum criteria can enter the Development Program (see next section) the other ones move to the next stage. Successful candidates after review of a sample of PHA report facilitated by the candidate, will move to the interview stage for final assessment.

The interview is conducted by a member of the PHA facilitation approval committee and an external professional recognized by the IChemE, thus providing an important external independent view.

During the interview, candidates are assessed on 5 topics:

1. *General knowledge of PHA methodologies:* how to select a method, purpose, typical documents required, nodes definition, general experience.
2. *Technical knowledge with a focus on applying Rio Tinto methodology:* how to properly define a cause, describe a consequence, a control, how to identify a critical control, etc.
3. *Report writing:* content of the report, quality check.
4. *Review preparation*
5. *Facilitation skills*

The interviewers provide the committee with a recommendation. A candidate can be:

- *Fully approved* for HAZID and/or HAZOP
- *Partially approved*. Meaning that gaps might have been identified during the interview, but candidate can facilitate and will generally have to provide his next report for review by the committee who will decide the way forward.
- *Not approved*. Feedback is provided to the candidate with proposed action to improve.

Once approved by the committee, a successful candidate is added to the official list of Rio Tinto Approved Facilitators for HAZID and HAZOP.

Some rules have been defined with the process:

- Members of the Rio Tinto PHA facilitators approval committee are approved by 3rd party assessor and qualifications are reviewed regularly (24 months)
- Approved facilitator performance is reviewed regularly by the PHA facilitator approval committee. Under performing facilitators will:
  - Have a performance review feedback session with a member of the committee.
  - Be removed from the list if no improvement observed.
- Candidates that are approved but have never applied a methodology like Rio Tinto's in the past will be required to provide the report for review by the PHA committee who will provide feedback and confirm approval.

3.3.2 Setting up a PHA facilitation approval committee

It became clear that the approval process needed some governance to be effective and the decision process required more transparency, clarity, and effectiveness to be accepted by the organization. A PHA facilitation approval committee was then set up to meet these objectives. A charter, approved by Leadership defined accountabilities of the committee:

- Review of proposed candidates (internal and external). The committee decides, informs, and documents any decision using the set of selection criteria defined by the process safety group.
- Monitor internal candidates under approval and development (progress of coaching). Collect feedback on difficulties faced and identification of actions to overcome these.
- Review periodically the performance of currently approved facilitators and take decision to maintain or remove them from the list.

- Review a set of performance indicators on topic to monitor performance of the facilitator approval process (duration of the approval cycle, number of approved facilitators per geographic area, etc.)

The committee meets monthly but can meet ad hoc if requested by one member (e.g., need for a quick approval) and is composed of:

- Chief advisor process safety (chair of the committee).
- Members of the Processing Center of Excellence (CoE) process safety team that have been approved.
- A maximum of one approved facilitator per business unit.
- An independent Rio Tinto employee from the in-house Technical Excellence team that has no direct interest in the process and who will provide an independent view and challenge on the proposed decisions.
- An independent expert recommended from a recognized institution (IChemE in our case).

All meetings of the committee are recorded, minutes of the meeting issued, and results of the discussions recorded in a master file to monitor progress of the various candidates.

3.3.3 Skills development (training, coaching)

As can be seen in figure.8 above, any internal candidate that does not meet the approval criteria can enter into a PHA facilitation development program. The overall program is presented in Figure.9 below.

Figure 7 Rio Tinto PHA facilitation development program

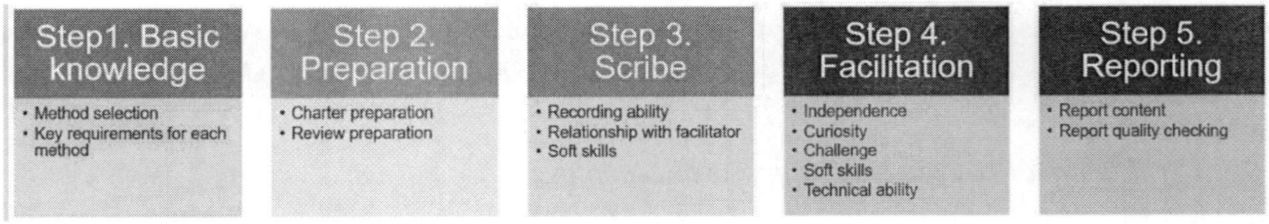

Each candidate is different, has different initial experience and might start from a different stage in the process. For each candidate, a coach (internally approved facilitator) is assigned. During the initial session, the coach will assess where the candidate sits within the process and will recommend a development program that suits the candidate's needs.

The Coach will then have an individual and personalized follow-up with the candidate, providing regular feedback and advice.

Let us take a fictitious example of a candidate. Andy is new to a process safety position and has never facilitated HAZID or HAZOP review in the past, has no knowledge of HAZID or HAZOP methodologies, has 10 years of experience overall. His proposed development program is provided in Figure.10 below.

Figure 8 Example of development program for a new candidate

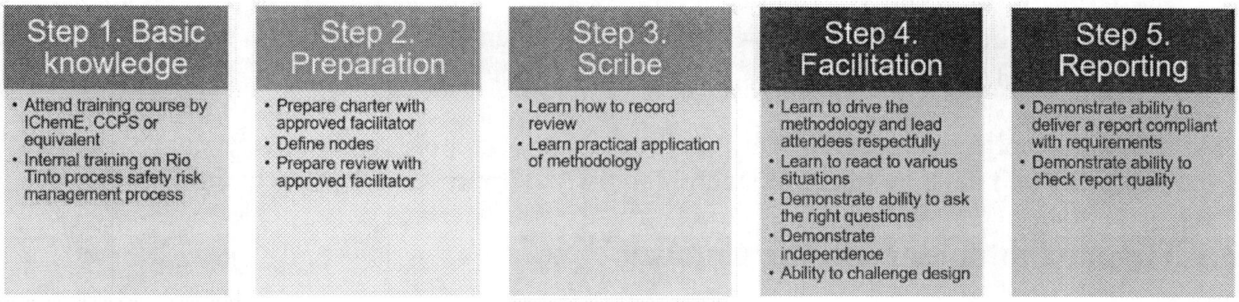

Steps 1, 2 and 3 are straightforward. Andy will be proposed to move into step 4 when scribing capabilities are satisfactory and method is properly understood (Step 3 might require several PHAs before moving into Step 4). Steps 3 and 5 are combined. Under Step 4, the candidate will facilitate under the supervision of an approved facilitator who will coach live. Once the candidate is comfortable with all the steps (this might require several PHAs), the coach will propose the candidate for final approval to the committee and will follow the approval process described in Figure.8.

In addition to this, we have developed an internal training package to:

- Better explain our methodology and clarify some methodological aspects (e.g., how to mitigate consequence in each scenario)
- Provide resources to facilitators, including:
  - Review report templates
  - Opening meeting templates and examples
  - PHA readiness checklist
  - PHA review charter template and examples
- Explain roles and responsibilities for the various participants.
  - PHA coordinator
  - PHA participants
- Provide examples of PHAs reviews to identify what good looks like

This training is currently in the form of Power Point materials but will turn into online sessions during 2024.

3.4 Assuring the process

To monitor the effectiveness of the process, an assurance program has been implemented that includes:

- Ad hoc reviews
  - Regular reviews of the PHA reports by an internally approved facilitator.
  - Review of PHA reports as a follow up for partially approved facilitator.
  - Attendance at PHA meetings by an internally approved facilitator to
    - Support a candidate during a review.
    - Monitor performance when concern is raised to the committee.

- First and second line assurance activities
  - Review of PHA reports during compliance audits, technical risk reviews.
- Incident investigations to check whether incidents scenarios were properly identified during PHAs.

All information collected is reviewed by the Processing Center of Excellence team and shared with the community and the committee to identify additional improvement opportunities.

3.5 Timeline to implement the changes.

Implementing these changes takes necessary time and doesn't appear overnight. We initially started with the first changes in 2020 (issue of the group procedure and the new methodologies) and we have defined a 5-year period to implement the proposed changes and regularly monitor the performance.

The 5-year period allows us to embed the changes into the PHA revalidation cycle.

4 Results observed.

The program has shown very interesting results to date.

- Successful development of internal candidates through the development program
  - Up to 10 internal candidates have been coached and developed through the program and are now on the list of approved candidates.
  - These internally approved facilitators are now coaching other candidates.
- Effectiveness of the PHA approval process
  - The committee has demonstrated response to specific needs, in particular, for Projects with short term needs.
  - Anticipation of needs in remote areas and areas of skills shortage (e.g., Mongolia, South Africa, New Zealand).
  - Decision making by the committee, in particular, to review PHA reports issued by already approved facilitators when concerns are raised by sites around the quality.
- Overall quality of PHAs has improved, resulting in:
  - a better understanding of risk
  - a better understanding of critical controls in place
  - definition and implementation of risk reduction measures that really reduce risk.
  - a drastic reduction of critical controls verification in the field (in some cases the application of the new methodology has resulted in a reduction of up to 40% in the number of critical controls identified).

5 Looking forward

Despite the encouraging results observed with the implementation of the changes, several challenges are ahead to us to seek consistent quality with the PHAs.

- Employee turnover, with people moving out of a position requiring PHA facilitation capability or exposure to PHA reviews.
- Put more focus on incident investigations to understand if the scenario was previously identified in a PHA.
- Improved connection with consulting companies / external facilitators to ensure the methodology is correctly understood and applied.

We will continue in the future to implement the plan and sustain this process over time. Keeping all stakeholders involved will be important, whether in operations or projects, and making sure they understand the importance of the PHAs and apply rigorously the process to ensure we design and operate safe facilities.

6 Conclusion

PHAs are an essential component of the Process Safety management system, deficiencies in the conduct and performance of these reviews can lead to scenarios not being identified and thus managed properly. Within Rio Tinto, the Process Safety management system started to be implemented in 2016 using the ICMM guidance on Critical Control Management as a reference. The initial focus was to identify key hazards and identify critical controls to be managed in the field. A combination of HAZID and Bow Ties was used to identify process safety risks.

This paper presented the improvement actions that Rio Tinto has implemented to enhance the critical control management process with a focus on reaching a more consistent quality of the PHAs that is the critical step to ensure:

- The right level of risk is assessed with an adequate number of critical controls.
- Real critical controls are identified and verified in the field.

An improvement plan has been developed and implemented since 2020, included the following and resulted in very encouraging results:

- Clarification of the PHA methodology:
    - Overall, the quality of PHA has improved.
    - a better understanding of risk
    - a better understanding of critical controls in place
    - definition and implementation of effective risk reduction measures.
- Revision of the methodology to identify critical controls:
    - drastic reduction of critical controls verification in the field (in some cases the application of the new methodology has resulted in a reduction of up to 40% in the number of critical controls identified).
- Development of PHA facilitation skills:
    - Successful development of internal candidates through the development program
    - Good effectiveness of the PHA facilitation approval process

We are conscious that despite these actions, there is still a way to go, and challenges are ahead of us in particular:

- Coping with turnover of personnel.
- Putting more focus in incident investigations on whether the incidental scenario was identified in a PHA.
- Connection with consulting companies / external facilitators to ensure the methodology is correctly understood and applied.

We trust that the current program brings us in the right direction and will give in the coming years a more consistent quality of PHA to better manage process safety risks.

7 Acronyms used.

AIChE – American Institute of Chemical Engineers
CADS – Control Activity Data Sheet
CCM – Critical control management
CCPS – Centre for Chemical Process Safety (of AIChE)
CoE – Centre of Excellence
HAZID – Hazard Identification - a hazard identification tool
HAZOP – Hazard and Operability Analysis – a hazard identification tool
ICMM - International Council on Mining and Metals
LOPA – Layer of Protection Analysis
MUE – Material unwanted event
P&ID – Piping & Instrumentation Diagram
PHA – Process Hazard Analysis
PPE – Personal Protection Equipment
PRV – Pressure Relief Valve
PSM – Process Safety Management
UK – United Kingdom

8 References.

Book:

[1] Health and Safety Critical Control Management. *Good practice guide*. International Council on Mining and Metals; 2015.

[2] Critical Control Management. *Implementation guide*. International Council on Mining and Metals; 2015.

[3] *Process Safety Management Standard (D6)*. Rio Tinto; 2015.

[4] CCPS, *Layer of Protection Analysis, Simplified Process Risk Assessment*, American Institute of Chemical Engineers, New York, New York, 2001.

[5] Center for Chemical Process Safety. *Guidelines for initiating events and independent protection layers in Layer of Protection Analysis*, New York, 2015

[6] ISO 31000, *Risk management guidelines*, Second edition 2018-02, Vernier, Geneva, Switzerland, 2018

Cooking up the Ideal Recipe for a PHA Facilitator

Alison Ballon
AcuTech Group, Inc.
1919 Gallows Rd, Vienna, VA 22182
aballon@acutech-consulting.com

Sam Aigen, Nicholas Francis, Alyse Keller, Amanda Lanson, and Nick White
AcuTech Group, Inc.

Keywords: PHA, HAZOP, LOPA, Facilitator, Process Safety Management

Abstract

As with any recipe, it is necessary to start with the right ingredients, and this paper will share case studies of attributes of effective PHA facilitators:

- **Operational Experience**: Operational experience helps the facilitator to draw a distinction in the group discussions between what is realistic and what is not. It is important a facilitator knows when they are not the "expert in the room" and when it is appropriate to ask "stupid questions" to drive the team to focused, beneficial conversations. This is part of leveraging operational experience to bring more flavor to your recipe.
- **Conducting a Certain Number of PHAs a Year:** It is beneficial to conduct a minimum number of PHAs annually to maintain and continually develop their skills.
- **Hazard Modeling Experience:** Understanding modeling and potential hazard extent helps the facilitator picture the variables that can impact the exposure, like tasting your cooking to know an ingredient is missing.
- **Management Experience:** Reports and recommendations are reviewed by management and ensuring the text is written to resonate with its audience, and that practical risk management solutions are identified to impact change is a crucial ingredient.
- **PHA Preparation and Organization:** Just like a chef cooking a meal, a PHA facilitator needs to make sure they are properly prepared for the PHA. With just one wrong ingredient, like missing process safety information, the study may be delayed or worse.
- **Essential Soft Skills:** Lastly, the true recipe for success requires PHA facilitators to have several soft skills such as understanding personalities, ability to be patient and engaging, and a general desire to have safe operations.

1 Introduction

Is it okay for an engineer with less than three years of experience to facilitate my PHA? What about an engineer with fifteen or more years of experience? Does a facilitator even need to be an engineer, or could they be a highly experienced operator? The answer is it depends. It depends on the kinds of experiences that they have up to now, their understanding of very basic technical process concepts, and the complexity of the study.

A further question is, how do I know I am ready to lead my first PHA? Some corporate PHA guidelines make it difficult for a new team leader to facilitate their first study. Can we help those people get their feet off the ground?

PHA Team Leaders can come from a wide range of fields of expertise, amount of training, years of industry experience, and other personal factors. An effective PHA Team Leader is an essential component of a successful PHA study team. Which begs the question: what are the factors that help create a truly effective PHA team leader? This paper will examine the recipe to become an effective and well-rounded PHA facilitator. The paper will also discuss how companies can help develop their team leaders to achieve the minimum required skills to begin facilitating PHAs, while ensuring efficiency and effectiveness.

As with any recipe, it is necessary to start with the right ingredients. For PHA facilitators and teams, you need to start with the right set of skills, experience, and expertise. There are a variety of skills that one can attain to become an effective PHA team leader. This paper will share insight into how these aspects can help make PHAs successful and suggest practical approaches to developing team capabilities.

2 Operational Experience

Have you ever tried to develop your own recipe? Have you thrown dinner together by looking at what's in the fridge and whipping up a meal? That task is much easier when you have more experience cooking, understanding flavor profiles, and how one "pinch of salt" can really make a dish. The same principle applies to operational experience in connection with process hazard analysis. It is important for a facilitator to understand the difference between the worst-case ***conceivable*** consequences and the worst-case ***credible*** consequences. A PHA full of the former often ends up in management's cross hairs for inconsistency and unrealistic recommendations, much like your first attempt at making a stromboli from scratch ends up in a trash can. A PHA full of the latter, is realistic, based in fact, and can easily be defensible and auditable. The Occupational Safety and Health Administration's (OSHA) Compliance Guidelines and Recommendations for the PSM Standard state that a PHA is focused on "analyzing potential causes and consequences of fires, explosions, releases of toxic or flammable chemicals and major spills of hazardous chemicals." However, OSHA does not give additional guidance and instead points toward additional "Sources of Further Information." [1] This highlights the importance of the PHA facilitator being able to delineate between the worst-case credible consequence instead of simply producing the worst-case conceivable consequence.

Operational experience gives the facilitator the perspective to understand when to challenge the team and when, or how, to ask follow-up questions. Facilitators do not need to be experts in the chemical process of every study; however, facilitators need to understand basic principles and be able to apply their own past operational experience to effectively guide the team. This experience will help drive conversations that adequately address the scenario under consideration. With support from the facilitator, the team will be able to address the fact that not all potentially hazardous scenarios end in credible catastrophic consequences. Similarly, the facilitator needs a basic background in operations/thermodynamics to understand when the team may be missing a potential consequence. For example, when examining the scenario of a tower level control valve failing open to downstream low-pressure storage, a team may arrive at the consequence 'operability issue only'. A facilitator with an understanding of operations can question the team to also consider if an overpressure scenario in the downstream low-pressure storage is credible due to gas blowby from the tower. This operational experience is further useful to know how to navigate the conversations when participants want to bring mitigation solutions into the cause/consequence discussion, as mitigation measures should only be addressed after the cause/consequence pairs have been defined.

Operational experience alone does not make you an expert in all processes and equipment. It is equally important for the facilitator to be aware of what they do not know. This means that they acknowledge when they do not have familiarity with a given piece of equipment or situation and when to ask for additional expertise from the team, or the need for additional subject matter experts. For example, it is usually a good idea to call the rotating equipment expert when discussing consequences of a major pump or compressor failure.

3 Conducting a Certain Number of PHAs a Year

Have you ever noticed that the second time you try a new recipe, it tastes different than the first? By the time you have tried the recipe 5 or 6 times, you usually reduce the variability between batches. PHA facilitation is often the same where continued opportunities to conduct and lead sessions, or a specific job assignment, can provide important opportunities to maintain and continually develop facilitation skills.

Conducting a minimum number of PHA per year can provide tangible benefits for consistency. For example, the language used to describe a potential overpressure scenario can vary widely, but it is important to establish and maintain consistency in describing these types of critical process safety scenarios. To continue with the storage overpressure example, it could be beneficial to establish personal "templates" to refer to when describing credibility of potential overpressure to the ASME designated Maximum Allowable Working Pressure (MAWP).[2] For this specific example, you may have a template that states "Scenarios <1.5x MAWP, assume no hazardous consequences, scenarios 1.5-2x MAWP assume leaks from flanges/gaskets/seals, and scenarios >2x MAWP assume vessel rupture." This language can be applied across studies to ensure consistency. This serves as one example of how a PHA facilitator can establish consistent language when describing specific scenarios. It ensures that future auditors will be able to easily see that the scenario aligns with established industry knowledge. The minimum number of PHAs per year may vary by company and role, developing facilitators should seek out guidance from more experienced facilitators to determine the frequency that will most benefit their growth

Regularly practicing facilitating PHAs will help a new facilitator process information, respond, and record the information for the study with the speed and thoroughness needed.[3] Those are not typical activities that occur when conducting other job functions, such as leading an audit or performing a dispersion analysis. Facilitators that regularly conduct PHAs can continually refine the unique skills and best practices that make a polished facilitator, like a chef making an award-winning recipe after thousands of attempts. Such facilitators can come out of each PHA with more knowledge of the unique circumstance studied. Even better, if this knowledge is shared properly, it can increase the skillset of other PHA facilitators within their company or organization. This can also help with company standardization and recognition of opportunities to update procedures and streamline workflows to help improve PHA processes.

4 Hazard Modeling Experience

Have you ever tried to spread flour across your kitchen counter to make homemade bread? Understanding how the flour will disperse helps to minimize potential catastrophes in your kitchen but is also needed to make a rich, tasteful dough. Knowledge of dispersion modeling provides similar benefits in PHAs regarding understanding the actual, credible consequence in a specific PHA scenario. PHA teams often consider a simplified ultimate consequence for loss of primary containment (LOPC) events for overpressure scenarios. However, it is possible for the severity of the consequence to be relatively minor based on pressure, equipment spacing, and other factors as determined by detailed modeling. This simplified approach to ultimate consequences can drive PHA teams to conservatively recommend additional safeguards or layers of protection to reduce risk or close identified gaps that may not exist.

A more effective closure of these gaps could be recommending hazard modeling to better understand the potential consequences. Normally in a HAZOP, risk is reduced only by lowering the likelihood of an event. Consequence modeling can lower potential severity by portraying a more accurate representation of the hazards, impacting the scenario risk level. However, due to cost and time constraints, it is typically not feasible to perform an advanced consequence modeling on every hazardous release scenario. A team can still minimize their dependence on assumptions/opinions by performing these detailed calculations in a more generic manner. This can be done by using group experience or industry databases to agree upon or calculate a hole size for a given scenario. These databases may include the United Kingdom Health and Safety Executive (UKHSE) Hydrocarbon Release Database (HCRD) or 'The Purple Book' by the Ministry of Housing Spatial Planning and the Environment (VROM) of the Netherlands. Then a series of modeling can be performed for various hole sizes at various operating conditions to identify the release quantity over a given period.[4] These results can provide more realistic consequences rather than relying on excessively conservative assumptions. Consequence modeling is very useful as it provides a more quantifiable way to assess the severity of different release scenarios.

5 Management Experience

Have you ever wondered why an organic ingredient is worth the higher cost or when a store brand ingredient will do the trick without compromising flavor? Understanding management's perspective in looking at hazard reviews and recommendations will help a facilitator calibrate the way a project is facilitated, increasing the chances that risk reduction measures will be accepted and ultimately implemented. Many recommendations developed in a PHA have an associated cost. Some are simple, such as the cost for drafting and drawing updates. They may be more complex, such as equipment rental to install car seals on already open valves to prevent mis-operation. And some are truly complicated like the addition of a relief valve or addition of safety instrumented systems, which also increase operational expenses (new inspections and testing for relief valves).

Facilitators with an understanding of how management may view the team's PHA results and recommendations can often write the recommendation in a realistic and practical way the first time. A list of feasible recommendations that show an emphasis towards responsible expenditure of resources (personnel efforts and funding) encourages management to accept the recommendations instead of rewriting. Understanding management's perspective allows the text of a recommendation to better communicate the risk and potential benefits. Much in the same way as you can proudly show off your guacamole recipe with organic avocados and store brand onions.

6 PHA Preparation and Organization

Have you ever started a recipe but realized halfway through that you are missing an ingredient because you did not check the pantry before starting? Similarly, in a PHA, a PHA facilitator must ensure the team is prepared for the study and has the necessary and expected information ready. It is also important to check the expiration dates of the food in the pantry and, in the same way, the accuracy of the process safety information (PSI). Having out of date PSI can make for incorrect results/assumptions or extend a PHA session as participants attempt to chase down the right information, which is just as frustrating as having to run back to the store in the middle of a recipe.

Before the first PHA meeting can begin, the PHA facilitator must ensure that all relevant PSI is available for the team to reference during the study, is up to date and reflects current facility operations. OSHA's PSM Standard states that PSI includes information on the hazards of the highly hazardous chemicals used or produced by the process, as well as the equipment and technology used in the process.[1] Important questions a facilitator should ask include: Do you have access to the process' drawings or diagrams, such as Piping and Instrumentation Diagrams (P&IDs) and/or Process Flow Diagrams (PFDs)? Have you confirmed that these drawings depict the current state of the process?

Another critical item to have prepared is the relevant relief system design and design basis documentation. If your process includes a relief device, a well prepared PHA facilitator ensures the team has access to information relating to each device's design basis. Without being able to confirm the design basis for a relief device, your team would not be able to credit those relief devices as effective safeguards. Confirming that you and your team have access to the necessary and current process safety information will ensure that the PHA runs smoothly, avoiding the need for study pauses and delayed time confirming data are accurate.

7 Essential Soft Skills

Have you ever tried to make a roux by following the recipe exactly and found it did not come out "just right"? The intangible knowledge and experience that comes from practicing how to make the perfect gumbo is a soft skill development which is the last piece of the recipe to elevate a PHA facilitator from good to great.

The role of a facilitator is to encourage the team to apply critical thinking and logic while being naturally curious about solving the puzzle that is a PHA. There are many variables when leading a team, each with different personalities, so it is essential that the facilitator is patient and understands operational constraints that distract people (i.e., their unit is in upset conditions). Additionally, it is beneficial for PHA facilitators to have communications training to help assess challenging situations, how people are acting (what their motivations are), and how to encourage participation from everyone in the room, not just the loudest person. It is helpful to be humble and have a sense of humor (as opposed to getting defensive), especially when a team member is more focused on spelling mistakes than the content of the documentation.

In addition to the interpersonal skills discussed here, a facilitator can build the skills discussed in this paper and improve their confidence over time through scribing opportunities. The scribe's role is to capture the information discussed during the PHA sessions and, in doing so, the scribe can observe how a facilitator effectively moves the team through the study. Scribing once or twice and then participating in a PHA Team Leader class is beneficial as you can bring real-world experiences to the class and case studies discussed.

Just as a gardener tending to the vegetables helps them grow better than on their own, having a mentor through the process of growing into a PHA facilitator is very beneficial. Consider having a senior facilitator attend a PHA workshop and provide feedback to the developing facilitator. Remote PHAs provide a great opportunity for the mentor to observe the developing facilitator privately and provide feedback to the facilitator via text or chat in a private and real-time manner. This can also be accomplished during in person PHAs by having someone senior sit in on the workshop. It is essential that the mentor allow the developing facilitator to lead the study, does not jump in to take over due to stylistic differences but still ensure critical safety cases are not overlooked. As with all communication, the way the mentor observes and communicates with the developing facilitator is key – it is not just about the words, but how they are delivered.

8 Conclusion

Following the guidance discussed in this paper will allow for a PHA team leader to become more effective and confident when preparing for PHAs, conducting team meetings, and presenting the results of a PHA to management. The continuous improvement lifecycle involves developing and maintaining professional skills and tools. Applying the elements outlined in this paper provides a clear path for individuals and companies to hone the skills required for an effective PHA facilitator.

The best chefs are known for their delicious culinary abilities around the world. These skills were not developed overnight, instead were refined over time and countless hours of practice. Of course, one can find food from a fry cook, or in a vending machine, which may sustain survival.

But, when one wants the best tasting foods, they know to seek chefs that have dedicated time to perfect their craft. Similarly, the craft of an effective PHA facilitator must be developed, practiced, and refined over time and experience. The guidance provided throughout this paper gives someone a framework of how to increase their abilities to ultimately create a well-seasoned PHA facilitator and leader.

9 References

[1] Occupational Safety and Health Administration. (1970). Occupational safety and health standards: Hazardous Materials (Standard No. 1910.119). Retrieved from https://www.osha.gov/laws-regs/regulations/standardnumber/1910/1910.119AppC. Accessed 1 January, 2024.

[2] Carter, Will J., and Bruce E. Ball. ASME Section VIII Div. 1, Pressure Vessels. CASTI Pub., 2000.

[3] Fields, Douglas. “How Does ‘Muscle Memory’ Work?” Psychology Today, Sussex Publishers, 2021, www.psychologytoday.com/us/blog/the-new-brain/202106/how-does-muscle-memory-work. Accessed 1 February, 2024.

[4] Aigen, Sam. (2018, October 23-25). Development of Company Specific Severity Model to Improve Efficiency and Consistency in Process Hazard Analyses [Conference presentation]. Mary Kay O'Conner Process Safety Center 2018 International Symposium, College Station, TX, United States.

Ensuring Success in the Process Hazard Analysis: The Imperative of Accurate and Accessible Process Safety Information

Morgan McVey
Morgan.McVey@NebulaSafety.com

John Johnson
John.Johnson@NebulaSafety.com
Nebula Safety & Environmental, LLC.
8583 Irvine Center Dr. #184
Irvine, CA 92618

Keywords: Process Hazard Analysis, PHA, Process Safety Information, PSI

Abstract

This technical paper discusses the importance of accurate and available Process Safety Information (PSI) in conducting an effective Process Hazard Analysis (PHA) in industrial settings, particularly in facilities dealing with hazardous chemicals. This paper will identify specific examples of how PSI is utilized during the PHA as well as real world incidents demonstrating the significant impacts that can occur when inadequate or inaccurate PSI is utilized.

1 Introduction

The Occupational Safety and Health Administration (OSHA) Process Safety Management (PSM) program was developed in response to a series of major industrial accidents, such as the Bhopal disaster in 1984 and the Phillips 66 Houston Chemical Complex explosion in 1989. These incidents highlighted the need for a systematic approach to managing the hazards associated with highly hazardous chemicals and processes. PSM was developed with two goals in mind. "The first goal of PSM is to take appropriate steps at the right time to prevent incidents in order to protect life, property, and the environment.... The second goal of PSM is to minimize the hazardous consequences of incidents that do occur.[1]" The likelihood of success of PSM implementation of the various prevention program elements including Process Hazard Analysis, is highly dependent upon available and accurate PSI documentation.

OSHA 1910.119 and EPA 40 CFR Part 68.67 both require that "The owner or operator shall complete a compilation of written process safety information before conducting a process hazard analysis required by the rule" [2, 3]. A PHA is a type of systematic risk assessment utilized to identify and characterize potential hazards associated with the storage, processing, or handling of hazardous materials [4]. Examples used for illustration in this paper will primarily be associated with qualitative studies utilizing the concept of severity, tempered by the likelihood (frequency) of an event to conduct the qualitative risk assessment, utilizing a risk ranking matrix. The PHA team first systematically identifies the causes for deviations from the intended process operation. For causes deemed credible the PHA team identifies the severity of possible consequences and the likelihood that such an event occurs, considering the process operations and any mitigating protections (safeguards) present in the process. The severity and likelihood rankings identify the overall scenario risk based on the risk ranking matrix. The company risk ranking matrix utilized by the PHA Team represents the level of risk deemed acceptable by the organization (risk tolerance) and guides the PHA Team throughout the qualitative risk assessment process.

Process Safety Information forms the backbone of any effective Process Hazard Analysis. Its importance is reflected in the regulations as PSI is explicitly required to be compiled prior to a PHA, emphasizing the reliance on PSI within the PHA and its impact on the accuracy and quality of the study. A well-known concept in computing "garbage in, garbage out" also applies to the PHA. Accurate and comprehensive PSI equips the PHA team with the necessary tools to provide the most accurate risk assessment possible, by correctly identifying potential hazards, and evaluating the effectiveness of safety measures to properly quantify risk within a process containing hazardous materials. As differences in process safety information can lead to different outcomes in PHA studies, the three common pitfalls of PSI (incomplete PSI, inaccurate PSI, and outdated PSI) must be carefully managed to avoid misrepresentation of risk, which could lead to significant process safety events.

2 Process Safety Information

Per the OSHA PSM and EPA RMP regulations, process safety information is categorized into three groups with each group intent on defining information about the chemicals used in the process, how the chemicals are processed, and the equipment used for processing the chemicals. The groups and associated components are listed below for reference [2]. Refer to Appendix A for a more detailed description of each PSI element and how it is used in a PHA.

i. Information pertaining to the hazards of the highly hazardous chemicals used in the process.
   a. Toxicity information
   b. Permissible exposure limits
   c. Physical data
   d. Reactivity data

e. Corrosivity data
f. Thermal and chemical stability data
g. Hazardous effects of inadvertent mixing of different materials that could foreseeably occur

ii. Information pertaining to the technology of the process.

a. A block flow diagram or a simplified process flow diagram
b. Process chemistry
c. Maximum intended inventory
d. Safe upper and lower limits for such items as temperatures, pressures, flows or compositions
e. An evaluation of the consequences of deviations, including those affecting the safety and health of employees

iii. Information pertaining to the equipment in the process.

a. Materials of construction
b. Piping and instrumentation diagrams (P&ID's)
c. Electrical classification
d. Relief system design and design basis
e. Ventilation system design
f. Design codes and standards employed
g. Material and energy balances (for processes built after May 26, 1992)
h. Safety systems (e.g., interlocks, detection, or suppression systems)

The above listed PSI is listed directly from the OSHA PSM standard; however, it is important to note, that in some cases the indicated PSI may not be applicable. For example, if the covered process is located outside, the ventilation design basis may not be applicable. Another example is for the storage and handling of sulfur dioxide cylinders, where the cylinders are never opened/used, process chemistry data may not be applicable. Lastly, another case of potentially inapplicable PSI would be for a facility with only a single chemical onsite. In this case the hazardous effects of inadvertent mixing of different materials may not be applicable. It should be noted that for this last example, chemicals not considered part of the process (e.g., cleaning chemicals) should be considered when identifying the hazardous effects of inadvertent mixing.

3 The Stage of Design Impacts the Availability of PSI to the PHA Team

A typical project developing a new chemical process will have several design stages and it is not atypical to have several PHA's as part of the design process, typically at the Front-End Engineering Design (FEED) stage, 30% stage, 60% stage, and 90% (detailed engineering design stage of the project). Throughout each stage, it's important to conduct regular reviews and updates to the process safety information as the project progresses, and new information becomes available. A common PHA methodology utilized at the early design stages is a Hazard Identification (HAZID) study. The HAZID study still requires a team with expertise in the engineering and operation of the chemical process but since at the early design stages P&ID's may not be available, block flow diagrams or PFDs are instead used to brainstorm the potential consequences of process deviations and identify potential safeguards to protect against those deviations that may result in a hazardous consequence. Each subsequent PHA will have access to a more complete set of PSI, and therefore each subsequent design stage will incorporate more design changes and safety systems identified by the PHA teams. It is generally assumed that at the 90% stage of a PSM covered process all regulatory required PSI will be available to the PHA team prior to the commencement of the 90% PHA. This PHA will then serve as the initial PHA for the purposes of subsequent revalidation PHA's for a PSM covered process.

4 Mock PHA Scenarios

PHA studies are used to identify and evaluate potential hazards within a process to determine whether the associated risk is acceptable or whether additional mitigations are required to achieve acceptable risk. Differences in process safety information can lead to different outcomes within these studies. The three common PSI deficiencies are (1) incomplete PSI, (2) inaccurate PSI, and (3) outdated PSI.

When the necessary process safety information is not available during the PHA, assumptions are sometimes made to facilitate study progress. Depending upon the assumption, this could lead to overstating or understating the risk. Both situations are undesirable as overstating the risk may lead to unnecessary equipment, operational, or administrative changes while understating or missing the risk entirely may result in latent hazards, unrecognized, in the process. The remainder of this section provides illustrative examples of how incorrect or missing PSI may result in inadequate PHA studies.

In Figure 1, a flow control valve FCV-1 controls discharge flow from a charge pump P-101 through the tube side of exchanger E-102 and on to downstream equipment. The process chemical is a hydrocarbon liquid with H2S content above the safe Immediately Dangerous to Life or Health (IDLH) limit. The temperature of the fluid is above the fluid autoignition temperature.

The PHA team identifies a potential no flow scenario due to FCV-1 malfunctioning. FCV-1 is a fail close valve that controls flow based on setpoint from downstream flow transmitter FT-1. The team notes that FT-1 misreading a high flow rate would throttle the flow control valve to maintain setpoint. Another potential fail mode would be a loss of instrument air resulting in the FCV-1 going to its fail position. In either case FCV-1 has closed or closed too far and now there are potential impacts on both the upstream and downstream equipment. The three scenarios detailed below are focused only on the upstream impacts. The three scenarios are meant to illustrate how varying PSI quality and availability may impact the PHA.

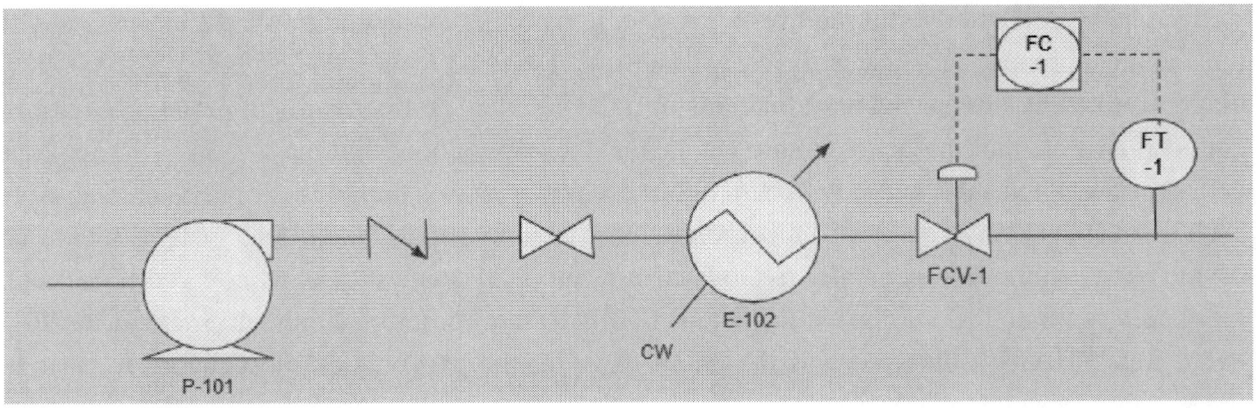

Figure 1. Scenario #1 and #2 schematic

Scenario #1 – missing pump data: The Team identifies, based on the P&ID symbology and operator feedback, that P-101 is a centrifugal pump and therefore will have a finite blocked-in discharge pressure. However, the pump data is not available and therefore the impacts on downstream piping and equipment, including E-102, cannot be determined. In the worst case, the team only identifies the potential for seal failure in the pump and ignores the impacts on downstream equipment. The most likely scenario is that the PHA facilitator assigns the team engineering representative a follow up item to procure the pump data sheets and or pump curves and this information is received before the conclusion of the PHA so the team can accurately assess impacts on downstream piping and exchanger. This scenario would most likely end up with a PHA recommendation to determine the maximum pump blocked discharge pressure and then revisit this scenario at a later date. The lack of PSI at the outset of the PHA has resulted in lost time in the future if the scenario is to be revisited and the potential to ignore overpressure impacts on the discharge piping/inline exchanger if the blocked-in discharge pressure cannot be determined. For illustrative purposes, we will assume that the pump maximum blocked-in discharge pressure is 100 psig and that the downstream piping/E-102 tube is rated to 300 psig, meaning that the weak point is pump seal failure due to deadhead.

Scenario #2 – inaccurate P&ID, incorrect equipment identified: In this next scenario, operations and engineering representatives are unfamiliar with P-101 but based on the P&ID symbol assume it's a centrifugal pump. Unknown to the team is that P-101 had been replaced previously due to an unrelated project. In the field P-101 is now P-102 and is actually a positive displacement pump rather than the original centrifugal pump. Without this knowledge, the team may mischaracterize the scenario as a pump seal failure and completely miss the more immediate concern that a blocked-in positive displacement pump can reach a theoretical infinite pressure, and without proper mitigation, result in overpressure of the discharge piping and E-102 exchanger. Fortunately for the team the P-102 pump was installed in compliance with RAGAGEP (Recognized and Generally Accepted Good Engineering Practices) and is equipped with a discharge pressure relief recirculation valve sized for blocked flow and set at E-102 tube Maximum

Allowable Working Pressure (MAWP) of 300 psig to protect the blocked-in equipment. The inaccurate PSI resulted in a complete mischaracterization of the scenario.

Scenario #3 - inaccurate P&IDs, missing equipment: Out-of-date P&ID's missing key process equipment ensures that the missing equipment will not be analyzed as part of the PHA. In this particular scenario an inline filter was missing from the process segment (see Figure 1 above). Fortunately the operator was able to identify that a filter, located immediately downstream of the E-102 exchanger, was missing from the P&ID. The PHA session was halted so that the operator and engineer could walk-down the process, field verify the filter, and update the P&ID's (see Figure 2). The engineering representative was able to locate the filter tag and identify the filter MAWP as 100 psig. The PHA team reevaluated the blocked flow scenario, due to FCV-1 malfunction and noted that the expected pressure increase in the line would be well above the MAWP of the filter (positive displacement pump, recall that E-102 tube MAWP is 300 psig), and therefore overpressure and catastrophic failure of the filter was the scenario of concern for the blocked-in flow case. Missing information on the P&IDs delayed the completion of the PHA as the team members had to perform an unexpected field walkdown. Additionally and more importantly, if not for the diligence of the operations representative, this PSI gap may have resulted in a hazardous scenario going unrealized by the PHA team.

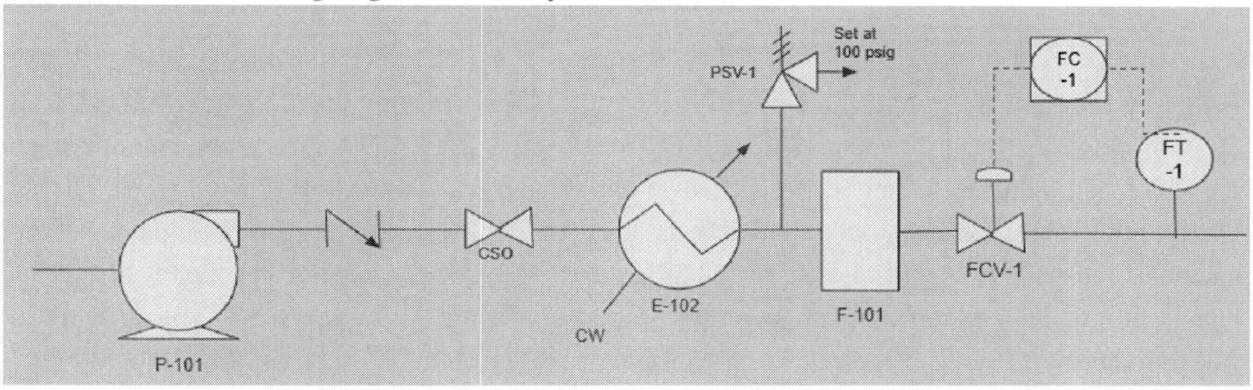

Figure 2. Scenario #3 schematic

Table 1 below captures the analysis of scenarios #1-3 above. Process safety information significantly impacts the scenario / consequence identified, along with the associated safeguards. Without PSI, the PHA team might be overly conservative and assume overpressure of E-102 and recommend a PSV to be sized for a blocked outlet case when in fact there may be no overpressure scenario as identified in Scenario 1. The PHA Team might also assume that P-101 maximum blocked-in discharge pressure is below the equipment MAWP and identify the seal failure consequence. The safeguards for seal failure would not protect against the overpressure consequences identified in Scenarios #2 and #3 and the overpressure protection for scenario #2 (relief valve set at 300 psig) would not provide overpressure protection for scenario #3 for the filter (MAWP 100 psig), and the risk would go unidentified and unmitigated.

Table 1. Summary

Scenario	Consequence	Missing or Inaccurate PSI	Concerns
#1	Blocked-outlet of centrifugal pump, leading to **P-101 seal failure**	Missing pump data	Potential mischaracterization of scenario if pump deadhead pressure can exceed safe operating limits of the downstream equipment
#2	Blocked-outlet of PD pump, leading to **E-102 overpressure**	Inaccurate P&ID's, incorrect equipment	Potential for the team to miss the overpressure risk due to blocked positive displacement pump discharge
#3	Blocked-outlet of PD pump, leading to **F-101 overpressure**	Inaccurate P&ID's, missing equipment	Potential for the team to miss the overpressure risk of the F-101 filter

Industry Incident Related to Incorrect P&IDs: An industry example of incorrect P&IDs leading to an unidentified hazard occurred at the Williams Olefins Plant in Geismar, Louisiana, on June 13, 2013. The explosion and fire resulted in the deaths of two Williams employees. Briefly, changes to a reboiler lineup that included installation of block valves along the reboilers relief path had occurred in 2001 as part of a project enabling the site to utilize only one reboiler at a time. The installation of the isolation valves were not properly captured in the P&ID's. Typical operation was that during routine maintenance of the primary reboiler a standby reboiler was brought online. At the time of the incident the standby reboiler had been isolated and filled with nitrogen over a period of 16 months. During this period flammable liquid propane had accumulated in the shell side of the standby reboiler. When the standby reboiler was brought online, the tube-side hot quench valves were lined up to the process resulting in significantly increased temperature in the shell side containing the liquid propane. The shell side was still isolated at this time and the resultant rapid increase in shell-side pressure resulted in an overpressure, release, and explosion in the plant. Among other deficiencies, the CSB report noted that the "PHA facilitator was under the incorrect impression that both propylene fractionator reboilers operated at the same time". It was not indicated in the PSI that the reboilers may be operated singularly and therefore would not have access to the relief system[5]. Had the isolation valves been depicted on the P&IDs, the potential for equipment to be isolated from its relief system might have been discussed and a recommendation made accordingly.

Scenario #4 - unavailable relief valve design basis information: In Figure 3 Feed Pump P-101 is pumping a hydrocarbon liquid to a Separation Vessel V-1. Vapor is flashed in the separation vessel and proceeds to downstream gas treatment. Liquid level is maintained in the vessel by LT-1 which opens LCV-1, when the setpoint is reached, to dump liquid to downstream oil processing equipment. The separation vessel is configured with an independent level alarm LAH-2 that will alert operations to high V-1 level.

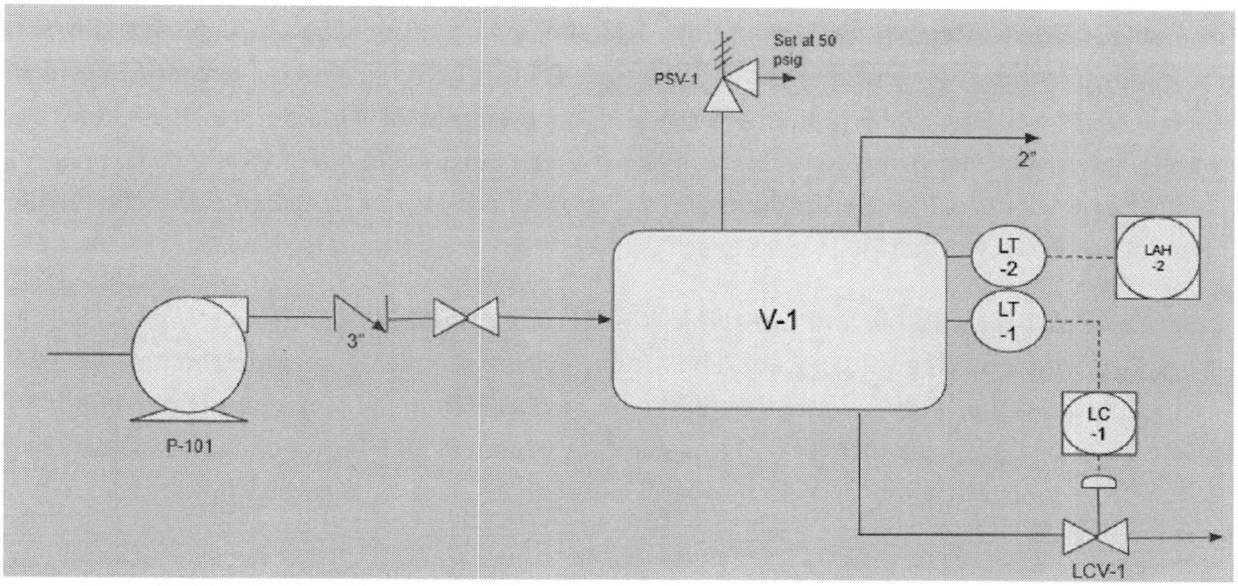

Figure 3. Scenario #4 schematic

The team identifies a potential cause for high level in the separation vessel due to LCV-1 failing to open when needed. It is noted by the team that the most likely cause would be a false low signal from LT-1, forcing the level control valve to stay closed. The Team identifies the most likely consequence of LCV-1 malfunction to be an increased level in the separation vessel. The team also noted that, without safeguards, liquid would continue to flow into the separation vessel until the vessel overfills. Finally, the team noted that it could not be verified that the 2" line at the vessel overhead could provide enough throughput to maintain the vessel within a safe pressure envelope and that overpressure of the vessel is the likely ultimate potential consequence of the level control valve failure. The maximum potential pressure achieved in the vessel would be based on the deadhead pressure of the upstream pump. The team identifies the independent level alarm, LAH-2, as a safeguard but based on the potential severity of an overpressure event, this single safeguard is not considered adequate to achieve acceptable risk. The Team identifies the relief valve on the vessel, PSV-1, as second safeguard but the relief valve sizing basis has not been provided to the team and no record of the PSV documentation is available for team review. The high risk scenario coupled with lack of adequate safeguards results in the development of a recommendation to calculate the relief requirements for this event and to determine if PSV–1 is appropriately sized for this case. Lack of relief valve design information has resulted in what could possibly be an unnecessary recommendation to develop documentation that should be in existence and available to the PHA team. In the worst case, an inexperienced team may have simply assumed that the relief valve was adequate protection and risk ranked the scenario as an acceptable risk. There is a high probability that the relief valve is only sized for the fire case and may not have the relief capacity to accommodate continuous flow of liquid from the upstream pump (2-phase flow). This could be the difference between an acceptable and unacceptable risk, that might go unidentified and unmitigated if unqualified assumptions are made.

Industry Incident Regarding Inadequate Relief Valve Sizing: The 2020 "Fatal Chemical Decomposition Reaction and Explosion at Optima Belle LLC" illustrates a scenario where the relief system was not sized properly for a sodium dichloroisocyanurate (NaDCC) dihydrate decomposition and was unable to relieve at a rate sufficient to prevent overpressure in the pressure-rated rotary cone dryer utilized for the process. The resultant overpressure and explosion resulted in the fatality of an Optima Belle employee[6].

Scenario #5 PSI is available and utilized by the PHA team: Figure 4 shows a reactor R-101 that is equipped with a cooling water jacket. The cooling system is under temperature control via TCV-1, which throttles the cooling water flow based on feedback from the temperature transmitter TT-1. The reaction is an exothermic continuous flow reaction and the cooling water system is necessary to control reactor temperature. Reactants are fed to a catalyst bed in the reactor from Feed Pump P-101.

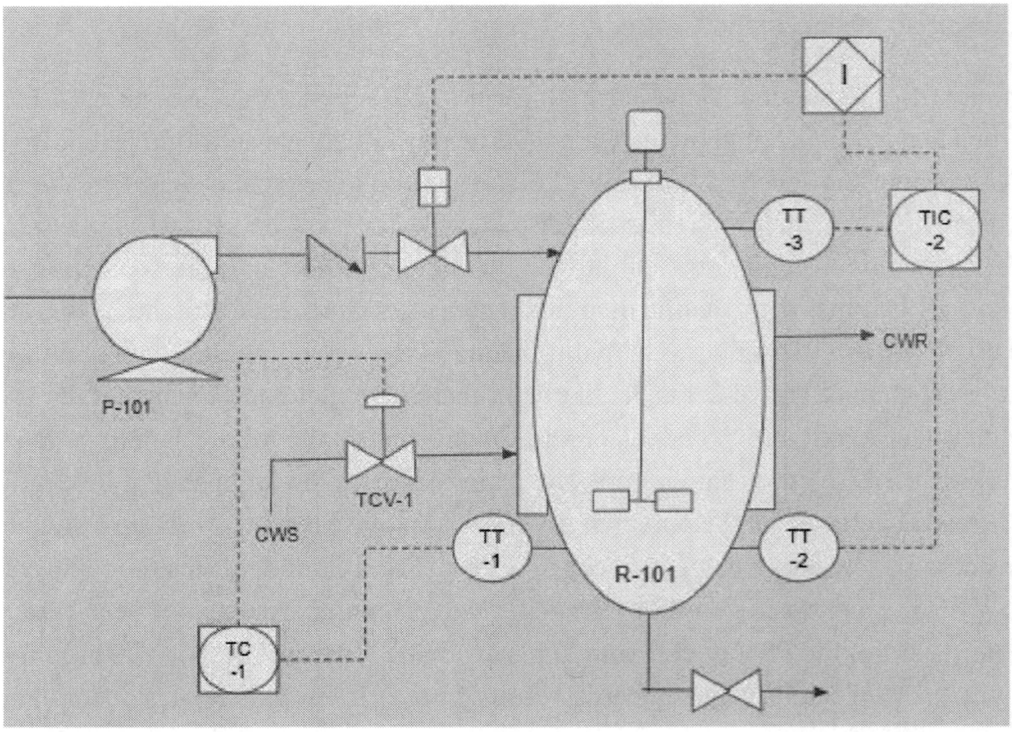

Figure 4. Scenario #5 schematic

The PHA team identifies a high temperature scenario due to unintentional closure of TCV-1, with the most likely cause of temperature transmitter TT-1 reading a false low temperature. For this example the team has all of the necessary PSI and is able to determine that the Reactor R-101 MAWP is 50 psig and the maximum allowable design temperature (MADT) is 200F. The team notes that the loss of cooling water may result in increased temperature, increased reaction rate and heat generation, runaway reaction and potential to exceed the vessel MAWT leading to hot spots, vessel damage and release of chemicals to atmosphere. This is identified as a high severity event. Reviewing the cause and effect matrix the team identifies TIC-2 as a potential safeguard. The cause and effect matrix identifies that TIC-2 receives inputs from both TT-2 and TT-3 and is configured to close the inlet feed from pump P-101. The team identifies that the setpoint that would initiate feed valve closure is below the MAWT of the reactor. However, as this is a high severity event, the presence of a single viable safeguard is not considered adequate for the identified risk. A recommendation is made to consider upgrading the TIC-2 instrumentation to a SIL-2 configuration, including addition of another temperature transmitter, two out of three voting and installation of a second inline feed shut off valve.

Industry Incident Regarding Incomplete Reactivity Knowledge: In May and June of 2017 there were two separate explosions that occurred at the Midland Resource Recovery facility in Philippi, West Virginia. The explosions took place during activities related to the tank decommissioning and deodorizing process via treatment with sodium hypochlorite. The US Chemical Safety and Hazard Investigation Board (CSB) identified that the "probable cause of these incidents was reactive, unstable chemicals that exploded when workers tried to drain the uncharacterized, chemically treated liquid from natural gas odorizer equipment". The initial explosion in May resulted in the deaths of two company employees and severe injuries to a third. The second June 2017 explosion resulted in the fatality of a contractor, onsite as part of the investigation and mitigation after the May 2017 explosion[7].

A primary function of the PHA team is to identify the risk of hazards on paper, so that the process is protected against the same hazards occurring in the field. Incomplete reactivity, thermal and chemical reactivity, and hazardous effects of mixing data limits the PHA teams ability to brainstorm and identify potential reactivity hazards associated with the chemicals used in a process. This information used in conjunction with process chemistry PSI provides the team with enough information to qualitatively assess potential chemical reactions both intended and to brainstorm unintended reactions during a PHA.

Scenario #6 - illustrating the importance of understanding material hazards for the PHA: The scenario below (Figure 5) depicts a powder transfer system from a feed hopper to a product hopper, including pneumatic conveyance with a blower providing motive force, and a dust collection system for a sugar processing facility. The dust collection system includes a cyclone for separation of fines and then a baghouse with intermittent backpulsing for fines collection. The material being conveyed is sugar.

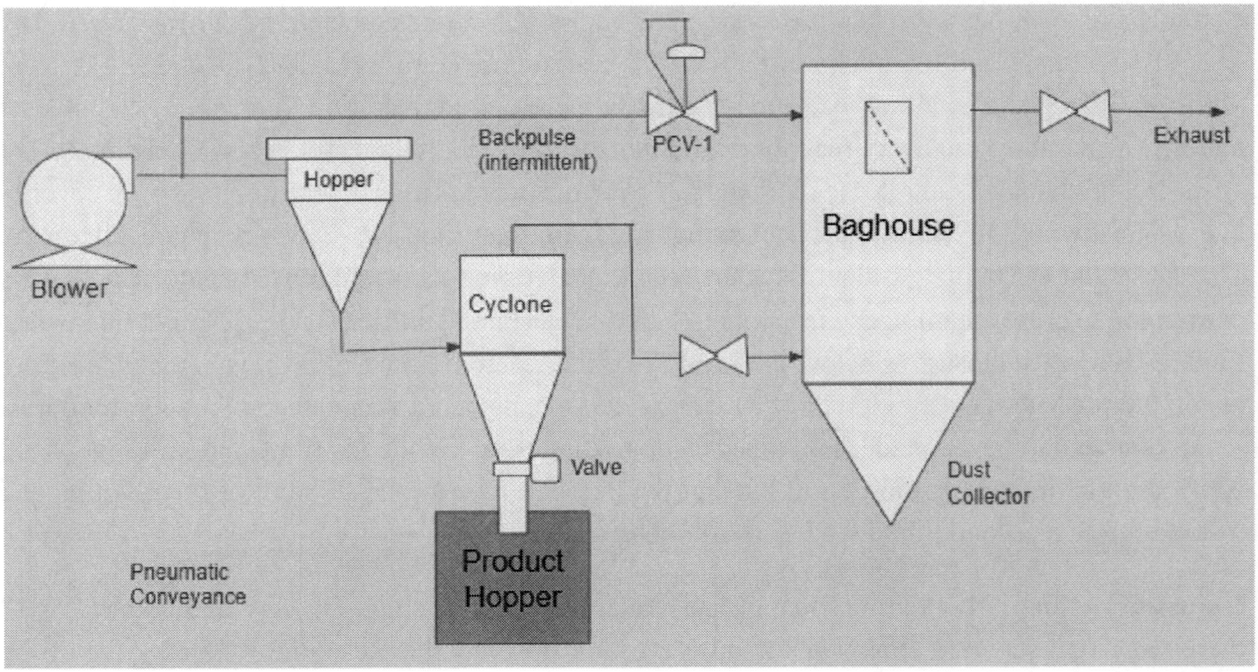

Figure 5. Scenario #6 schematic

The PHA Team is reviewing hazards associated with inadvertent closure of the block valve on the baghouse exhaust. The blower maximum blocked-in discharge pressure is 200 psig; while the equipment is rated to 80 psig. The Team analyzes the overpressure consequence and identifies a loss of containment of sugar and an asset impact. The overpressure ratio isn't significant enough for a catastrophic equipment failure with shrapnel impact. Additionally, the Team did not identify significant consequences associated with loss of containment of sugar, as material hazards were unknown / had not been identified.

Unfortunately there wasn't proper combustible dust testing to identify whether the fines could create a combustible dust explosion. Years later, due to accumulation of powder, lack of hazard recognition, and the occurrence of an ignition source, a dust explosion occurs resulting in 14 fatalities, numerous injuries, and extensive damage to the facility. This was the Imperial Sugar Company combustible dust explosion that occurred on February 7, 2008, at the company's sugar refinery in Port Wentworth [8]. The industry at the time, including the sugar processing sector, had insufficient awareness of the potential hazards associated with combustible dust, leading to a lack of recognition of the explosive nature of sugar dust.

Scenarios 1-6 illustrate the importance of reviewing applicable process safety information when identifying a credible cause of an event, determining the severity of the event, identifying the associated likelihood, and listing appropriate safeguards designed to mitigate the event, and finally, developing PHA recommendations appropriate to the process under review. Some of the scenarios identified PSI deficiencies, some of the scenarios noted that all necessary PSI was available. This is not unlike what the PHA team encounters every day. In the next sections we discuss some common errors in maintaining PSI and some tips for PHA teams as well as facilities on gathering, organizing, and providing the proper PSI for conducting a PHA.

5 Common Errors in PSI Management & Solutions

Many companies have experienced difficulties when it comes to PSI Management and ensuring that information is kept current. Some of these difficulties can be attributed to improperly managed changes, inadequate document control management, lack of knowledge needed to identify which PSI docs would be impacted by a given change, etc. Some of these common issues are listed below with potential solutions discussed.

1. Missing or Incomplete Safety Information: for older processes, there may be missing PSI such as pump curves, equipment MAWPs, metallurgy, etc. As discussed above, this information is critical for the PHA and therefore would need to be somehow located or recreated. Some solutions are listed below:
    a. Contact the Original Equipment Manufacturer (OEM) of the equipment to request historical data, manuals, or specifications, including pump curves.
    b. Use engineering principles and available data to recreate pump curves or other performance data based on known parameters such as pump size, speed, and system characteristics.
    c. If the specific equipment data is not available, data from similar equipment in the same or similar operating conditions may be used as a basis for estimation.
    d. Review historical records, maintenance logs, or operating data as this may provide insights into the performance and characteristics of the equipment.
    e. Use process simulation software to model the system and estimate the performance of the equipment based on known parameters.
    f. Conduct field tests or performance tests on the equipment under controlled conditions to help recreate missing information. Rerating and hydro testing equipment in place may be an option.
    g. Consult with subject matter experts or experienced engineers familiar with the equipment or process to provide guidance for recreating missing information.
    h. Some regulatory agencies or industry organizations may have historical data or guidelines that can be used to recreate missing information.

2. Inaccurate Process Safety Information: Errors in the PSI, possibly due to P&ID redline errors, swapping U1A forms or pump curves for equipment, etc. Some solutions are listed below:
    a. Implement a periodic PSI audit to verify existing PSI through representative sampling. For example, pick 3 relief valves and cross reference the PSV

datasheet setpoint and orifice with the P&ID depiction with the poptesting records and finally against the physical equipment tag in the field to ensure that they are all aligned. This should be conducted prior to a PHA with sufficient time to allow for required updates before the study begins.

b. Ensure that PHA recommendations, identifying PSI discrepancies, are prioritized similarly to process safety recommendations. The PHA is where many PSI gaps are identified. The PHA can provide an additional tool for identification of PSI inaccuracies.

3. Failure to Update Safety Information: Changes in the process, equipment, or operating conditions may not be adequately documented or updated in the process safety information. This could be due to insufficient assessment or errors in the management of change (MOC) process. It could also be due to inadequate training programs and awareness of what is classified as process safety information and the associated importance. Some solutions are listed below:
    a. Review the MOC process and ensure that there are appropriate triggers for PSI updates based upon the change.
    b. Ensure training programs clearly delineate what is classified as process safety information and to provide training on which PSI could be affected based upon the type of change.
    c. Ensure training programs emphasize the importance of accurate PSI to ensure teamwork and shared responsibility on keeping PSI current.
    d. Ensure that MOC coordinators/owners are familiar with the various PSI documents and what each entails so they can better identify required PSI updates associated with each MOC.
    e. Train MOC reviewers to ensure that they confirm the required PSI documents to be updated before approving MOC
4. Not easily accessible PSI: PHA Teams may make assumptions if the PSI isn't readily available at the time of the study (e.g., if pump curves are stored in a vault and not accessible online). Another common issue is inadequate training for PHA team members on how to access required PSI for the study. Potential solutions are listed below:
    a. Have a list of required PSI and require Teams to come prepared to the meetings and familiar with where to locate these documents.
    b. If the PSI storage protocol for the facility is hard copy equipment folders containing equipment datasheets and pump curves, ensure that this information is compiled beforehand including MAWPs/MAWTs, hydrotest pressures, pump maximum blocked-in discharge pressures, etc.
    c. Scan the PSI and store it electronically so that this information can be easily accessible during the study.

d. Ensure proper training for the PHA team members on how to access PSI so it is readily available when needed.

6 Tips for PHA Teams / PSI Management

From a PHA facilitator's perspective, based on real world experience, procuring adequate PSI for a PHA requires constant focus, and quite a bit of badgering of those persons who can provide the PSI. Many PHA facilitators are contracted consultants and as such have limited access to servers where a company's internal documentation may be stored. Furthermore as consultants there is always pressure to keep the client happy, and in those cases try and complete the PHA with what has been provided, even if it is inadequate. As a rule of thumb every PHA should have a PSI preparation checklist where all tasks required to conduct a successful PHA are listed and checked off as they are completed. As part of that checklist all items identified in section 2 should be entered in as a line item and noted with the date the PSI item has been received. As some information may not be easily electronically transmissible, the PHA facilitator should contact the client to identify the location of where the PSI element is located (e.g., onsite library containing relief valve sizing studies). Information gathering should be conducted well in advance of the PHA, typically it is suggested about two months in advance to allow for updates if gaps are identified. Of course any contracted PHA facilitator knows that projects can sometimes materialize and the preparation time will be severely limited. The preparation checklist should be generic in nature so that as soon as a contract is awarded the information request can be submitted to the client.

Another tip for PHA facilitators is to make contact with the PHA team engineering representative well in advance of the PHA. The engineering representatives typically know where to find all of the PSI and if they do not, they know who to contact in order to procure the necessary PSI. The PHA facilitator should make partners of the PHA team members well in advance of the PHA. Remember that the PHA team members have jobs and tasks outside of the PHA and so out of respect for their time give them as much advance notice as possible when soliciting their help in PSI procurement.

For those companies storing, handling, or processing highly hazardous materials, consider the following for organizing and maintaining chemical process PSI:

- Develop a PSI matrix to clearly identify what is classified as PSI at the facility
- Create electronic folders for each equipment (with hardcopy backup folders), with scanned U1A forms and pump curves to facilitate ease of reference
- Designate a document management system owner to preserve PSI integrity
- Ensure the PSI matrix is integrated in the MOC system to ensure PSI is updated accordingly. And ensure there is a system to clearly archive old PSI to limit confusion

- Designate single point ownership for PSI class and clear roles and responsibilities for PSI management

7 Conclusion

Accurate PSI is indispensable for conducting a thorough and effective PHA. It enables the identification of potential hazards, assessment of risks, and the implementation of appropriate control measures. Inadequate or inaccurate PSI can lead to overlooked hazards, insufficient understanding of process risks, and consequently, increased potential for accidents. The availability and accuracy of Process Safety Information is paramount in conducting effective Process Hazard Analysis. It forms the foundation for identifying potential hazards and implementing safety measures. Investing in robust PSI management systems is not just a regulatory requirement but a fundamental aspect of proactive safety management in the chemical industry.

If the process chemistry is not well known, if the relevant drawings and procedures are not accurate, or if the process knowledge available from a study team does not reflect the way the system is actually operated, then the results of the hazard evaluation may be invalid. This could lead managers to make poor risk management decisions[9].

8 References

[1] (1995). Process Safety Documentation Overview. In Guidelines for Process Safety Documentation, (Ed.). https://doi.org/10.1002/9780470938072.ch2

[2] "1910.119 - Process Safety Management of Highly Hazardous Chemicals." *Occupational Safety and Health Administration*, U.S. Department of Labor, www.osha.gov/laws-regs/regulations/standardnumber/1910/1910.119. Accessed 1 Feb. 2024.

[3] "The Federal Register." *Federal Register :: Request Access*, Code of Federal Regulations, www.ecfr.gov/current/title-40/chapter-I/subchapter-C/part-68. Accessed 1 Feb. 2024.

[4] "Process Hazard Analysis." *AIChE*, 18 Dec. 2014, www.aiche.org/ccps/resources/glossary/process-safety-glossary/process-hazard-analysis.

[5] "Williams Olefins Plant Explosion and Fire." *CSB*, U.S. Chemical Safety Board, www.csb.gov/williams-olefins-plant-explosion-and-fire-/. Accessed 1 Feb. 2024.

[6] "Fatal Chemical Decomposition Reaction and Explosion at Optima Belle LLC." *CSB*, U.S. Chemical Safety Board, 6 July 2023, www.csb.gov/assets/1/6/optima_report_for_publication.pdf.

[7] "Midland Resource Recovery Explosion." *CSB*, U.S. Chemical Safety Board, www.csb.gov/midland-resource-recovery-explosion-/. Accessed 1 Feb. 2024.

[8] "Imperial Sugar Company Dust Explosion and Fire." CSB, U.S. Chemical Safety Board, www.csb.gov/imperial-sugar-company-dust-explosion-and-fire/. Accessed 21 Feb. 2024.

[9] CCPS Guidelines for Hazard Evaluation Procedures III Center for Chemical Process Safety, New York, 2008

[10] "Process Safety Information (PSI)." *AIChE*, 15 Aug. 2018, www.aiche.org/ccps/resources/glossary/process-safety-glossary/process-safety-information-psi.

[11] *Safety Data Sheet*, Airgas, www.airgas.com/msds/001003.pdf. Accessed 1 Feb. 2024.

[12] *Safety Data Sheet*, Airgas, www.airgas.com/msds/001195.pdf. Accessed 1 Feb. 2024.

[13] "Gaseous Hydrogen Embrittlement of Materials in Energy Technologies." Edited by Richard P Gangloff and Brian P Somerday, *ScienceDirect*, Woodhead Publishing, 2012, www.sciencedirect.com/book/9781845696771/gaseous-hydrogen-embrittlement-of-materials-in-energy-technologies.

[14] "What Does 'Electrical Classification' Mean?" *EPA*, Environmental Protection Agency, 23 May 2023, www.epa.gov/rmp/what-does-electrical-classification-mean.

[15] "What Does 'Relief System Design Basis' Mean?" *EPA*, Environmental Protection Agency, 12 Oct. 2023, www.epa.gov/rmp/what-does-relief-system-design-basis-mean.

[16] Fairfax, Richard E. "Process Safety Information for Ventilation System Design." *Occupational Safety and Health Administration - Standard Interpretations*, U.S. Department of Labor, www.osha.gov/laws-regs/standardinterpretations/2003-12-22-2. Accessed 1 Feb. 2024.

[17] "Recognized and Generally Accepted Good Engineering Practices in Process Safety Management Enforcement." *Occupational Safety and Health Administration*, U.S. Department of Labor, 11 May 2016, www.osha.gov/laws-regs/standardinterpretations/2016-05-11-0.

Appendix A: Process Safety Information

Process Safety information is the "physical, chemical, and toxicological information related to the chemicals, process, and equipment" [10]. Per the OSHA PSM and EPA RMP regulations, process safety information is categorized in three groups with each group intent on defining information about the chemicals used in the process, how the chemicals are processed, and the equipment used for processing the chemicals. The following information is a brief description of each PSI element and how it is used in a PHA.

1 Information pertaining to the hazards of highly hazardous chemicals in the process

Information pertaining to the hazards of highly hazardous chemicals assist the PHA team in identifying what occurs, within the process, under specific process conditions, both normal operating conditions and upset conditions.

1.1 Toxicity information and permissible exposure limits

Toxicity information and exposure limits provide the basis for evaluation of the severity of a release. For example, consider an aqueous ammonia release versus a release of anhydrous ammonia. Anhydrous ammonia is characterized as an LC50 inhalation gas with the dose identified as 7338 ppm while aqueous ammonia (20-30%) is characterized as an LD50 Oral with a lethal dose of 350 mg/kg [11, 12]. When presented to a PHA Team this information would be used to identify the severity of a consequence based on the potential for inhalation exposure of the anhydrous form of the chemical and oral exposure of the aqueous form. The PHA team will typically assign the higher risk to the inhalation of the chemical versus ingestion. Toxicity information including the permissible exposure limits enables the PHA team to make an informed assessment as to the severity of a scenario upon a release to the atmosphere. Ammonia was used as an example for this section as it is a widely understood chemical; however, there are many chemicals in industrial use that are not as well characterized. Incorrect or missing chemical toxicity information will limit the PHA teams' ability to properly evaluate the severity upon a release scenario, leading to personnel exposure.

1.2 Physical Data

Chemicals have inherent and specific physical properties. The vapor pressure, boiling point, freezing point, and other physical characteristics provide the PHA team with information as to how the chemical will interact with the expected, as well as unexpected, process conditions. Boiling points and melting points are useful for PHA teams to know whether the LOC scenario would result in solid, liquid or gas release based upon the local ambient temperatures. For example, the team may identify a PHA scenario that results in an increased hydrocarbon separator temperature. The increased temperature will typically lead to a corresponding increase in the separators internal pressure. Utilizing the chemical physical data (i.e., the chemical vapor pressure) the team can calculate a theoretical maximum pressure for the separator and then identify if the achieved pressure exceeds separator design limits. Incorrect or missing physical data will inhibit the PHA teams' ability to accurately assess risk for some PHA scenarios.

1.3 Reactivity Data, Thermal and Chemical Reactivity Data, and Hazardous Effects of Inadvertent Mixing of different materials that could foreseeably occur

In a perfect world a chemical process would proceed as designed and provide a calculable amount of product from a specified quantity of reactants. The expectation is that only those constituents identified in the reaction chemistry are going to interact. In the real world there are numerous examples, for a variety of reasons, of chemical constituents input into a process in a way not identified in the initial design of the process equipment. This can be due to new chemical formulations intentionally contrived to provide more of an existing product or to produce a newer product. This can also be due to the use of cleaning chemicals, when the process is operational or shut down for maintenance, with trace cleaning chemicals not properly removed. Accurate reactivity data should be provided to a PHA team so that the team can "brainstorm" all possible chemical interactions that may occur in the process, both intentional and accidental.

1.4 Corrosivity Data

Corrosivity is managed in a PHA in many ways. Some simpler PHA's utilize a corrosion deviation within each node and poll team members as to the types of corrosion encountered during regularly scheduled inspections. The team then attempts to identify potential "bad actors" in the process that show a tendency for increased rates of pitting, depositing, weathering, or any other signs that piping or equipment is degrading over time. Other PHA's utilize internal company checklists to review operating conditions under which certain instances of corrosion, based on the chemicals being processed, may occur at a higher frequency. Still other companies perform Risk Based Mechanical Integrity (RBMI) studies outside of the PHA and provide those studies to the PHA team for review with the intention of identifying any potential corrosion issues or operating conditions not properly identified in the RBMI study.

Regardless of the approach, corrosion is dependent on a wide range of factors including the constituents of the chemicals being processed, the material selection of the piping and processing equipment, and the operating parameters being utilized for the process. Up to date corrosivity data, in the form of generalized material incompatibilities, safe operating ranges, and known attack vectors allow the PHA team to make qualitative assessments of the potential corrosion issues that may be present in the process under scrutiny. Incorrect corrosivity data may result in potential serious corrosion concerns to be overlooked during the PHA.

1.5 Safety Data Sheets (SDS)

Per 1910.119(d)(1)(vii) "safety data sheets meeting the requirements of 29 CFR 1910.1200(g) may be used to comply with this requirement to the extent they contain the information required by the subparagraph" [2]. Often during a PHA the PHA facilitator will request the SDS for those chemicals used in the process. The SDS provides a repository of the PSI elements discussed above. It is important that the SDS utilized in the study include the most up to date information possible so that the PHA team can properly assess hazards associated with the chemicals utilized in the process.

2 Information pertaining to the technology of the process

The information pertaining to the technology of the process includes an overview of process design and operation, the utilization of chemicals in the process, the quantities of chemicals used in the process and the safe operating parameters for the equipment in the process.

2.1 Block flow diagram or a simplified process flow diagram

A block flow diagram simplified Process Flow Diagram (PFD) provides a schematic overview of how a chemical gets from point A to B in a chemical process. The block flow diagram or PFD is often utilized in team discussions when analyzing PHA scenarios and if these documents are incorrect then the associated analysis may also be incorrect. Block flow diagrams are used to show the major process equipment and interconnecting process flow lines. Process Flow Diagrams (PFD's) are more complex and will show main flow streams, including valves, along with control loops and key utility systems. It is common for PFDs to become outdated if the changes to the process are not properly managed by updating all relevant PSI. Fortunately, this deficiency is easily identified by the operations representatives in the PHA meetings. Unfortunately, this may occasionally result in loss of session time as the relevant PFD's are updated for accuracy.

2.2 Process Chemistry

Process chemistry is intended to provide an overview of the reactions occurring in the process. The process chemistry documentation should accurately reflect all of the interactions between the different chemicals in the process. This information should include the normal reactions, along with abnormal situations, including the resulting primary, secondary, side and intermediate reactions. It is critical to understand potential undesired reactions that may adversely impact process safety (e.g., exothermic runaway reactions, hot spot formation). Kinetic data may also be important as this may impact process safety systems, such as pressure relief requirements.

Depending on the complexity of the process there may be several cascading reactions occurring or no reactions at all. Understanding the reaction chemistry provides the PHA team with insight into what chemical constituents are being consumed and what products are being produced. It may be a case where the non-hazardous reactants are being processed into a hazardous product are vice versa. The reaction chemistry alerts the team to what segments of the process may contain the hazardous materials and what segments of the process contain materials of lesser concern. The process chemistry also provides information as to the quantity of materials consumed or produced as well as whether the reaction is endothermic or exothermic. All of this information provides the basis for determining whether a scenario is hazardous, operational, or not credible. Inaccurate process chemistry may result in a misrepresentation of which segments of the process are areas of concern and which segments present operational concerns only.

2.3 Maximum intended inventory

The maximum intended inventory provides the PHA team with an estimate of the worst possible case if a loss of containment were to result in a complete loss of containment of the highly hazardous chemical. Intuitively the team understands that the larger the quantity of a highly hazardous chemical that is released the more significant the severity ranking should be. The maximum intended inventory should include the volume from all interconnected sources, not just the maximum volume within a single vessel or tank. This includes the volume of hazardous materials within the process piping. Additionally, the maximum intended inventory should specify if administratively controlled maximum allowable fill limits are used in the calculation. An inaccurate maximum intended inventory may result in the PHA team underestimating the full impact of a catastrophic release scenario. Accurate maximum intended inventory will set safe upper operating limits for tanks and vessels.

2.4 Safe upper and lower limits for such items as temperatures, pressures, flows, or compositions

The safe upper and lower limits identify the range the equipment and/or process can be operated safely. These should not be confused with operational limits or parameters used to ensure product quality. From the PHA perspective, operating outside of the safe upper and lower limits represents a hazardous condition and should be evaluated as such. Typically, the safe upper and lower limits may be found in the unit operating procedures. Ideally a table of the safe upper and lower limits, specifying the consequences of deviation outside of the safe operating parameters, is provided to the PHA Team.

2.5 An evaluation of the consequences of deviations, including those affecting the safety and health of employees

Often a facility PHA is referenced as an evaluation of the consequences of deviation. This leads to a circular "chicken or the egg" scenario whereby the PHA team is required to have, in hand, the consequences of deviation prior to performing an initial PHA but the site is waiting for the PHA to develop the consequences of deviation. If the site is operational then a review of procedures may provide some insight into the consequences of deviation. Many times, the consequences of deviation are encapsulated as warnings in the procedures and equipment manuals. Availability of the consequences of deviation, including those affecting the safety and health of employees provides the PHA team with confirmation of their own assessments and may also identify hazards missed in the PHA Team evaluation.

3 Information pertaining to the equipment on the process

The information pertaining to the equipment in the process comprises information about the design, materials, operation, safety systems, and codes and standards employed for a chemical process. These materials provide the PHA team with an overview of the process conditions, process interactions, and potential safeguards to be used for PHA scenario analysis.

3.1 Materials of construction

Piping and equipment utilized in a chemical process may encounter a variety of chemical constituents at a range of operating parameters (temperatures, pressures, flow rates, etc.). When using a PHA to analyze a process, an important consideration is if the equipment can safely operate under all expected process conditions (both intentional and unintentional). For example, the ideal gas law shows that a decrease in pressure has a corresponding decrease in temperature. This is known as the Joule-Thomson effect and some materials are more susceptible to associated embrittlement impacts than others. Similarly High Temperature Hydrogen Attack (HTHA) may occur at high temperatures due to the interaction between molecular hydrogen in the process stream and the carbon atoms in the piping or equipment material [13]. Material selection and identification is an important tool for identifying potential failure modes of equipment undergoing PHA evaluation. These considerations should include normal, non-routing, and unintended process conditions. The material selection may have considered Joule-Thomson effects under normal operating conditions, but impacts should also be considered when evaluating process upsets, for instance in the case of a control valve failure resulting in a significant drop in pressure can the materials selected withstand the corresponding significant decrease in temperature. The PHA team should be provided with the materials of construction in order to properly calibrate the risk of different process deviations.

Another consideration for the use of the materials of construction information in a PHA is the identification of potential material incompatibilities (e.g., copper piping in anhydrous ammonia service). Incorrect materials of construction information may result in a PHA team incorrectly evaluating or missing altogether potential hazards inherent in a process design.

3.2 Piping and instrumentation diagrams

Typically, PHA's are conducted as a meeting, in an office setting. Piping and instrumentation diagrams (P&IDs) are the schematic illustration of a process being evaluated by a PHA team. A fundamental assumption is that the P&IDs are an accurate representation of the process as it exists and operates at the time of the PHA. P&ID's that are out-of-date or missing information may lead to inaccurate evaluation of scenarios. For instance, for an overpressure case the PHA team will identify the potential maximum pressure that may be developed in the equipment under scrutiny. The maximum pressure is then compared to the MAWP of the equipment. If the maximum potential pressure is below the MAWP of the vessel then physical overpressure of the vessel may not be considered credible (although there may be other impacts associated with the increased pressure). If the maximum potential pressure achieved is over the MAWP of the vessel then the PHA team will often characterize an overpressure scenario considering other factors such as the percentage over the MAWP, the hydrotest pressure of the equipment, etc. to accurately characterize the severity of the overpressure event. If the equipment information is incorrect then the identified hazards may also be incorrect. Risk may be either overestimated or underestimated meaning on the discrepancy.

A more obvious case is when the P&ID's do not accurately depict the process as it is operating at the time of the PHA. PHA facilitators are very familiar with the case where the PHA team's operations representatives identify that the lineups depicted on the P&ID's do not match what is occurring in the field. This often results in a cessation of PHA activities while the P&IDs are walked down and "redlined". In the worst case, the discrepancy may not be identified, and potential hazards completely ignored.

3.3 Electrical classification

A typical PHA team assumption, for the purposes of identifying the severity of a given scenario, is that any uncontrolled atmospheric release of a flammable chemical, at ground level, will find an ignition source and proceed directly to a fire and/or explosion. Accurate and up-to-date electrical information is used by a PHA team to inform the likelihood that ignition occurs and that the identified severity is attained.

The electrical classification information provided to the PHA team should identify the class, group, and operating parameters of electrical equipment in the areas where flammable materials are stored and handled [14]. Typically, a plot plan with area electrical classification overlays is provided to the PHA team. An inaccurate or out-of-date electrical area classification document may result in misinterpretation, by the PHA team, of the actual flammability hazards in the area of concern.

3.4 Relief system design and design basis

At the heart of a risk assessment is "brainstorming" whereby the team identifies potential causes. Over the life of a process, operations experience, lessons learned, and changes to the process may add additional scenarios not identified during the initial design of the process. Overpressure scenarios are often encountered during a PHA evaluation and may occur due to a wide range of causes. The relief system and design basis, if accurate and available to the PHA team, enable the PHA team to verify that the relief system will respond as intended to the overpressure case under scrutiny. An often-noted example is that ASME coded vessel will have a fire-case relief valve that may not be viable for blocked vapor line or liquid overfill cases. Utilization of the fire-case relief valve for protection against the blocked vapor line or liquid overfill would be an egregious misrepresentation of the actual risk in the scenario.

The relief system design basis provides a "description of the overpressure scenarios considered, the scenario that creates the largest load to be relieved, the assumptions used, and if the device meets a certain code" [15]. Additionally, if there is a flare system, a flare design basis should be included in the PSI provided to the PHA team so that the team can accurately assess that the flare system will respond accordingly even in the event of simultaneous release from multiple relief valves (e.g., in the event of external fire).

3.5 Ventilation system design

For a covered facility that includes ventilation systems as part of the process safety management program the ventilation system must be treated as any other segment of the process safety management program, including the requirements for compiling written process safety information for the system, including the design basis for it [16].

During a PHA, for scenarios examining a release of a hazardous substance into an enclosed space configured with a ventilation system, the ventilation system design, calculations, number of air changes, etc. provide basis for the determination of the likelihood of developing a hazardous atmosphere, within the enclosure, in the event of a release. Inaccurate or missing ventilation system design information may result in inaccurate evaluations and misrepresentation of the actual risk for the scenario.

3.6 Design codes and standards employed

A PHA is a qualitative study relying on the expertise of the PHA team. A list of the design codes and standards employed provides the team with comparisons between how the process is currently operating and how the process was intended to operate. It is often assumed that these items would be equivalent, but over time, small tweaks, projects, and updates can move the process further away from the intended design. A robust Management of Change (MOC) program is typically used to navigate these changes safely, however looking at both the current operating philosophy and the design codes and standards employed in the process design inform the PHA team if there have been deviations from the original design intent. An inaccurate list of design codes and standards employed may result in a team misidentifying the original design intent of the process. The basis for PHA's, such as a HAZOP, is the deviation from intended operation. A misunderstanding of those deviations may result in an inaccurate evaluation of the risks of the process under evaluation.

3.7 Material and energy balances

The material and energy balances are utilized by a PHA team to identify the quantity of a regulated substance input into a chemical process, the amount of the regulated substance consumed by the process, and the amount of a regulated substance exiting a process. Similarly, the material and energy balance identifies the energy input, consumed, and output from a process. As the PHA team works to identify the potential risk of a given scenario, understanding the current quantity of both energy and of the regulated substance is used to inform both the severity of an event, more energy and/or more of a regulated substance typically generates a higher risk ranking. A misrepresentation of these values may result in the PHA team either overestimating or underestimating the risk of a given scenario.

3.8 Safety systems

Safety systems is a broad category that includes all equipment in a chemical process designed and installed to provide mitigation against hazardous process upsets. This includes fire suppression systems like sprinklers or deluge, leak detection systems, process interlocks/trips, relief systems, fixed secondary containment, etc. The risk ranking developed as part of the PHA exercise depends on characterization of both the severity of an event and the likelihood of occurrence. The likelihood of occurrence is directly impacted by the presence of mitigating safeguards in the process therefore accurate information pertaining to the safety systems in the process should be provided to the PHA team. Inaccurate information may result in A PHA team misrepresenting a scenario's overall risk due to either crediting an ineffective safeguard or ignoring the presence of a safeguard and identifying PHA action items that may be unnecessary.

Since the quality of the PHA is so dependent on the safety systems information it is useful for the safety systems information to be provided to the team in an easily readable format. For very simple processes with only a limited number of alarms, interlocks, and safety instrumented functions a list of the safety systems may be sufficient. For more complex processes alarm tables and cause & effect diagrams (summary table of safety instrumentation associated with a chemical process) as well as an up-to-date control narrative may be required to provide the team with the necessary safety systems information.

3.9 Compliance with recognized and generally accepted good engineering practices

Recognized and generally accepted and good engineering practices, commonly referred to as RAGAGEP provides guidance as to the construction, operation, and maintenance of a chemical process. PHAs typically make an assumption that that a PSM covered process, complies with the

PSM standard 29 CFR 1910.119 in that the all equipment in PSM-covered processes complies with RAGAGEP, inspections and tests are performed on process equipment subject to the standard's mechanical integrity requirements in accordance with RAGAGEP, and that inspection and test frequency follows manufacturer's recommendations and good engineering practice, and more frequently if indicated by operating experience. However, due to the importance of producing an accurate PHA it is recommended to have written documentation that the process is in compliance with RAGAGEP so that the team can compare that documentation to what is actually occurring in the field. If the PHA team identifies deficiencies in the process's compliance with RAGAGEP, PHA recommendations can be made to bring the process into compliance[17].

AUTHOR INDEX

AIChE
120 Wall Street, FL 23
New York, NY 10005-4020

ISBN 979-8-3313-1698-3